NUMERICAL CONTROL PROGRAMMING IN APT

Irvin H. Kral

CAD/CAM
General Dynamics
Pomona Division
Pomona, California

PRENTICE-HALL
Englewood Cliffs, New Jersey 07632

Library of Congress Cataloging-in-Publication Data

KRAL, IRVIN H., (date)
    Numerical control programming in APT.

    Bibliography: p.
    Includes index.
    1. Machine-tools—Numerical control.   2. APT
(Computer program language)   I. Title.
TJ1189.K73   1986   621.9′023   85-28307
ISBN   0-13-626599-5

Editorial/production supervision and
    interior design: Reynold Rieger
Cover design: Diane Saxe
Manufacturing buyer: Gordon Osbourne

Printed in the United States of America

10   9   8   7   6   5   4   3   2   1

ISBN   0-13-626599-5   025

Prentice-Hall International (UK) Limited, *London*
Prentice-Hall of Australia Pty. Limited, *Sydney*
Prentice-Hall Canada Inc., *Toronto*
Prentice-Hall Hispanoamericana, S.A., *Mexico*
Prentice-Hall of India Private Limited, *New Delhi*
Prentice-Hall of Japan, Inc., *Tokyo*
Prentice-Hall of Southeast Asia Pte. Ltd., *Singapore*
Editora Prentice-Hall do Brasil, Ltda., *Rio de Janeiro*
Whitehall Books Limited, *Wellington, New Zealand*

Contents

3   Points, Lines, and Circles

60

PART II
TRANSITIONAL INTERLUDE

7 Coordinate Transformations, COPY, and TRACUT   202

Contents

PART III
THREE-DIMENSIONAL AND MULTIAXIS PROGRAMMING

12   Geometry of Three Dimensions   368

13   Three-Axis Contouring   411

14 Multiaxis Part Programming   448

Appendix: APT Reserved Words   485

Bibliography   489

Index   491

Preface

The use of digital computers for the computational effort related to digital control of machine tools is necessary for machine operations depending on computations arising from complex geometry or from simpler geometry that requires a large number of computations. When using digital computers in this application, the choice is either to write computer programs with the computational algorithms needed by using a general-purpose scientific programming language or to use a high-level computer language that has the computational algorithms embedded within the language implementation. One widely known and extensively used high-level language is APT (Automatically Programmed Tools). Other languages are APT-like in form and content and may be subsets. This book is about APT.

A high-level language is designed to fulfill various needs in an application area, one of the needs being to serve as the vehicle by which persons skilled in the application area are given an opportunity to apply the computer without becoming skilled in programming at the assembly language level or in another high-level language not designed for the application area. This, in turn, implies the need for tutorial material for persons who have no computer programming knowledge. Clearly, the need is for a textbook. This book fills that void with respect to the computer language APT. The lack of a proper textbook has made it difficult for instructors to teach the subject and for workers in research and industry to learn about and apply the language.

This book is written for persons with a technical background, such as engineers and technicians in industry and students of industrial technology, engineering technology, engineering, and computer science. It is written at the undergraduate level and is suitable for a two-semester course. No prior computer programming experience is assumed or required for using the book. It is appropriate for industrial technology, engineering technology, engineering, and computer science courses in universities and colleges, junior colleges and trade schools, and in courses offered by computer system manufacturers and numerical control machine tool manufacturers. It is also suitable for instruction in the many APT-like languages and subsets, such as ADAPT, EXAPT, NELNC, MINI-APT, UNIAPT, and so on.

New material is introduced incrementally and its understanding depends only on the material introduced in previous chapters. Each chapter is self-contained with respect to the new material being introduced. This approach differs from the encyclopedia/dictionary format that is used in other material on this subject and that is completely unsuited for tutorial purposes because of the discontinuities it introduces in the learning process. Continuous path part programming techniques are introduced first, with the conceptually easier point-to-point part programming language features deferred until later. This is done to emphasize the power of the language for those applications that require constrained tool motion and to avoid the tendency of students to use sequences of simple statements as a substitute for more sophisticated language features. Initially, at least, this ordering of material may be more mentally demanding but it will lead to simpler part programs and a more disciplined approach to part programming. Unfortunately, APT contains few structured programming features. However, we follow the structured programming concept to the extent practical. All examples were run with the IBM APT-AC system, Version 1, Mod 2 or Mod 3.

The book is divided into three parts. Part I (Chapters 2 through 6) includes material fundamental to learning and applying APT to applications of two-dimensional geometry. Part II (Chapters 7 through 11) includes features of APT best presented with a limited coverage of three-dimensional geometry and features topics pertinent to completing the part program. It is organized for convenience when making the transition from two-dimensional to three-dimensional geometry, yet permits instruction to be terminated without loss of continuity because of time or computer program limitations (such as course length or because only a subset of APT is available). Part III (Chapters 12 through 14) completes the instruction on APT by extending the coverage to three-dimensional geometry and multiaxis programming.

Chapter 1 contains an overview of numerical control part programming in APT. The language is placed in perspective relative to modern trends in numerical control programming. Part programmer educational prerequisites are identified and the programming philosophy of this book is explained. The system of notation for the statement formats used in the book is explained.

Chapter 2 deals with the continuous path tool motion commands for parts of simple two-dimensional geometry and includes a complete example for commanding the tool. It is assumed in this chapter that the geometric definitions have already been prepared and the requirement now is to describe and command the tool. This approach, and using the exercises provided, allows the student to begin immediately preparing part programs for parts of simple geometry. Interpretation of the computer output is discussed and the mechanics of preparing a part program is presented. Program debugging and verification procedures are introduced.

Chapter 3 presents simple geometric definitions—points, lines, and circles. The concept of definition nesting is explained and the canonical form discussed. This material, together with that of Chapter 2, now enables the student to prepare complete part programs for parts of simple two-dimensional geometry.

Chapter 4 introduces the more powerful features of the language, including its computational capability. Because computations are frequently made to support decision making, the branching and looping features of the language are also introduced. The student should now be able to prepare complete part programs with computational segments, that selectively execute portions of the part program as determined by conditions established during the computer run, and that may contain repeatedly executed segments.

Chapter 5 defines arrays and introduces the use of subscripted variables. This material enables the student to prepare some part programs more conveniently than the material of Chapter 4 required, and it enlarges slightly the class of problems for which part programs may be written.

Chapter 6 explores general conics, loft conics, and tabulated cylinders, geometric entities that the student has thus far been unable to handle. These features greatly enlarge the class of problems for which part programs may now be written, still of two-dimensional geometry.

Chapter 7 includes a detailed presentation of coordinate transformations and the use of matrices in conjunction with the COPY, REFSYS, and TRACUT features for applying the language to part geometries that contain certain distinguishing characteristics. The COPY feature is normally used as a convenience feature of the language but coordinate transformations in general represent a significant extension of the capabilities of the language. The student should now be able to prepare complete part programs for two-dimensional problems requiring cutting and noncutting tool motions and which apply to part geometries complex enough to be described with fewer geometric definitions but with more coordinate transformations.

Chapter 8 deals with the definition of plane surfaces, initially for z-axis control, and with part surface specification. These additional language features permit three-dimensional tool motion control where z-axis motion is permitted only along a planar surface but is subject to program redefinition according to the problem requirements. This limited three-dimensional application is appropriate for APT subsets and APT-like languages with restricted capability. This material results in a considerable enlargement of the class of problems that the student should now be able to handle.

Chapter 9 introduces the concept of point-to-point part programming, a concept much easier to understand and learn than continuous path programming. The material is deferred until this chapter for reasons mentioned earlier. Point definition via the pattern feature is also presented.

Chapter 10 introduces the use of postprocessor commands. The set of postprocessor commands actually available is dependent on a particular controller/machine tool combination and its associated postprocessor, but a typical set of such commands is defined and incorporated in various examples. The student now has been introduced to as much material as required to prepare a complete part program and should be quite capable of structuring a part program for a reasonably complex problem.

Chapter 11 deals with the subject of macros. This language element is perhaps used more as a convenience feature by the part programmer but it is much more important in that it serves as a vehicle by which some structured programming techniques may be employed. Also, powerful language features may be more readily identified and appreciated when macros are used. The pocketing command, a convenience feature to reduce the programming effort for a certain class of problems, is described.

Chapter 12 contains definitions of objects in three-dimensional space and completes the coverage of geometric definitions. Vectors and planes are again discussed with a view toward exploiting these features for multiaxis control.

Chapter 13 covers vector mathematics and its application to numerical control in a three-dimensional context. This is followed by the tool shape description allowed for multiaxis contouring and a discussion of problems in maintaining proper tool-to-surface relationships.

Chapter 14 is devoted to multiaxis part programming, both contouring and point-to-point, and completes the material for programming in APT. The coverage is now enlarged to include the widest class of problems that APT is able to handle.

This book is intended for use in the classroom but it may also be used for self-instruction. It evolved from material prepared for use by industrial technology, engineering, and computer science students and its organization reflects the experience obtained from this use. To appreciate the contents fully, the student should have a substantial mathematical background. For students having less mathematical ability, the subject matter may be covered in less depth and the description of the more mathematically oriented examples supplemented through classroom discussion. The material should be covered in chapter order with frequent discussions of proper program construction and how to cope with errors.

Surely the greatest progress in learning a computer programming language comes from running programs on a computer. Thus students should have the opportunity to become involved in the mechanics of part program preparation, program debugging, and part program output verification whether or not a numerically controlled machine tool is available for instructional use. In support of the programming effort, the reference manual for the particular language implementation should be available. The author has specified this manual among the required course materials. If a machine tool is available for control, the applicable postprocessor reference manual should also be made available for student use. A programming problem assigned from each chapter will thoroughly involve the student in the subject matter, but it is essential that sufficient time be allowed for problem analysis, part program preparation and debugging, and program output verification.

This book does not cover part program verification with computer graphics output devices such as plotters, nor does it include a discussion of interactive computer graphics techniques when applied to APT. Also, postprocessors are not covered in detail and the content of the machine control tape is not covered at all. However, these subject areas provide sufficient material for use in term projects, senior projects, and independent study courses.

It is a special privilege to acknowledge the freedom given me by Dave Stephens to experiment with the APT language and the ideas and suggestions on APT offered by

Fred Smith, Tom Furois, and Mike Girouard, all of General Dynamics. I also appreciate the patience of students who used early versions of the manuscript, and the helpful comments and suggestions of the anonymous reviewers.

IRV KRAL

Upland, California

CHAPTER 1

Introduction

An introduction to numerical control and the task of developing a computer program to augment the various activities for controlling a machine tool are presented in this chapter. A brief description of the APT computer language is included to introduce capabilities that are discussed in detail later. Also, because people must describe the tasks to be performed to the machine, we identify the educational prerequisites needed to function as a part programmer. It is not necessary that one know how a computer works in order to use it, and knowledge of computer design or assembly language programming is not an essential prerequisite to learning APT.

1.1 NUMERICAL CONTROL OF MACHINES

Originally, powered machines used to create parts were completely manually operated. With time, some mechanical automatic control was incorporated, for example automatic tool feed of lathes. Later, some of these functions were continuously controlled electronically. Now, we find machines controlled in an incremental manner by computers. Those using the digital computer to directly or indirectly affect the operation of a machine while it produces a part may consider manual control of machines a page from history. This viewpoint could surely be justified when recognizing the role that the computer has played in the development of what has become known as **numerical control** (NC).

Numerical control represents a significant departure from the way in which machines historically have been controlled. Figure 1.1 illustrates the evolutionary process of machine control. Briefly, the operator of manually controlled machines exercised complete control over the machine, as depicted by Figure 1.1(a). Some of man's limitations could

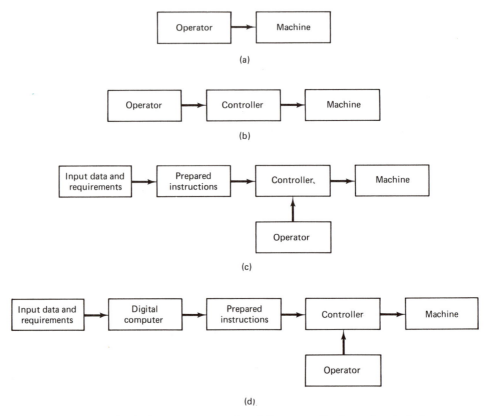

Figure 1.1   Block diagram representation of the preparation for and control of a machine process.

be overcome with equipment that substituted for human physical behavior, so machine control systems were developed to partially replace the human operator with equipment known as the controller. As shown in Figure 1.1(b), the operator now interacted with the **controller**, an assembly of electrical and mechanical components that caused physical actions to be performed during machine operation.

A logical step was to remove the operator further from direct machine control through independent preparation of controller instructions. Now the operator only initiated operations, as depicted by Figure 1.1(c). The instructions for the controller were prepared by the operator and also by others who possessed skills in areas other than just machine operation. Under this concept the machine control instructions were completely prepared in advance by people who used any means available, including the digital computer, to assist them in performing their task. Later [Figure 1.1(d)] the computer was used exclusively for this purpose.

Manual part machining depends on the skills of the operator, such as knowledge of the machine tool, ability to interpret engineering drawings, talent for deriving quantities mathematically, mental alertness and concentration, physical conditioning, and job attitude. In numerical control, part machining depends not on operator characteristics but on a clearly defined machine process plan whereby the functions and control of the machine are determined through a conceptual documentation procedure.

1.2 PREPARING A MACHINE CONTROL TAPE

Figure 1.2 is an expanded version of Figure 1.1(d). It shows a sequential ordering of functions that are performed while using a numerical control computer program for preparing the controller instructions. The computer is used to assist the **part programmer** (the person preparing the instructions for the computer) in computing the machine tool coordinates and in converting the instructions from an internal computer representation to a form recognized by the controller.

The part programmer prepares the **part program**, sometimes called a **manuscript**, which is a set of instructions in the NC program language that describes the part geometry and motion commands to direct the machine during production of the part. The instructions are written in statement form and entered into the computer for processing to produce

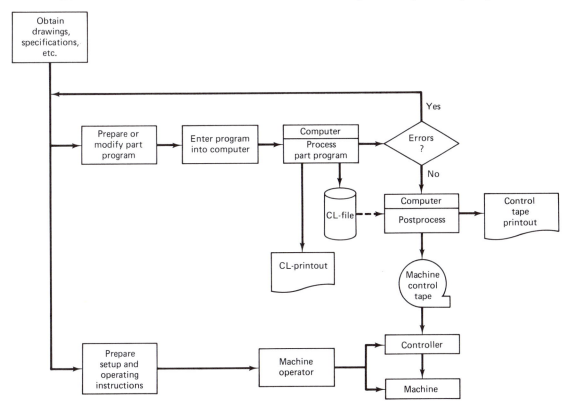

Figure 1.2   Functions performed while preparing a machine control tape.

the **CL-data**, which is recorded on the **CL-file** (CL stands for cutter location). The CL-data is specially formatted machine tool coordinates and other pertinent information that must be further processed before it can be used to control a machine. The CL-file, which may be on a magnetic tape or on a magnetic disk, is a data storage structure. A **file**, in a computer context, denotes information storage locations. An edited version of the CL-file data, the **CL-printout**, is printed for use by the part programmer in verifying correctness of the part program. Verification is done by checking the syntax of the part program statements (they must be corrected when diagnostic messages are printed for statements in error), by determining that the correct part will be produced (usually by inspecting, and making calculations with, the machine tool coordinates printed out by computing the tool position with respect to the tolerances of the desired part boundary, and by noting auxiliary action sequencing, such as spindle and coolant control, etc.).

When the part program is verified to be correct, its output, the CL-file, becomes the input to another program called the **postprocessor**. The postprocessor is written for a specific controller and machine combination. It converts the CL-file data into instructions for the controller and records them onto a **machine control tape**. These instructions are codes usually punched as holes in a tape (Mylar or paper) or recorded onto a magnetic tape or onto a magnetic disk. A readable edited version of the information on the control tape is also printed. The control tape is used as input to the controller. When ready to produce the part, the machine operator mounts the control tape onto a tape reader, readies the controller and machine combination, and initiates its automatic operation.

Some computer installations have auxiliary equipment such as graphic display terminals, hard-copy plotters, and other keyboard and special data-entry devices to aid in part program entry and program verification. Modern computer installations will also have interactive graphics devices to aid the part program design process.

When programming the computer to perform the foregoing functions in the manner described, we are following an off-line programming methodology. This is because the computer is not electrically connected to the machine being controlled by the instructions it prepares. If the machine were directly controlled by the computer with the instructions we programmed into it, we would be following an on-line programming methodology.

The part programmer usually has other functions to perform besides writing the part program and producing the machine control tape. These include selecting designs appropriate for production on numerical control machines, planning setup and machining operations, specifying tools and fixtures, and verifying the first part produced by the machine control tape.

1.3 THE NC PROGRAMMING PROCESS

In this discussion, numerical control programming means that computers are used to assist in producing a set of coded instructions for the machine controller. Determining the coded instructions without use of the computer is not considered in the remainder of this book.

In principle, the part programming effort can be reduced to that of performing a sequence of several clearly defined events. The events include describing the part geometry, defining selected machining parameters, specifying toolpath constraints, and translating the resulting toolpath information to the input format for a selected machine tool controller.

The complexity of the part determines the effort required to describe the geometry that will be referenced while computing the toolpath. The geometry may be described simply with lines and arcs, or it may require the calculation of complex curves and surfaces and specification of their interrelationships.

Defining machining parameters requires knowledge of tool application, machining processes, material characteristics, and part production details. Decisions must be made with respect to tool selection, machine spindle speeds and tool feed rates, coolant application, tolerance selection, and so on.

With few exceptions, the toolpath is determined a priori. Motion of the tool is usually described explicitly with reference to geometric constraining surfaces or coordinate specification. Decisions must be made with respect to machining sequences, tool location reference points, part clamping procedures, and so on.

The computed toolpath information must eventually be converted to the input requirements of the controller. Very little knowledge of NC machines and related considerations is required here since conversion is done by the postprocessor.

Characteristics of the NC machine influence the manner in which the part program is written. One characteristic allows for classification into **continuous path** (contouring) or **point-to-point** (discrete positioning) machines. A continuous path machine operates with its tool in continuous contact with the part in a machining mode while the machine control system guides the tool along a predefined path. A point-to-point machine operates with the tool not in contact with the part while it travels from one location where a machining operation is performed to another. Continuous path machines typically perform operations such as milling, cutting, turning, grinding, and sawing, while point-to-point machines perform operations such as drilling, punching, riveting, tapping, and spot welding. Continuous path part programming is more demanding of the part programmer.

The part programmer assumes responsibility for directing the machine via the part program. The part program documents the machine control motions. The machine control motions must be precise so that the machine/controller combination can produce the proper part. Consequently, the part programmer must be able to visualize the part geometry and the desired tool motion and translate these into program statements for computer processing. Ideally, one should just have to produce the part. But because some engineering drawings are not drawn in an easy-to-understand manner, skill may also be needed in interpreting the design intent.

The requirement for computer preparation of controller instructions has resulted in development of several computer programming languages suitable for this application. The purpose of the programming language is to provide a vehicle by which the requirements for a machine control operation may be conveniently described to the computer. A programming language developed for an application area permits those persons using the language to employ terminology closely resembling that used in the application area. One programming language for numerical control is APT (**A**utomatically **P**rogrammed **T**ools).

1.4 THE APT SYSTEM

APT is the most widely used of the more than 50 numerical control programming languages available. Many of the new NC languages are APT based. Versions of the program generally conform with the requirements of X3.37-1980, a Standard of the American National Standards Institute, Inc. (ANSI). This Standard establishes the form and interpretation of programs expressed in the APT language. Differences from the requirements of the Standard may arise from extensions to the language, slight modifications necessitated by computer equipment characteristics, or some combination of the two. Although part programs written in APT should, in principle, run on any implementation of APT, such may not occur because of the slight differences mentioned above.

With APT, machine control instructions can be produced for machines operating in either a continuous path or a point-to-point manner. For continuous path operation, the geometry of the part must be described in terms of curves. For point-to-point operation, the geometry of the part must be described in terms of points.

The principal output of an APT part program is a sequence of tool-end coordinates that describe the path of the tool as it is commanded to perform a designated function. This tool may be a milling cutter of some shape, a drill, a lathe cutter, a drafting machine pen, a sewing machine needle, the flame of a welding torch, a metal stamping punch, an inspection tool, or some other suitably controlled element. APT computes tool-end coordinates from the part geometry description, tool motion commands, tolerance specifications, and the tool description. A pair of tool-end coordinates determines the point locations between which the tool path is established. While contouring, a tool follows a straight-line path between the points. For point-to-point work, the path of the tool between the points is determined by the characteristics of the controller/machine combination.

Given the starting tool-end coordinates, a vector to the next tool-end location is required for contouring purposes. This is referred to as a **cut vector** since it has both a magnitude and a direction and because the tool usually cuts material during its movement. Tool-end coordinates to produce the part are computed from a series of cut vectors. The phrase "cut vector" is often used in place of the phrase "tool-end coordinates." Tool-end coordinates are computed from the part program statements that are processed by the APT system, also referred to as the **processor**, which is organized into sections. Each section is designed to operate on the input statements or on intermediate results such that collectively they can produce the CL-file and, ultimately, the machine control tape. Figure 1.3 is a simplified flowchart of this organization and processor operation. With APT we follow the off-line programming methodology.

Figure 1.3 shows Section 0 as the executive control routine, a program that controls the overall operation of the APT system and links the other sections so that information flows smoothly throughout the system from the time the part program statements are read

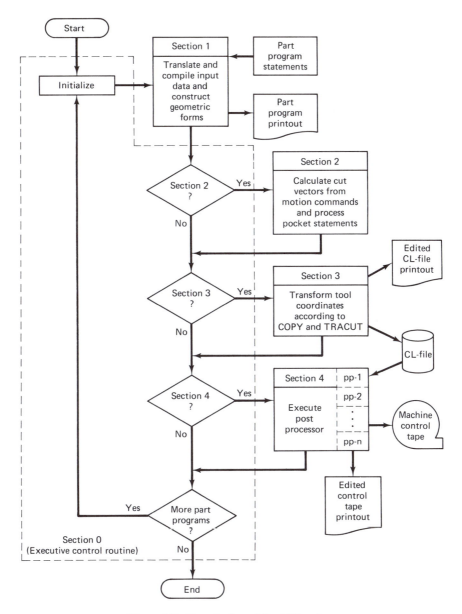

Figure 1.3 Organization of the APT system.

until the machine control tape is produced. Section 1 translates and compiles the input statements and constructs geometric forms. The part program is printed along with any error diagnostics.

Section 2 computes the cut vectors from those APT statements that cause tool motion. Section 3 performs geometric transformations of tool-end coordinates as directed by the part programmer and edits and prints the information from the CL-file. This CL-printout includes Section 2 and Section 3 error diagnostics, if any. Section 4 initiates the execution of postprocessors which, in turn, prepare the machine control tapes and print an edited version of each tape. The postprocessors are not considered part of the APT system but their execution is controlled by APT through the linkage provided by Section 4. Only printouts and tapes are shown in the figure. The use of intermediate tapes and the flow of control when fatal errors occur are not shown.

1.5 AN IMPORTANT APT CAPABILITY

The machining capability realized from an NC controller/machine tool combination is a function of (1) the capability of the machine, for which the principal limitation that we are concerned with now is the number of axes along which the tool may be simultaneously controlled; (2) the limitations of the programming language; and (3) the programmer's ability to program for the complexity of the part. The first two factors are discussed briefly below so that the power of the APT language can be appreciated for the applications for which it is intended. This book is written to help others overcome the limitations of the third factor.

Regardless of the manner in which the machine tool operates, the outcome of the machining operation is determined by the relative motion between the tool and the part and the orientation of the tool with respect to the part. The number of simultaneously controllable machine axes limits the complexity of parts that may be produced on a machine but clever programming may circumvent some such limitations.

First, consider the coordinate system within which we view the machine tool cutter and the part. The standard three-dimensional Cartesian, or rectangular, coordinate system shown in Figure 1.4 serves as the basis reference system. Tool position coordinates and tool orientation can be precisely located within this reference system. This coordinate system contains three mutually perpendicular **principal axes** and, when labeled as shown

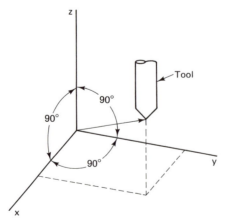

Figure 1.4   Cartesian three-dimensional coordinate system.

in the figure, is known as a right-handed coordinate system. This nomenclature arises from the fact that when the *x*-axis is rotated through the smaller angle toward the *y*-axis, a right-handed screw would progress in the direction of the *z*-axis. Alternatively, if the curved fingers of the right hand indicate the direction through which the positive *x*-axis is turned toward the positive *y*-axis, the thumb shows the direction of the positive *z*-axis.

If the machine is able to simultaneously control the tool along only two of the axes, it is referred to as a **two-axis** machine. The tool is parallel with and independently controlled along the third axis, for example. This machine allows for tool motion as shown in Figure 1.5(a), where the *z*-axis control plane would be parallel with the *xy*-plane. It is called a **two-dimensional** system. This means that the tool may be guided along a contour whose equation describes the contour in two dimensions only with

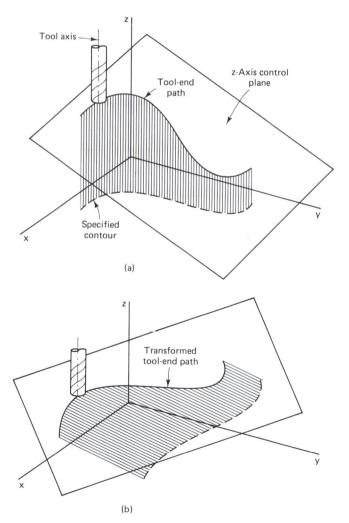

Figure 1.5   (a) A representative tool-end path for a 2½-dimensional system; (b) the same path transformed.

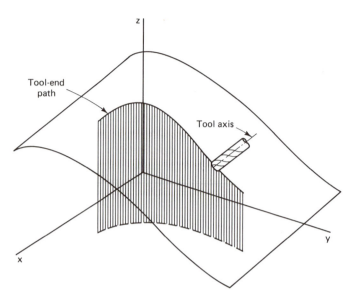

Figure 1.6   Tool axis orientation and a tool-end path permitted in the APT system.

movement along the third axis independently specified. If the tool is controlled to follow an inclined z-axis control plane, we have a **2½-dimensional** system.

If simultaneous tool control is along all three axes, we have a **three-axis** machine. This tool motion is shown in Figure 1.5(b). It is a **three-dimensional** system. If the tool axis orientation varies with tool motion in three dimensions we now have a **multiaxis** machine, perhaps a four-, five-, or six-axis machine. Each additional axis represents a rotation about one of the principal axes. For example, a six-axis machine will not only be able to move the tool simultaneously along each principal axis but will also be able to simultaneously rotate it about each principal axis. Tool rotation for cutting purposes is not considered to be one of these axes. Representative tool motion is shown in Figure 1.6. To effect simultaneous control along multiple axes, the NC programming language must contain features for expressing such motion with resultant computation of tool position and tool axis orientation. The APT system performs multiaxis computations for machine control in a single coordinate system. Control among multiple coordinate systems is not available.

With APT we obtain tool-end coordinates within the coordinate system located by the part programmer with respect to features of the workpiece. This coordinate system may not coincide with orientation of the machine tool axes of control. Differences between the two systems are resolved by the postprocessor.

Other important features of the APT system to support the most complex of computations are listed below. Limitations of a programming language are not necessarily evident from a listing of its features however. Neither are its strengths. Detailed investigation of each feature is required to fully appreciate the capability of the language. For now, the reader is asked to accept the claim that APT provides full capability for multiaxis part programming. The following additional features of APT are discussed in detail in this book.

Language features:
    Symbolic name references
    Subscripted scalar variables and geometric entities

Geometric definitions:
    Analytic representations: point, line, circle, plane, general conic, quadric surface, etc.
    Discrete representations: tabulated cylinder, patterns
    Z-surface specification for coordinate assignment
    Nesting: within geometric definitions, motion commands, and modal commands

Motion statements:
    Multiaxis contouring
    Point-to-point positioning
    Direction specification
    Multiple intersection specification
    Feed rate specification

Tool and modal condition specification:
    Tool shape description
    Tool-to-part tolerances
    Part surface specification
    Tool axis orientation

Computation:
    Arithmetic statements: addition, subtraction, multiplication, division, and exponentiation operators
    Built-in mathematical functions: square root, sine, cosine, arctangent, etc.
    Nesting: within arithmetic statements, geometric definitions, motion commands, and modal commands

Repetitive programming:
    Looping and branching
    Tool-end coordinate copying
    Coordinate transformations
    Macros

Program and diagnostic printouts:
    Part program listing with syntax error diagnostics
    Part programmer annotation of listings
    Edited CL-file printout with computational error diagnostics
    APT system error printouts

Postprocessors:
    Standard vocabulary words
    Automatic processing

1.6 APT AND MODERN NC TECHNOLOGY

Rapid advancements occurred in computer technology during the past decade. Advances include smaller, less expensive, and faster computers with larger high-speed memories. Less expensive peripheral devices, such as interactive terminals, graphic display terminals, bulk storage units such as disks, and hard-copy printers are readily available. With these advances we also see an expansion in computer applications that have affected numerical control. In this section we place in perspective the computer advances and their effect on the APT programming language. In particular, we consider whether or not its application will expand or diminish in the changing environment.

APT has been in existence since the early 1960s. It has been improved substantially during this period and has inspired the development of a number of programs possessing APT-like characteristics. Its basic operational philosophy has not changed, namely, the preparation of the CL-file for postprocessing.

The problem of verifying the accuracy of the data on the CL-file was addressed through development of programs to display the computed cutter path graphically on plotters or on graphics display terminals. Despite all the advances, the verification problem remains. Other improvements involve part geometry description via computer graphics terminal entry. From a graphics description, APT geometric definition statements can be developed and the desired tool path described to the computer for cut vector computation. Previously expensive graphics terminals and the heavy computational load inhibited application of this approach. Now, with the available computational capability with less expensive graphics terminals, this approach is receiving more attention.

Since the late 1970s small, inexpensive, reliable, easy-to-operate computers are being incorporated in the controller. This computer is used to control the machine tool with CL-file data stored within its memory. The computer becomes part of the controller and is dedicated to the machine control task. This computer/controller/machine tool combination is referred to as **computer numerical control** (CNC). It is depicted in Figure 1.7(a). Since the program and data are local to the controller, the operator can alter (edit) memory contents to adjust feed rates or coolant control, optimize noncutting cutter paths at run time, correct minor errors in the cutter path, and so on, all without recomputing the CL-file via APT reruns.

Instead of dedicating a computer to one controller with its machine tool, it is possible to control several machines with one computer. This configuration, called **direct numerical control** (DNC), also requires data derived from the CL-file. When operated in this manner, the computer can be used to collect statistics, such as machine utilization, for management report purposes. Figure 1.7(b) shows one computer in a DNC configuration. In reality, multiple computers may be interconnected hierarchically to control many machines.

A logical next step is to program the computer in a CNC installation to accept APT statements directly, thus eliminating the CL-file transfer operation. Indeed, this is being accomplished with a high-level language with limited features for parts of simple geometry. The computer executes a limited-capability APT system.

Finally, it is becoming possible to describe the part graphically and identify the

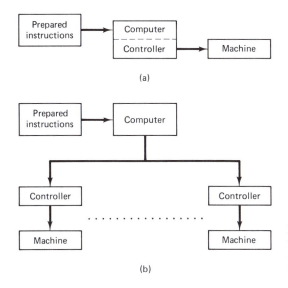

Figure 1.7   **Representation of (a) a computer numerical control (CNC) system and (b) a direct numerical control (DNC) system.**

surfaces to be machined, then via a toolpath generating program compute the required cutting motions. The part programmer works with bounded geometry parts, instead of with the infinite entities on which APT is based. The toolpath generating program produces an APT file although the part programmer no longer needs to know APT.

Computer databases now contain part descriptions obtained from engineering drawing capture with computer-aided design programs. Databases also contain manufacturing process control data. Thus, the possibility exists for a highly automated manufacturing process through advanced numerical control concepts. Considerable effort is being directed toward the development of a **flexible manufacturing system** (FMS), a fully self-sufficient DNC system in which distinct and often complex tasks are performed on a variety of parts, possibly moving in random order through the operation.

Because complex interrelated decisions involving material processes, machining parameter selections, and so on, must be made in a highly developed computer-integrated manufacturing system, we can expect to find artificial intelligence techniques incorporated in advanced production systems. With time, the role of the part programmer will diminish in importance.

Technological changes within an industry evolve with time at a speed dependent on the product. The need for funding and intellectual talent to participate in state-of-the-art development activities prohibit all but a few companies from engaging in this activity. Also, large capital investments in machines functions as an inertial force that causes most companies to resist adopting the new developments immediately, small companies especially. On this basis, we can propose at least three scenarios that would counteract the threat to the future of APT:

1. Small shops with NC machines will buy computers with the APT system because of decreasing computer equipment cost. Jobs previously farmed out or not bid on

because of their complexity or because of time constraints will be feasible for local production without tedious and lengthy manual part programming. This scenario assumes an increase in the use of current numerical control equipment.

2. A general downward migration of technological advances will spread use of numerical control with local programming capability. Inexpensive CNC machines will accelerate this migration, and thus the use of APT. This scenario is technology based and probably most affects small shops.

3. The APT language may not remain static. Advances in computer-aided design can be expected to force changes in the APT language that would accelerate the conversion from part conception and design to part production. Initially, the changes would be APT extensions. Later, they could appear in a revision of the ANSI APT Standard. The changes may be programmed routines stored in a library for access by the part programmer, they could be new features closely related to the geometric descriptions developed by computer-aided design systems, or they may be language extensions appropriate for off-line programming of automated manufacturing processes, perhaps using robots.

Should one of the scenarios become true, we invoke the capital investment argument again and await the next cycle for insertion of more new developments. The demise of APT will eventually follow.

1.7 PART PROGRAMMER EDUCATIONAL PREREQUISITES

The educational prerequisites for NC programming in a high-level language such as APT provide a baseline from which programmers can measure their ability and place in perspective the substance and method of presentation of material of this book. Important prerequisites include knowledge of machine tools and cutting tools, an ability to read and understand engineering drawings, and knowledge of procedures used in machine shops. Also, the programmer must have an ability to visualize spatial relationships derived from orthographic projection views as well as possess the numerous characteristics discussed previously.

The prerequisite of concern here is the mathematical background of the part programmer. In the context of part programming in a high-level language such as APT, mathematical ability is often needed for the understanding and visualization of concepts, rather than for derivation and computation purposes. Indeed, using APT allows the programmer to avoid the computations required for manual part programming. The immediate effect of using APT is to raise the level of ability of the part programmer. However, to use the full power of the language, mathematical skill, in levels varying from algebra through differential equations, may be required depending on the complexity of the part to be produced.

Some concepts are best understood symbolically through mathematical representation and manipulation. But in numerical control, such concepts ultimately must be reduced to numbers (cut vectors) through some transformation process. Fortunately, high-level numerical control languages facilitate this transformation. It is demonstrated throughout this book that mathematical prerequisites are dependent on part complexity. This is analogous to showing that the ease or difficulty of program development is application dependent. For example, writing programs for applications in business and engineering requires equal skill in applying a programming language such as FORTRAN or Pascal. However, payroll tax, inventory, and similar business computations are mathematically less demanding than are computations for the dynamic motion of particles subject to time-varying forces as encountered in engineering.

Most part programmers probably have the option of writing their part programs in a mathematically oriented language such as FORTRAN. However, assuming that they have access to APT, they would not choose FORTRAN over APT. FORTRAN has input and output statements with special formatting capability, multidimension arrays, and function and subroutine subprogram defining capabilities not found in APT. It is, therefore, a more versatile general-purpose language. But the special features of APT make it a more powerful language for numerical control applications. The full power of the language can be appreciated when applied to part complexities demanding of mathematical skill. We adhere to the following philosophy in the mathematically oriented presentations of this book:

1. The mathematical capability inherent in APT, when APT statements or standard procedures are available for the purpose, is preferred over programmer-developed procedures.
2. The mathematical level employed is appropriate for the application being described. However, the level at which equations are derived for part production may differ from the level needed for verification or justification purposes.
3. The sequence of presenting the material is not governed by the order of increasing mathematical prerequisites. Later material builds on that presented earlier but may be at a higher or lower mathematical level.

1.8 THE PROGRAMMING PHILOSOPHY ADOPTED FOR THIS BOOK

This book teaches part programming in APT. We discuss features of the language so that programs can be written to perform a predetermined function. We will not write just any program that does the job. We will write programs that reflect the principles of good programming. These principles lead to good design (the programs work and are easy to maintain and modify), correct results (unreasonable verification procedures are not required), ease of use (the programs are well documented and conveniently used), and operational efficiency (the cost of running the programs is reasonable).

Contributions from the techniques for improving the programming process can be adopted for improving the program design. These include top-down design (specifying the task in general terms, then expanding it into successively more specific and detailed actions until the entire program is developed) and structured design (controlling the design of a program so that its parts can be integrated into a whole without tight coupling between the parts). Both techniques are used in this book to the extent possible. APT does not contain the limited set of flow control statements associated with the structured programming concept, such as the IF-THEN-ELSE and WHILE-DO constructs, and the block organization.

We adopt the view that computer hardware costs (run time and consumable supplies) are much less than computer programmer costs. Consequently, it is important that programmers accomplish their jobs quickly and without substantial inconvenience. This means that the time for diagnosing program errors be minimized. Program efficiency may be sacrificed to achieve this goal. Since APT programs are never production programs (they produce the CL-file from which the machine control tape is produced which is then used for part production purposes), inefficiency here is of little concern. It is also unimportant for jobs to be run only a few times for, if they cannot be completed quickly, they may be canceled. The CL-file created from an inefficient APT part program may not cause a machine tool to run inefficiently however. As a rule, we do not advocate sacrificing machine tool efficiency for expedient program development. There are legitimate reasons for running a machine tool inefficiently (maybe only one part is to be produced), but multiple part production almost always requires efficiently run machines.

We follow this recommended approach to part program development:

1. Emphasize the writing of readable and understandable code, saving efficiency considerations for later. Write the simplest (not simplistic) and clearest code to do the job. Because of this, the APT synonym feature is not used in this book.
2. Write, check, and test the code in stages. Decompose the task into manageable steps and concentrate on the central, most important issues first.
3. Invest extra effort in the creative part of the task, which is the program design and coding process, to reduce the testing and debugging effort.

Principles of good programming are discussed throughout the book. Separate chapters on good program design, efficiency, testing, or documentation are not included. These subjects are pertinent in varying degrees to all programs and therefore are discussed when appropriate.

To learn programming, it is essential that one write, read, revise, reread, and test programs. Because it is difficult to keep mentally organized in short-term memory more than about seven things at a time, it will be necessary to partition the programs into manageable modules to control their complexity.

It is not difficult to follow the programming philosophy described above when writing modest-size programs as given in this book. Huge programs cannot illustrate the APT language features or part program characteristics more effectively than the well-chosen and well-designed programs of this book. But those included here serve as a bridge to understanding the huge production programs. We aid this transition by limiting the scope of material in the following way:

1. No programming tricks are used (we use APT statements for their intended purpose and do not disguise our programming intent through an abstract interpretation).

2. A pseudostandardization of material is effected by selecting APT features and statement formats and syntax common to several widely used implementations. Extensions to the language are avoided unless they strongly reinforce the philosophy stated above. Most of what we have chosen adhere to the ANSI Standard.

3. The context in which the features are described is kept simple; postprocessor commands are included in a single chapter, reference to workpiece clamping is omitted, and part mounting and machine preparation procedures are omitted.

4. Language features are illustrated in applications even though APT may no longer be appropriate for the application. For example, we use NC drafting machines in examples even though they have been replaced by computer graphics equipment.

5. Tool dynamics and material characteristics are excluded from the presentation.

6. The topic of writing postprocessors is not covered.

1.9 SYSTEM OF NOTATION FOR STATEMENT FORMATS

Throughout this book, statement formats for essential elements of the APT language are displayed in a consistent manner utilizing the following conventions:

1. Words having a preassigned meaning in the APT system (reserved words) are printed entirely in capital letters.

2. Words printed in italics represent information that the part programmer must supply.

3. Punctuation and special characters are essential where shown and must be included.

4. Choices in a required portion of the format are indicated by enclosing the choices within braces {}. One and only one of the choices must be selected, the others omitted.

5. Optional portions of the format are enclosed within brackets []. These may be included or omitted as required for the part program. When choices are available within the optional portion, one and only one of the choices must be selected, the others omitted. All punctuation must be included.

Examples:

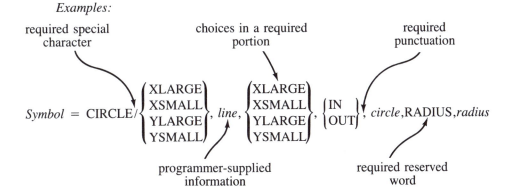

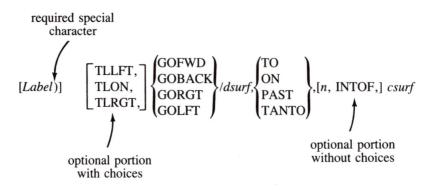

CHAPTER 2

Continuous Path
Part Programming

Numerical control part programming in APT requires that the geometry of the part be described and the motion commands be prepared so that the part can be produced. This chapter describes the motion commands to direct the tool in a continuous manner, that is, to move within a specified tolerance along a curve. Part programming in this manner is referred to as continuous path part programming. Continuous path machining is also referred to as contouring.

All APT continuous path motion commands are introduced in this chapter. We include statements for describing the shape and size of the tool and for specifying the tolerances to be used in computing the tool position. We assume that the geometric description of the part has already been prepared and that the surfaces we are concerned with have been given a symbolic name. We will use these symbolic names while referencing the surfaces in motion commands. Making this assumption allows us to focus our attention on the motion commands, tolerance considerations, and the tool description. We show these geometric descriptions in the examples and provide them for the problems. As a result we will be able to write complete APT part programs and run them on the computer to get the CL-printout for program verification. The manner in which elementary part geometry is described is presented in Chapter 3.

Use of an NC programming language such as APT requires that the part program be prepared for and run on a computer. The mechanics of preparing a part program in computer-compatible form are included.

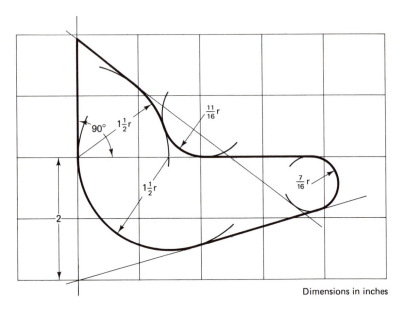

Figure 2.1   Part to be machined.

2.1 EXAMPLE

Example 2.1: Part Machining Problem

An example APT part program is now developed for machining the part in Figure 2.1. Assume that the dimensions are in inches and that the part will be milled from ½-in.-thick aluminum stock.

The simplicity of the drawing is evident. Although a machinist would consider this drawing inadequate because several dimensions would have to be computed to provide a minimum set from which the machine could be set up and controlled, a complete APT part program can easily be written from the information supplied. The APT system will be used to locate the centers of the circles, the points of tangency, the tool radius offset, and the portion of the tool path around the circular curves which, when connected by straight-line segments, define the center of the tool path. These items would all be difficult for the machinist to compute from the information given.

The APT part program will be written from Figure 2.2, which is Figure 2.1 with the circles and lines symbolically labeled. Two points of special interest in this drawing have been labeled ORIG and SETPT. ORIG has special significance in describing the geometry of the part. As stated earlier, we are not developing the geometric definitions at this time. SETPT represents that point over which the longitudinal axis of the tool is to be aligned prior to beginning the machining operation. The machinist will be given written instructions on how to set up the machine and install the workpiece. The part programmer usually prepares the instructions.

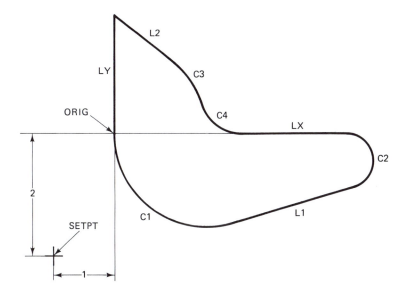

Figure 2.2   Part with symbolically labeled curves and points.

Certain practical problems must be resolved before writing the part program. Those that concern this part, but which will be ignored so as not to detract from the APT part program explanation, are the following:

1. A scheme must be developed for moving clamps holding the workpiece.
2. Tool feed rates and coolant control must be determined.
3. A method for mounting the workpiece on the machine must be determined so that after machining no burrs or uncut edges remain on the bottom of the workpiece and the tool did not cut into the machine bed. We would normally use spacing buttons.

The APT part program of Figure 2.3 will produce the CL-file for this part. It consists of a series of instructions, called statements, that specify tolerances, describe the tool, describe the part geometry, and specify the general path of the tool. A ½-in.-diameter tool is specified and directed counterclockwise around the part beginning from SETPT. Beginning with the first statement, the APT system processes the statements sequentially and produces the part program output listing (CL-printout) of Figure 2.4. The CL-printout shows the tool-end coordinate values which, when plotted and connected by straight-line segments, will trace the center of the tool path during the cutting operation. Because the tool is ¼ in. in radius, the material will be cut ¼ in. on each side of the straight-line segments. The result is a part whose periphery at all points is within the specified tolerances of the programmed part outline. Figure 2.5 is a plot of the tool path. In this figure the centers of the octagons define the location of the plotted points. The octagons are plotting symbols; they are not drawn to the scale of the tool cross-sectional area. The allowance of the tool radius is called the tool offset.

```
ISN 00001 PARTNO   APT PART PROGRAM.
ISN 00002          CLPRNT
ISN 00003 REMARK      * * *   GEOMETRIC DEFINITIONS   * * *
ISN 00004 SETPT    = POINT/-1,-2,1.5
ISN 00005 ORIG     = POINT/0,0
ISN 00006 LX       = LINE/XAXIS
ISN 00007 LY       = LINE/YAXIS
ISN 00008 C1       = CIRCLE/TANTO,LY,XLARGE,ORIG,RADIUS,1.5
ISN 00009 L1       = LINE/(POINT/0,-2),RIGHT,TANTO,C1
ISN 00010 C2       = CIRCLE/YSMALL,LX,YLARGE,L1,RADIUS,.4375
ISN 00011 C3       = CIRCLE/CENTER,ORIG,RADIUS,1.5
ISN 00012 C4       = CIRCLE/YLARGE,LX,XLARGE,OUT,C3,RADIUS,.6875
ISN 00013 L2       = LINE/LEFT,TANTO,C2,RIGHT,TANTO,C3
ISN 00014 REMARK      * * *   TOOL AND TOLERANCE SPECIFICATION   * * *
ISN 00015          CUTTER/0.5
ISN 00016          INTOL/.0005,.005        ++ TOLERANCES IN INCHES
ISN 00017          OUTTOL/.0005,.005
ISN 00018 REMARK      * * *   MOTION COMMANDS * * *
ISN 00019          FROM/SETPT              ++ TOOL ALIGNMENT LOCATION
ISN 00020          INDIRV/1,0,0
ISN 00021          GO/PAST,L1              ++ FOLLOWED BY RAMPING CUT
ISN 00022          TLRGT,GOFWD/L1,TANTO,C2 ++ COUNTERCLOCKWISE CUTTING
ISN 00023          GOFWD/C2,TANTO,LX
ISN 00024          GOFWD/LX,TANTO,C4
ISN 00025          GOFWD/C4,TANTO,C3
ISN 00026          GOFWD/C3,TANTO,L2
ISN 00027          GOFWD/L2,PAST,LY
ISN 00028          GOLFT/LY,TANTO,C1
ISN 00029          GOFWD/C1,TANTO,L1
ISN 00030          TLLFT,GOBACK/L1,TO,LY   ++ MOVE TOOL CLEAR OF PART
ISN 00031          GOTO/SETPT              ++ AND RETURN TO SETPT
ISN 00032          FINI
```

Figure 2.3   APT part program listing.

The CL-printout is an edited version of the information in the CL-file. From Figure 2.4 we note that the information printed is keyed to the part program statements by referencing internal sequence numbers (ISNs). The tool description, ISN 15, and tolerance specifications, ISNs 16 and 17, are printed in the order in which they are processed by the APT system. The first motion command, ISN 19, causes the coordinates of SETPT to be printed. Subsequent motion commands result in the computation of additional tool-end coordinate values as shown. These are the points plotted in Figure 2.5. The straight lines there trace the path of the cutter endpoint. The CL-printout also contains the center coordinate values and radius of each circle whose symbol is referenced in a motion command. For example, the first circular part of the periphery referenced is labeled C2 (ISN 23). Its coordinate values ($x = 3.7946$, $y = -0.4375$, $z = 0$) and its radius (.4375) appear on the CL-printout and, of course, in the CL-file. The tool at this time cuts along an arc of C2. The circle geometry descriptions appear in the CL-file for use by postprocessors for controllers that employ circular interpolation techniques.

Now, given the description of the example and a brief interpretation of the CL-printout, it should not be difficult to infer the purpose of most statements of the part program listing of Figure 2.3. The remarks in the program and a commonsense interpretation of abbreviated words and phrases in the statements should avoid the need for detailed explanation of the program on first exposure to APT. Except for the geometric definitions of this part program (these are discussed in Chapter 3 for reasons mentioned previously), the various other APT statements are discussed in the sections to follow.

```
ISN
0001 PARTNO/  APT PART PROGRAM.
0015 CUTTER/     0.50000000
0016  INTOL/     0.00050000         0.00500000
0017 OUTTOL/     0.00050000         0.00500000
0019  FROM/         SETPT
              -1.00000000        -2.00000000         1.50000000
0021
0021  GOTO/          L1
               0.89285714        -2.00000000         0.0
0022  GOTO/          L1
               3.98714286        -1.09750000         0.0
0023 SURFACE         C2                                   CIRCLE    DS(IMP-TO)
               3.79464286        -0.43750000         0.0
               0.0                0.0                1.00000000         0.43750000
0023  GOTO/          C2
               4.09120935        -1.06328316         0.0
               4.27176807        -0.93940415         0.0
               4.40462240        -0.76534327         0.0
               4.47648915        -0.55850364         0.0
               4.48018288        -0.33956571         0.0
               4.41533428        -0.13041954         0.0
               4.28842710         0.04802381         0.0
               4.11214990         0.17792305         0.0
               3.90412740         0.24629045         0.0
               3.79464286         0.25000000         0.0
0024  GOTO/          LX
               2.07665597         0.25000000         0.0
0025 SURFACE         C4                                   CIRCLE    DS(IMP-TO)
               2.07665597         0.68750000         0.0
               0.0                0.0                1.00000000         0.68750000
0025  GOTO/          C4
               1.98505508         0.25458485         0.0
               1.81755450         0.32879026         0.0
               1.69446615         0.46448163         0.0
               1.66132477         0.55000000         0.0
0026 SURFACE         C3                                   CIRCLE    DS(IMP-TO)
               0.0                0.0                0.0
               0.0                0.0                1.00000000         1.50000000
0026  GOTO/          C3
               1.60216509         0.71630441         0.0
               1.42644336         1.02239148         0.0
               1.19303106         1.28712932         0.0
               1.05454669         1.39657842         0.0
0027  GOTO/          L2
              -0.25000000         2.38163259         0.0
0028  GOTO/          LY
              -0.25000000         0.0                0.0
0029 SURFACE         C1                                   CIRCLE    DS(IMP-TO)
               1.50000000         0.0                0.0
               0.0                0.0                1.00000000         1.50000000
0029  GOTO/          C1
              -0.24569296        -0.18050229         0.0
              -0.17182790        -0.53386934         0.0
              -0.02722321        -0.86464690         0.0
               0.18200248        -1.15883888         0.0
               0.44699627        -1.40399721         0.0
               0.75654554        -1.58974856         0.0
               1.09755241        -1.70823328         0.0
               1.45358794        -1.75443796         0.0
               1.81550266        -1.72640756         0.0
               1.99000000        -1.68000000         0.0
0030  GOTO/          L1
               0.25000000        -2.18750000         0.0
0031  GOTO/         SETPT
              -1.00000000        -2.00000000         1.50000000
0032 ***** FINI *****
```

Figure 2.4   APT part program output listing (CL-printout).

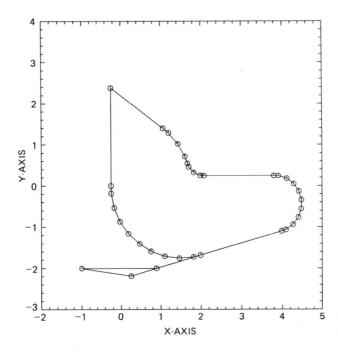

Figure 2.5   Tool-end path plotted by connecting the CL-printout coordinate values with straight-line segments.

2.2 THE TOOL DESCRIPTION

An appropriate tool description is required in the part program so that proper tool path coordinates may be computed by APT. This tool description is given in the CUTTER statement. For two-dimensional applications, it is of the form

CUTTER/d[,r]
where d = diameter of the tool
r = corner radius

This statement is shown in the system of notation adopted for this book (see Section 1.9). As a reminder, the square brackets mean that r is optional unless needed to further describe the tool. The cutter statement of Figure 2.3 is CUTTER/0.5. The parameters are usually given as numeric literals, although they may also be scalar variables as described in Chapter 4. The interpretation of the parameters d and r for a straight-sided tool is shown in Figure 2.6. Parameter r may be omitted when $r = 0$. Note the tool-end location, that point of the tool that conforms with the coordinates being output by the APT system.

The height of the tool defined by the foregoing form of the CUTTER statement is assumed to be 5 units. Since APT assumes quantities to be dimensionless, they must be interpreted with respect to the conversion necessary to associate them with the application. Lengths often are in units of inches or millimeters, so the tool height as defined above

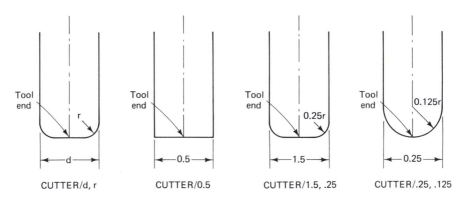

Figure 2.6   Examples of tool specification.

would be, say, 5 in. The tool height used by APT for calculation purposes bears no relation to the actual tool length, which is specified with postprocessor statements as described in Chapter 10. For two-dimensional applications the height is usually irrelevant, so we will not be concerned about it until we discuss three-dimensional applications.

Although Figure 2.6 shows the tool configuration assumed by the APT system (straight sided with diameter d and possibly a corner radius r), the actual tool used may have a quite different appearance. A tool with $d = 0$ may be used for the pen of a drafting machine for which no tool offset is desired. It may also be used for the needle of a sewing machine, the flame tip of a welder, or the end of an engraving tool. The end of a lathe cutter may be used with its tool end located as shown in Figure 2.7. Other interpretations are used for tools such as fly cutters, grinders, and those used for wire wrapping and component insertion. For offset and clearance purposes, the shapes shown in Figure 2.6 are used by the APT system for two-dimensional work. An expanded form of the CUTTER statement is defined in Section 13.2 in connection with three-dimensional machining.

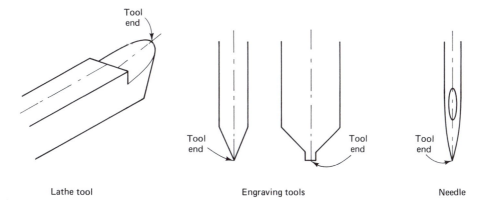

Figure 2.7   Tool-end identification.

2.3 COMMANDING THE TOOL

A part program in the APT language is written so that the general path of the tool is described to perform a predetermined function on a stationary workpiece. Describing the general path of the tool is referred to as **commanding**, or **directing**, the tool. Inasmuch as a desired result can be achieved either through movement of a tool with respect to a stationary workpiece, through movement of a workpiece with respect to a stationary tool, or by some combination of these two actions, the essential ingredient is the relative movement between the tool and workpiece. The assumption of a movable tool only in no way limits the scope or power of the APT language to those machines operating in this manner since it is really the function of the postprocessor to correctly transform tool movement to actual machine commands so as to reproduce the desired relative movement. Also, APT in no way limits the size of workpiece which may be accommodated, and thus the tool for all practical purposes has unlimited travel in both directions for each of the three axes.

2.3.1 Part, Drive, and Check Surfaces

The part programmer formulates a set of instructions to uniquely command the tool to accomplish the desired results. Implicit in the commands is the concept of controlling surfaces. There are part, drive, and check surfaces. Continuous path commands function such that the computed tool path results in the tool being within specified tolerances of the drive and part surfaces during tool movement, and within specified tolerances of the drive, part, and check surfaces when completing a commanded action. Tolerances, discussed in Section 2.4, specify the degree to which the part produced by straight-line cutting motions will approximate the desired part contour.

Drive, part, and check surfaces are illustrated in Figure 2.8. For two-dimensional contouring the **part surface** PS controls movement of the tool along the tool axis consistent with the specified tolerances. It controls the depth of the machining operation. Some part of the tool is always within tolerance of the part surface. The **drive surface** DS controls movement of the tool perpendicular to the direction in which the tool is moving, that is, the drive surface is always tangent to the direction of tool movement. It guides the tool for a motion command and must be referenced explicitly in the command. The **check surface** CS terminates movement of the tool for a given command. The drive and part surfaces are shown as intersecting surfaces as are the part and check surfaces. The drive and check surfaces may or may not intersect. For two-dimensional machining the part surface is the *xy*-plane unless another part surface is explicitly defined.

Figure 2.8(a) shows a flat-end tool at position A with the tool-end center point within tolerance of the part surface, since the part surface is parallel with the *xy*-plane. The periphery of the tool is tangent to, and within tolerance of, the drive surface. For the given direction of tool motion, the tool is to the right of the drive surface. When the tool reaches position B, its periphery will be tangent to, and within tolerance of, the check surface also. Figure 2.8(b) shows the tool-end center point at position A within tolerance of both the part and drive surfaces. We assume that the tool is commanded to

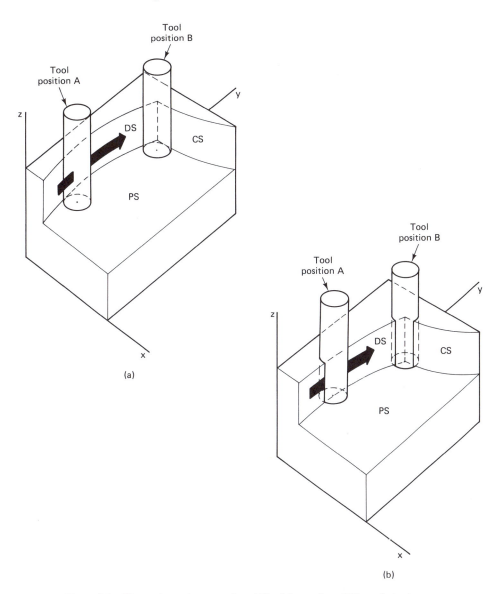

Figure 2.8   Illustrations of part surface (PS), drive surface (DS), and check surface (CS).

move **on** the drive surface, so therefore its tool-end center point is also on the drive surface. When the tool reaches position B, the tool-end center point will be within tolerance of all three surfaces since we assume that it is commanded to move **on** the check surface. This figure shows that the tool periphery does not necessarily have to be tangent to the drive and check surfaces.

2.3.2 Continuous Path Commands

After startup conditions are established, movement of the tool along a predetermined path is specified by a motion command in the following general form:

Statement-label)　　　　Major-section/minor-section

Specifically, using the system of notation for statement formats, we have

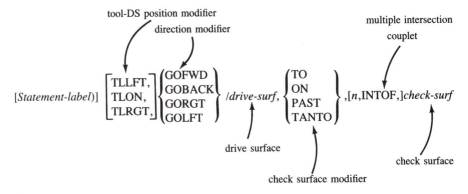

The purpose of the statement label is discussed in Chapter 4. It will not be required until then. The major section always contains a direction modifier and may contain a tool position modifier. These are followed by the slash, which, in turn, is followed by drive and check surfaces references and the modifier to specify the tool-to-check-surface relationship. This motion command generates an **elementary cut sequence**, a series of cut vectors from the tool location at the beginning of the command until it reaches the check surface. During this command the tool-to-surfaces relationship is maintained by APT according to the tolerance specifications.

A moving tool is always considered to be moving forward. The initial direction of tool movement for a subsequent motion command is relative to the direction the tool was moving at the end of the previous motion command. It is given by one of the four direction modifiers GOFWD (go forward), GOBACK (go back), GORGT (go right), or GOLFT (go left). These are interpreted in a general sense as shown by the regions delineated in Figure 2.9. Overlap of the regions permits an alternative in the choice of the direction modifier. Its selection is best made by first determining the most dominant direction of the desired tool path relative to the current forward direction and choosing the corresponding modifier. Later we will learn that this modifier may be chosen relative to an initially specified direction. In any case, the actual tool path is determined by the need for APT to maintain tool-to-drive-surface and tool-to-part-surface relationships.

The minor section is used to specify the drive and check surfaces for that motion command together with a check surface modifier for the tool-to-check-surface relationship when the motion command is completed. These modifiers apply when viewing from the tool to the check surface and are interpreted as follows.

Modifier	Terminal position of the tool
TO	The tool is tangent to the check surface on the near (tool) side of the check surface
ON	The tool is centered on the check surface
PAST	The tool is tangent to the check surface on the far (opposite the tool) side of the check surface
TANTO	The tool is at the point where the drive surface is tangent to the check surface

These modifiers are illustrated in Figure 2.10, where the tool is commanded along drive surface L2 to check surface C1. The comparable motion commands for the part program of Figure 2.3 appear in ISNs 22 through 30.

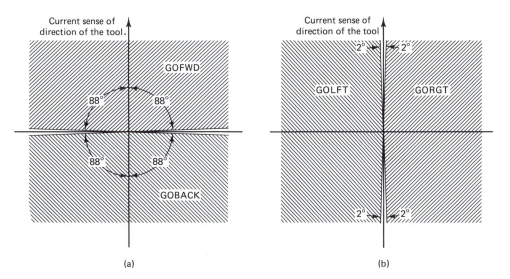

(a)   (b)

Figure 2.9   Regions appropriate for specifying tool motion: (a) in a forward or backward direction; (b) in a left or right direction.

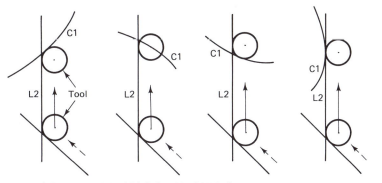

GORGT/L2, TO, C1   GORGT/L2, ON, C1   GORGT/L2, PAST, C1   GORGT/L2, TANTO, C1

Figure 2.10   Check surface modifiers in a motion command.

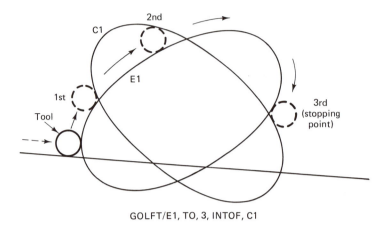

GOLFT/E1, TO, 3, INTOF, C1

Figure 2.11   The multiple intersection couplet.

Occasionally a tool may multiply intersect the check surface as it moves along the part and drive surfaces on its way to the desired check surface intersection. The use of the *n*,INTOF, multiple intersection couplet, where *n* is a positive integer, will result in selection of the *n*th intersection of the drive and check surfaces as the tool stopping point. Omission of this couplet results in selection of the first intersection as the stopping point. The use of this couplet is illustrated in Figure 2.11. Figure 2.12 illustrates the meaning of the check surface modifier when used in conjunction with the tool initial positions given.

The results of a particular motion command are dependent not only on the tool-to-check-surface and tool-to-part-surface specifications but also on the tool-to-drive-surface relationship. The tool-to-drive-surface relationship may be established implicitly, as in ISNs 23 through 29 in Figure 2.3, where the tool is always to the right of the part, or it may be established explicitly with a tool position modifier included in a continuous path motion command, as in ISNs 22 and 30 of Figure 2.3. Once specified, the modifier remains in effect with subsequent motion statements until countermanded by another modifier. The tool position modifiers are:

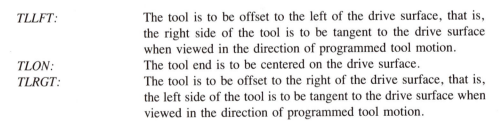

TLLFT:	The tool is to be offset to the left of the drive surface, that is, the right side of the tool is to be tangent to the drive surface when viewed in the direction of programmed tool motion.
TLON:	The tool end is to be centered on the drive surface.
TLRGT:	The tool is to be offset to the right of the drive surface, that is, the left side of the tool is to be tangent to the drive surface when viewed in the direction of programmed tool motion.

Figure 2.13 shows the tool-to-drive-surface relationships for the position modifiers defined above. Tool position modifiers are not always required in a motion statement since the APT system can often determine the correct modifier from the context. If the modifier

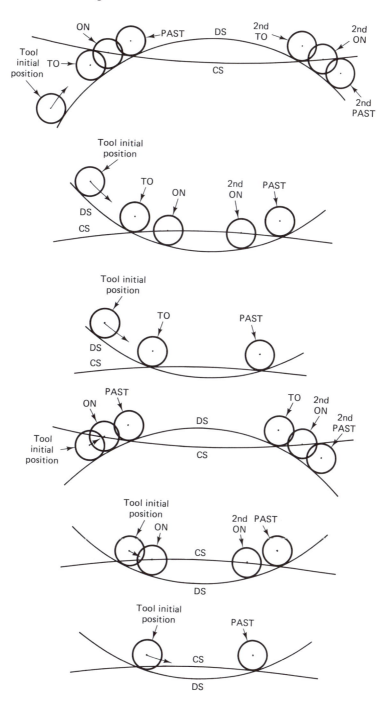

Figure 2.12   Interpretation of the check surface modifier.

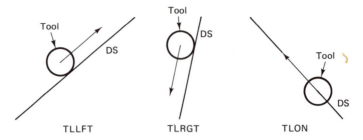

Figure 2.13 Tool-to-drive-surface relationships.

is used when not required, its redundancy does no harm. However, it is necessary to establish the tool position with a modifier for the first command of a sequence of commands in which the tool position does not change. An example is shown in Figure 2.14. For this example, the following part program segment would result in APT Section 2 error printout since the tool position modifiers are not supplied as required.

    GOFWD/L1,ON,C1
    GOLFT/C1,TO,C2
    GORGT/C2,PAST,L2
    GORGT/L2, . . .

However, the part program segment will work properly when the tool position modifiers are included as follows.

    GOFWD/L1,ON,C1
    TLON,GOLFT/C1,TO,C2
    TLRGT,GORGT/C2,PAST,L2
    TLLFT,GORGT/L2, . . .

Whenever two surfaces are specified in a motion command the first is always interpreted as the drive surface and the second as the check surface. This is explicit check

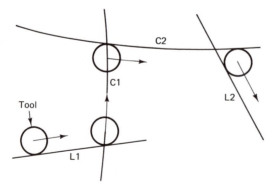

Figure 2.14 Example for tool position modifier usage.

surface specification. If only one surface is specified, it is the drive surface for that motion command and its check surface is the first surface of the following motion command. This is implicit check surface specification and is a feature included for the convenience of the part programmer (it results in less part program coding). This feature lacks the self-documenting characteristic of explicit check surface specification and the restrictions on its use lead to error-prone coding. It is not discussed in this book. Among the allowable drive and check surfaces are lines, circles, general conics, tabulated cylinders, and planes.

2.3.3 Startup Procedure

The need for initial alignment of the tool with respect to the workpiece was discussed in connection with Example 2.1. In that example we assumed that the tool was to be positioned over the point symbolically labeled as SETPT. The tool was commanded from that point to begin the machining operation. Once the tool is brought to within the specified tolerances of the part surface and a check surface, it can be commanded as discussed in Section 2.3.2. The sequence of instructions necessary to position the tool such that it can be commanded in this manner is referred to as the startup procedure. This procedure may require no more than a few simple obvious commands or several commands which collectively position the tool in a complex manner.

It is necessary that the tool have a specific coordinate location before the first motion command is given. This point serves as the beginning point of the first cut vector. Thereafter, each cut vector will have a starting point that is the terminal point of the previous cut vector. Additionally, a sense of direction must be established for the initial move. The process of assigning a specific coordinate location and establishing a sense of direction is termed initialization.

The FROM statement is used to assign a specific coordinate location so as to initialize an APT tool path calculation. It is usually required only at the start of a part program and when used results in no movement of the tool but tool endpoint coordinates are written to the CL-file. It may be used elsewhere in the part program but will then probably result in movement of the tool since it will produce the endpoint of a cut vector. FROM statement formats for two-dimensional machining are

$$\text{FROM/} \begin{Bmatrix} x\text{-}coord,y\text{-}coord[,z\text{-}coord] \\ point \end{Bmatrix}$$

where *x-coord*, *y-coord*, and *z-coord* are tool endpoint coordinate values as numeric literals or scalar variables (see Chapter 4) and *point* is a symbolically named point. Tool axis specification, omitted from the format, is discussed in Chapter 14. Examples of FROM statements are

FROM/2.5, − 4.0,1.5
FROM/ − 6,3
FROM/SETPT

A sense of direction may be established implicitly with the GO command, which is used as a startup motion command to position the tool within tolerance of the part surface and a surface that becomes the drive surface for the following motion command. The format of the GO command is

$$GO/\begin{Bmatrix} TO \\ ON \\ PAST \end{Bmatrix}, drive\text{-}surf \left[,\begin{Bmatrix} TO \\ ON \\ PAST \end{Bmatrix}, part\text{-}surf \left[,\begin{Bmatrix} TO \\ ON \\ PAST \\ TANTO \end{Bmatrix}, check\text{-}surf \right] \right]$$

This command has two optional parts. Thus three separate statements can be derived from it. These are referred to as one-, two-, and three-surface startup statements.

Usually, the tool is not within tolerance of any constraining surfaces until the GO command is executed, after which it will be within tolerance of a drive surface and a part surface. Interpretation of the surfaces of the GO command is with respect to the final tool position, the only tool-end coordinates written to the CL-file by this command.

The one-surface startup statement requires that *drive-surf* only be specified. This surface becomes the drive surface for the following motion command. In the absence of a part surface specification, discussed in Chapter 8, the *xy*-plane becomes the part surface. The resulting tool movement is along the minimum distance route from its current position TO the part surface and to *drive-surf*, terminating there according to the specified modifier TO, ON, or PAST interpreted in the same manner as for the continuous path commands of Section 2.3.2. This statement is suitable for two-dimensional programming.

The two-surface startup statement requires that both *drive-surf* and *part-surf* be specified. It is explicit in the specification of the part surface, which applies to later motion commands, and is suitable for two- and three-dimensional programming. Modifiers for both surfaces must be included. Minimum-distance tool movement results. Care must be used when the modifier PAST is used with *part-surf* since now the length of the tool is considered in determining final tool position (see Section 2.2).

The three-surface startup statement requires that *check-surf* also be specified. This third surface is used to further identify the final tool location. It is not otherwise associated with subsequent motion commands. The modifier TANTO specifies that the final tool position is within tolerance of the part surface at the tangent point of *drive-surf* and *check-surf*.

Symbolic names are generally used for surfaces *drive-surf*, *part-surf*, and *check-surf*. Figure 2.15 illustrates the GO command.

A sense of direction may be established explicitly with the INDIRV (in direction of vector) or INDIRP (in direction of point) statements. These statements establish the forward direction for the next motion command. Their formats are

$$INDIRV/\begin{Bmatrix} x\text{-}comp, y\text{-}comp, z\text{-}comp \\ vector \end{Bmatrix}$$

$$INDIRP/\begin{Bmatrix} x\text{-}coord, y\text{-}coord, z\text{-}coord \\ point \end{Bmatrix}$$

The vector may be given symbolically as *vector* or in terms of its components *x-comp*, *y-comp*, and *z-comp*. Similarly, the point may be given symbolically as *point* or in terms of its coordinates *x-coord*, *y-coord*, and *z-coord*.

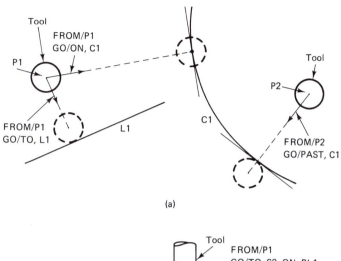

(a)

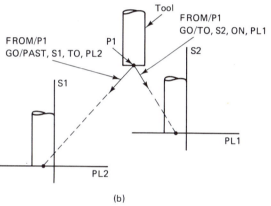

(b)

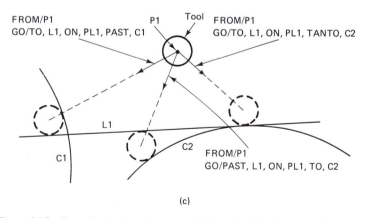

(c)

Figure 2.15 Example startup statements: (a) top view showing one-surface start-up statements (part surface assumed to be the *xy*-plane or a plane parallel with it); (b) side view showing two-surface startup statements for part surfaces that are planes parallel with the *xy*-plane; (c) top view showing three-surface startup statements (PL1 assumed parallel with the *xy*-plane).

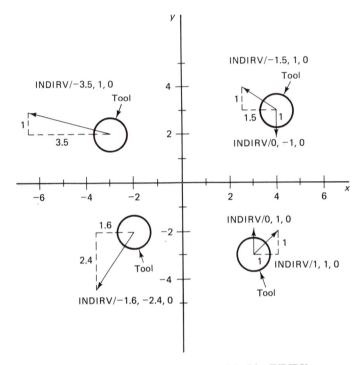

Figure 2.16   Sense of direction established by INDIRV.

The INDIRV statement defines a directed quantity originating at the last computed point (coordinates established by a FROM statement or terminal point of the last cut vector) and extending in a direction determined by the vector components along each axis. For two-dimensional machining only the x- and y-components are needed to establish a direction to the check surface. Figure 2.16 shows the sense of forward direction established by various INDIRV statements.

The INDIRP statement defines a directed quantity originating at the last computed point and extending toward the point specified in the minor section of the statement.

For startup purposes, the INDIRV and INDIRP statements immediately precede the GO command. For the one-surface GO, the tool moves exactly in the specified direction without regard to the minimum distance to the check surface. For the two-surface and three-surface GO commands, the vector establishes for tool movement a general direction only. Actual movement is governed by the need to satisfy the surface constraints of the GO command. Examples of tool movement resulting from INDIRV and INDIRP statements are shown in Figure 2.17.

The OFFSET command may be used instead of the GO command for startup purposes. The format of this statement is

$$\text{OFFSET/}\begin{Bmatrix} \text{TO} \\ \text{ON} \\ \text{PAST} \end{Bmatrix}\text{,}drive\text{-}surf\left[\text{,}\begin{Bmatrix} \text{TO} \\ \text{ON} \\ \text{PAST} \end{Bmatrix}\text{,}part\text{-}surf\right]$$

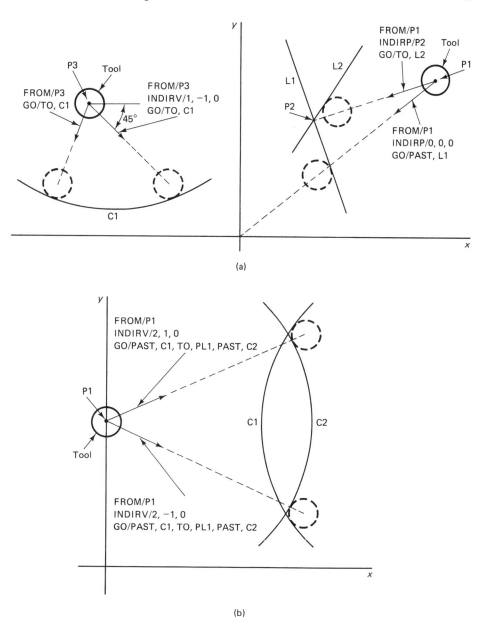

Figure 2.17   **Startup with INDIRV and INDIRP applied to two-dimensional problems: (a) one surface; (b) three surfaces.**

This command must be preceded by an INDIRV or INDIRP statement. The effect of this command differs only slightly from that of the GO command. The interpretation of the modifiers and of *drive-surf* and *part-surf* are as described for the GO command. A normal to *drive-surf* at the point where the vector defined by the INDIRV or INDIRP statement intersects it is computed and the tool is positioned on this normal according to the modifier associated with *drive-surf*. Tolerance within *part-surf* is determined as for the GO command. Figure 2.18 illustrates use of this command.

2.3.4 Ending the Tool Motion Process

Once the tool has been set in motion by the startup procedure, continuous path commands are used until a sequence of operations on the workpiece has been completed. At this time, the tool will be within tolerance of the machined part because of the last motion command. Although this could be the terminal position of the tool, it is more likely that operations such as retracting the tool from the workpiece and returning it to its starting position are the desired final operations. Such sequence ending operations may also be desired for tool changing, clamp repositioning, and so on. Movement of the tool at the end of a sequence of operations is now considered.

The final operation of a sequence of continuous path commands will leave the tool within tolerance of the part, drive, and check surfaces. To clear the tool of the workpiece, it may be sufficient just to select a set of drive and check surfaces, issue another motion

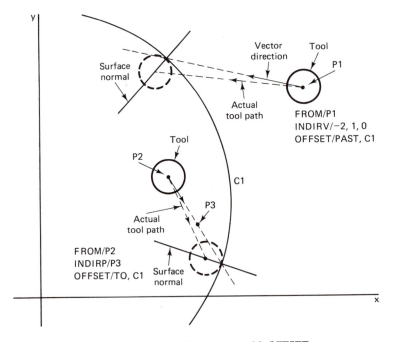

Figure 2.18   One-surface startup with OFFSET.

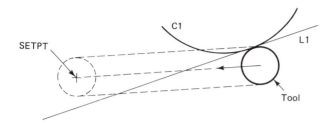

Figure 2.19   Cutting into the part by using an improper sequence of ending commands.

command, and displace the tool from the part being machined. This simple technique suffices when it is not necessary to return the tool to the initial alignment location. Of course, the initial alignment point may coincide with surfaces suitable for use as drive and check surfaces, in which case the ending procedure is just a continuation of the continuous path process.

It may be desired to return the tool to a point that is not within tolerance of any surfaces suitable for use as drive and check surfaces. Such is the case in Example 2.1, where the tool was returned to SETPT. An instruction suitable for this purpose is the GOTO command. Its format is

$$\text{GOTO/} \left\{ \begin{array}{l} x\text{-}coord, y\text{-}coord[,z\text{-}coord] \\ point \end{array} \right\}$$

This is a point-to-point motion command that causes the tool end-point to be positioned on the coordinates of the referenced point. The tool path is a straight line from the end of the last cut vector to the referenced point. There is no requirement for keeping the tool within tolerance of any part, drive, or check surfaces for this command.

An examination of the part program of Figure 2.3 shows that the part is completed when the command of ISN 29 is executed. At this time the tool is tangent to the part at the point where L1 is tangent to C1. The tool is then displaced from the part by another continuous path command, ISN 30, before the GOTO command is issued in ISN 31. The sequence of instructions is necessary to avoid cutting into the part. The effect of omitting ISN 30 is shown in Figure 2.19.

2.4 TOLERANCE SPECIFICATION

The APT system computes cut vectors for directing the tool. They are interpreted as successive straight-line segments. Allowing for the tool radius when following these straight-line segments, the profile of the machined part lies entirely within specified maximum allowable deviations from the shape of the part as programmed. We must specify the accuracy (tolerances) to which the part is to be machined. Factors that enter into the selection of tolerances include the machine tool positioning error, tool tolerances, tool and workpiece deflection, and temperature. The tolerance specifications suitable for two-dimensional machining are

INTOL/*part-tol,drive-tol,check-tol*
OUTTOL/*part-tol,drive-tol,check-tol*

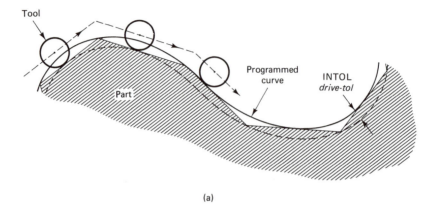

(a)

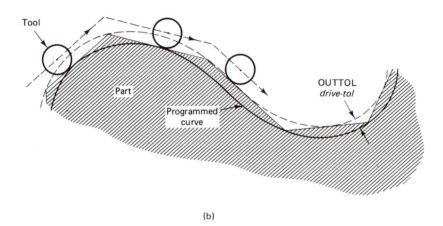

(b)

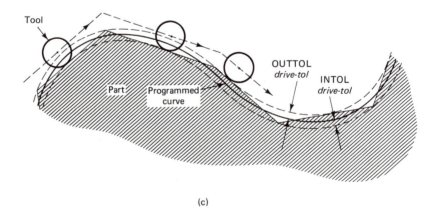

(c)

Figure 2.20   Drive surface tolerance specification interpretation: (a) INTOL > 0, OUTTOL = 0; (b) INTOL = 0, OUTTOL > 0; (c) INTOL > 0, OUTTOL > 0.

where *part-tol*, *drive-tol* and *check-tol* are the tolerance values for the part, drive, and check surfaces, respectively. They are positive numeric literals or scalar variables (see Chapter 4). A zero-valued tolerance is permissible, but both tolerances for a given surface may not simultaneously be zero. INTOL and OUTTOL may be used individually or in combination. Examples of both statements are (missing specifications cause default values to be assigned):

    INTOL/.0005,.001,.001
    INTOL/TOL1
    OUTTOL/T1,T2,T2
    OUTTOL/.0003

The values for INTOL specify the maximum allowable deviation (distance) permitted between the straight-line segments and the programmed curve on the part side of the curve. This is illustrated in Figure 2.20(a) for the drive surface with the OUTTOL value zero and where, relative to the tool position, the segments are tangential or chordal to convex and concave portions of the curve, respectively. The straight-line segments always lie within the programmed curve (on the part side), undercutting results, and the tool diameter results in fillets where adjacent chordal segments meet.

The values for OUTTOL specify the maximum allowable deviation (distance) permitted between the straight-line segments and the programmed curve on the tool side of the curve. This is illustrated in Figure 2.20(b) for the drive surface with the INTOL value zero and where the segments are also tangential or chordal to the curve. The straight-line segments always lie external to the programmed curve (on the tool side), excess stock results, and fillets also occur.

When both INTOL and OUTTOL values for a given surface are nonzero, the straight-line segments will lie alternately on the part and tool sides of the programmed curve, as shown in Figure 2.20(c).

INTOL and OUTTOL are modal commands that remain in effect for all subsequent motion statements until countermanded by another tolerance specification. The tool will always be within tolerance of the programmed curve for any tolerance increase. However, when the tolerance is decreased, it may be necessary to issue new startup commands since the tool probably will not be within tolerance of the programmed curve.

2.5 EXAMPLE 2.1 REVISITED

Let us return to Example 2.1 for further discussion of the motion commands and tolerance specifications. We will consider the lattitude of design provided the part programmer by the APT language features. As before, the analysis excludes discussion of the geometric definitions.

1. Frequently, the part programmer is confronted with selecting from among choices the appropriate tool motion direction modifier. However, in writing a program to follow the contour of Figure 2.2 there is only one opportunity for making a choice. This occurs when moving the tool along LY after machining L2. The example program shows a GOLFT as the choice, ISN 28, whereas GOBACK could just as well have been chosen. No other choice is available because of the tangent conditions occurring at the intersection of all other curves. A GOFWD modifier must always be used for such tangent conditions since the initial motion from a tangent intersection is at $0°$ relative to the direction of motion upon completion of the previous motion command.

    This is within the $\pm 2°$ exclusion region for GOLFT or GORGT as shown in Figure 2.9.

    2. A program for contouring the part in a clockwise direction with the same startup procedure is as follows:

TLRGT,GOLFT/L1,TANTO,C1	22
TLLFT,GOBACK/C1,TANTO,LY	23
GOFWD/LY,PAST,L2	24
GOBACK/L2,TANTO,C3	25
GOFWD/C3,TANTO,C4	26
GOFWD/C4,TANTO,LX	27
GOFWD/LX,TANTO,C2	28
GOFWD/C2,TANTO,L1	29
GOFWD/L1,PAST,LY	30

The lines of the original program are replaced number-for-number with those above. No other program changes are required.

    3. The startup procedure could be as follows:

FROM/SETPT
GO/PAST,L1

This procedure will command the tool to cross L1 along a path that is perpendicular to L1. As a result, the total tool path will be slightly longer than as originally programmed.

    4. It is possible that the tool, when located at some SETPT, will be within tolerance of a suitable drive surface. In this case a forward direction would be established with an INDIRV statement and the tool commanded immediately along the drive surface. Let us suppose that the tool, when at SETPT, is within tolerance of L1 and located above L1. The motion commands to get started are

FROM/SETPT
INDIRV/1,0,0
TLLFT,GOFWD/L1,TO,C1
TLRGT,GORGT/C1,PAST,L1
      etc.

Actually, the tool is not within tolerance of L1, as shown by the following computations.

The equation for C1 is

$$(x - 1.5)^2 + y^2 = 2.25 \qquad (2.1)$$

and the equation for L1 is

$$y = mx - 2 \qquad (2.2)$$

where m is the slope of the line. At the point of tangency the slopes of C1 and L1 are equal and given by

$$\frac{dy}{dx} = \frac{-(x - 1.5)}{y} = m$$

from which

$$x - 1.5 = -my \qquad (2.3)$$

Substituting this expression into equation (2.1) gives

$$(-my)^2 + y^2 = 2.25$$

from which

$$y^2 = \frac{2.25}{1 + m^2} \qquad (2.4)$$

Similarly, substituting equation (2.3) into equation (2.2) gives

$$y = m(1.5 - my) - 2$$

from which

$$y = \frac{1.5m - 2}{1 + m^2} \qquad (2.5)$$

Eliminating y from equations (2.4) and (2.5) gives

$$\left(\frac{1.5m - 2}{1 + m^2}\right)^2 - \frac{2.25}{1 + m^2}$$

$$(1.5m - 2)^2 = 2.25(1 + m^2)$$

from which we obtain the slope as

$$m = \frac{1.75}{6} = 0.291667 \qquad (2.6)$$

L1 intersects LX at [solving for x from equation (2.2)]

$$x = \frac{y + 2}{m} = \frac{0 + 2}{0.291667} = 6.85$$

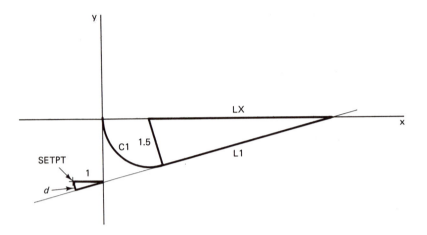

Figure 2.21   Geometry for the tolerance computation of paragraph 4 of Section 2.5.

By similar triangles from the construction shown in Figure 2.21, where d is the perpendicular distance from SETPT to L1, we have

$$\frac{d}{1.5} = \frac{1}{6.85 - 1.5}$$

from which

$$d = \frac{1.5}{5.35} = 0.291$$

For a tool whose diameter is 0.5, the periphery of the tool is short of L1 by $0.291 - 0.250 = 0.041$, which exceeds the specified drive surface tolerance of 0.005. One would obtain a Section 2 diagnostic indicating an out-of-tolerance condition for this sequence of statements.

5. An alternative startup procedure is the following:

FROM/SETPT
GO/TO,C1

In this case, the tool will be directed toward the center of C1 until it is within tolerance but outside C1. From this point a command to go right would be given for a CCW path. Of course, the tool now cannot be commanded to stop at this same point after having gone around the part since there is no check surface to reference. Overlap of tool paths will occur around part of C1.

6. An alternative startup procedure is the following:

FROM/SETPT
INDIRP/ORIG
GO/TO,C1

In this case, the tool will be tangent to C1 near ORIG. A GORGT instruction for a CCW direction will again result in overlap of tool paths since part of C1 will have been left uncut during startup and no suitable check surface exists for only part of C1.

The check surface for the foregoing startup procedure must be carefully specified so that the tool will be within tolerance of the next drive surface. For example, the following incorrect check surface specification will result in a Section 2 out-of-tolerance diagnostic:

FROM/SETPT
INDIRP/ORIG
GO/TO,LY
GOBACK/C1,TANTO,L1

The computation to verify an out-of-tolerance condition can be made with reference to Figure 2.22. By similar triangles, and using the coordinates of SETPT and a tool diameter of 0.5, the y-coordinate of the tool at the point of tangency to LY is computed from $y/2 = 0.25/1$, from which we observe that the point of tangency is $(0, -0.5)$. The distance from this point to C1 is

$$d = \sqrt{1.5^2 + 0.5^2} - 1.5 = 1.581139 - 1.5 = 0.081139.$$

This distance is substantially in excess of the 0.005 drive surface tolerance.

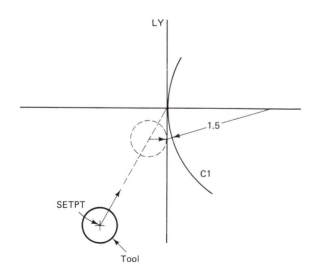

Figure 2.22   Geometry for computing an out-of-tolerance condition to surface C1.

7. Another suitable startup procedure is the following:

```
FROM/SETPT
INDIRP/ORIG
OFFSET/TO,LY      or      OFFSET/TO,C1
TLRGT,GORGT/C1,TANTO,L1
        .
        .
        .

GOLFT/LY,PAST,LX
GOTO/SETPT
```

The tool is positioned such that it is tangent to LY at ORIG. A sense of forward direction has been established so a continuous path command can be specified next. The final continuous path command uses PAST so that as the tool passes by ORIG the result is a smooth cut at ORIG.

2.6 THE MECHANICS OF PREPARING A PART PROGRAM

A complete part program is a collection of statements for which the most important requirement is to convey to the APT system the geometry of the part and the motion commands to produce the part. Among these statements are others that describe the tool environment and that provide instructions to the postprocessor. From these statements it is possible, perhaps with a great deal of effort, to determine the path that the tool will follow as it produces the part.

Certain APT language features are available that (1) aid in preparing a part program such that it is easily converted into machine-readable form (via punched cards or directly from a video terminal), (2) provide a permanent record of the part program suitably annotated and thus reduce the requirement for additional supporting documentation, and (3) aid program debugging and verification efforts. These features are primarily for the convenience of the part programmer.

2.6.1 The APT Character Set

The following characters are recognized by the APT system and are used individually or in combination to form the geometric definitions, tool motion commands, arithmetic statements, and so on.

Alphabetic characters:	A through Z
Decimal digits:	0 through 9
Special characters:	/ * + − = ( ) $. ,
	and blank (no visible character)

The meaning of special characters will be given in context as they are introduced.

2.6.2 Statement Preparation and Coding

In the context of punched card usage, or line display on a video terminal, column positions 1 through 72 are used for the APT statements, while columns 73 through 80 are used for line identification. With the exception of those APT reserved words that must begin in column 1, all statements are free-format, that is, they may begin in any of the statement columns. The free-format concept allows a part programmer to lay out the part program such that, when it is printed, indentation, alignment of particular program elements, character spacing, and comments will help in reading and understanding it. Blanks are ignored by the APT system and, with few exceptions, may be included anywhere in the statement to enhance program readability.

Line identification is usually optional, but recommended. When controlled by the programmer, identification schemes may consist of (1) strictly numeric integers arranged in numerically increasing order, (2) a combination of alphabetic and numeric characters whereby the alphabetic portion denotes a specific program part (e.g., geometric definitions, motion statements, etc.), (3) a real-number scheme in which the integer part may represent a particular statement type and the decimal part a sequence number for that statement type, or (4) some other representation. Line identification may not have to be strictly numeric nor in ordered form since ordering is not checked by the APT system. The line identification concept (1) assists the part programmer in keeping the statements in order (when an ordered scheme is employed), (2) is used for identifying statements containing errors since the line identification is printed together with diagnostic information for statements in error, and (3) is used for identifying tool center coordinates associated with a specific statement since the line identification is printed with the tool center coordinates on the CL-printout.

A statement need not be completed on one line (columns 1–72). The line continuation symbol $ signifies that the statement is continued on the following line (card). This symbol may appear in any column, but all characters following it on the same line are treated as comments only.

Figure 2.23 is an example of a part program as it appears on a coding form. It applies to the part geometry of Figure 2.24. Notice that all characters are printed in capital letters and that certain printing conventions are employed to avoid misinterpreting some alphabetic characters as numerals: for example, the slash through the letter O will distinguish it from the numeral 0, the bars on the letter I will distinguish it from the numeral 1, and so on.

2.6.3 Identifying and Terminating a Part Program

It is convenient to identify a part program by supplying it with a name, an identifying number, or some other meaningful phrase. This is done with the word PARTNO, which must appear in columns 1–6. The remainder of the statement through column 72 is treated as text and serves the purpose of part program identification. It also appears on the CL-printout. When used, the PARTNO statement is usually the first statement in the part program. The following are examples of its use.

```
PARTNO    APT PART PROGRAM CODING EXAMPLE.                                        10
          CLPRNT                                                                  20
          OUTTOL/.0005,.01                                                        30
          CUTTER/.25                 $$ DIA = .25                                 40
          RESERV/C,63                                                             50
          N = 5                      $$ N = NUM OF SMALL CIRCLES, (ODD NUM)       60
          R = .5                     $$ R = RADIUS OF SMALL CIRCLES               70
          LX = LINE/XAXIS                                                         80
          LOOPST                                                                  90
          J = 0                                                                  100
LP1)      J = J + 1                  $$ LOOP TO DEFINE                           110
          C(J) = CIRCLE/(R*(2*J-1)),0,R   $$ N CIRCLES                           120
          IF(N - J), LP2, LP2, LP1                                               130
LP2)      FROM/(SETPT=POINT/-1,0,1)                                              140
          GO/TO,(LY=LINE/YAXIS)                                                  150
          J = 1                                                                  160
          INDIRV/0,-1,0                                                          170
LP3)      TLRGT,GOFWD/C(J),ON,LX     $$ MOTION ALONG THE                         180
          IF(N - J), LP5, LP5, LP4   $$ N CIRCLES                                190
LP4)      J = J + 1                                                              200
          JUMPTO/LP3                                                             210
LP5)      LOOPND                                                                 220
          TLRGT,GOFWD/LI,TANTO,(CK=CIRCLE/YLARGE,LX,XSMALL,          $           230
             (LI=LINE/(2*R*N),0,(2*R*N),1),,RADIUS,(R*N))                        240
          GOFWD/CK,TANTO,LY                                                      250
          GOFWD/LY,PAST,LX                                                       260
          GOTO/SETPT                                                             270
          FINI                                                                   280
```

Figure 2.23   Part program coding example.

PARTNO   FOLLOWER PER DWG B-83309
PARTNO   NO. 228 ASSEMBLY TOOL HANDLE
PARTNO   PERFORATED DISK

The last statement of each part program must contain the single word FINI in free-format. It signifies to the APT system that the physical end of a part program has been reached.

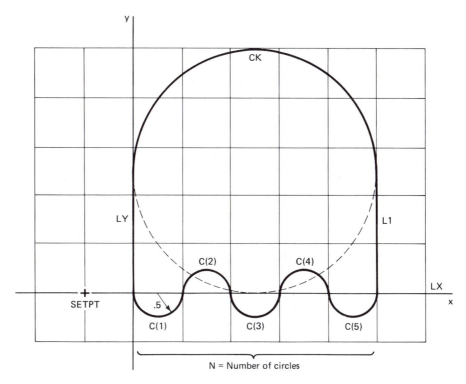

Figure 2.24   Part geometry for the coding example of Figure 2.23.

2.6.4 Annotating the Part Program

Documentation for production of a part will include the problem description, drawings of the part, the part program, setup and special instructions, and so on. Preparation of the part program documentation includes appropriate comments within the program. These comments assist the person reading the part program in understanding certain of its aspects. Comments may be inserted with the REMARK statement or with the $$ symbol.

The word REMARK in columns 1–6 causes that statement to be printed as part of the program listing and initiates no other action. Its contents are not interpreted by the APT system and cannot be executed. There is no limit on the number of REMARK statements that can be used in a part program.

The $$ symbol accomplishes the same purpose as REMARK but, additionally, it may be inserted anywhere in the statement. The entire portion of a statement following the $$ is treated as a comment and is printed with the part program. If the symbol appears in columns 1 and 2, the entire statement is treated as a comment. There can be no spaces between the $ signs. This symbol is often used to append a comment to a statement.

Examples of the REMARK and $$ appear in the part programs of Figures 2.3 and 2.23.

2.6.5 CLPRNT

The computations performed by the APT system result in the writing of a file that contains the coordinates of the tool center as it is directed along the path to produce the part, postprocessor commands, and other instructions produced by the APT system. This file, known as the CL-file, serves as input to the postprocessor. An edited printout of its contents is obtained with the reserved word CLPRNT placed in free-format anywhere within the part program. An example of this output is shown in Figure 2.4. This output is used in program verification. For this purpose some, or all, of the tool-end points would be plotted and the result inspected for proper program output. It may be sufficient just to inspect the printout visually. Omitting CLPRNT from the part program will not inhibit listing of the part program, its execution, or the writing of the CL-file.

2.7 INTRODUCING PART PROGRAM DEBUGGING AND VERIFICATION PROCEDURES

The material of this chapter is limited to the preparation of tool motion statements given that the geometry is described. Therefore, we will limit the discussion of part program debugging and verification to those factors associated with these statements and the output resulting from their execution. Debugging and verification procedures will be expanded in succeeding chapters as new concepts are introduced.

A sure way to quickly produce a functional part program is never to make errors while developing it: logical, syntax, or other. Since this is expecting a lot from the part programmer, the next best thing to do is to follow part programming practices that help in avoiding errors. Such practices must be learned with the introduction of programming features. Thus, in the context of the material of this chapter, we now present recommendations for preparing good programs and for debugging and verifying them.

2.7.1 Preliminaries

The part program must be understandable. This implies that it must be readable. Contributions toward this end include adequate and meaningful comments in the program and motion statements constructed to avoid ambiguous and contradictory interpretation.

The number of comments is adequate if collectively they reveal the intent of the part program and if they clearly delineate logical parts of the program. They are meaningful if they briefly describe important aspects of the program logic, if they reveal the task being performed by a program segment, if they are not too brief or cryptic so as to require study and assumptions for understanding, or if they avoid highlighting trivial or redundant program statements. Because textual interpretation accompanies examples in this book, comments are not liberally included.

Ambiguous and contradictory interpretation of motion statements can be avoided if a consistent tool-to-part viewing philosophy is adopted and if proper tool position modifiers are included in the motion statements. These factors reduce awkward motion statement construction and thereby eliminate "embedded errors."

2.7.2 Debugging

Debugging—the process of removing errors from the program—requires knowledge of the function of the part program, information regarding the part specifications, and familiarity with the statements of the APT language. Based on the elements of the language presented thus far, we choose to discuss the debugging process in the logical order that follows.

General Suggestions

1. Given a choice, set print options for full diagnostic message printout rather than code number printout only. Some APT systems print out the code numbers only. The part programmer must then consult a table for the meaning of the code number. Some APT systems print out full messages each time, while other systems allow the part programmer to select the form of printout. Full message printout is recommended to avoid interrupting concentration associated with discovering the error.

2. Note the severity of the error—fatal, warning, serious, and so on. Fatal errors inhibit further processing of the program. Serious errors indicate that processing was continued but that erroneous results may have been produced. Warning messages identify departure from expected program operation. Fatal and serious errors must be diagnosed and corrected. Warning messages must be disposed of by identifying the condition and correcting it to avoid future errors through misunderstanding of the code.

3. Liberalize tolerances for trial runs to reduce the amount of printout. It is important not to conceal critical printed values, diagnostic messages, auxiliary APT printout, and so on, by copious amounts of printout that is rendered meaningless because of errors. Tolerances can be tightened up after initial runs are proven satisfactory.

Compilation-Time Errors

1. These errors may be syntactic (e.g., inserting a comma instead of a period) or semantic (e.g., misspelling a reserved word). The diagnostic printout localizes the error, but further analysis of the statement is required.

2. Consider that the error reported may be a secondary error—one produced by errors elsewhere in the part program. The original errors may or may not be reported.

3. Simplify the part program to reduce the possibility of compilation-time errors. It is important to complete compilation and go into execution as soon as possible since it is at that time that most is learned about proper functioning of the part program.

Execution-Time Errors

Common errors involving motion commands are described below. They may be found by comparing the CL-printout values with a ''walk-through'' of the desired motion commands until departure from expected performance is observed.

1. The cutter is not within tolerance of the drive surface. This may be caused by inconsistent geometric definitions, by improper choice of check surface, by improper check surface modifier, and so on.
2. The check surface could not be found. An improper direction modifier was specified.
3. The tool is on the wrong side of the drive surface. This may be caused by specifying an improper tool position modifier, by implicitly specifying the tool position modifier but changing the tool-to-drive-surface relationship with a check surface modifier, by improper choice of check surface modifier, and so on.

2.7.3 Verification

The part program is not finished until the first part produced by it passes inspection. At this time the part program may be considered to have been verified as correct. For the simple part programs of this chapter, this definition of verification will suffice. However, before the part is produced (machined), the following actions can be taken to ensure a high probability of correct part production.

1. Perform a walk-through of the CL-printout values by comparing them to the expected position of the tool on a part drawing superimposed on grid paper. Allow for tool offset and, if necessary, reduce the cutter diameter to zero to facilitate locating the tool with respect to the part.
2. Compare CL-printout values against selected hand calculated values for key features of the part. As above, it may be advisable to reduce the cutter diameter for this purpose.

2.7.4 Examples of Errors in Motion Statements

Figure 2.25 shows motion commands of Figure 2.3 with errors deliberately inserted to force Section 2 errors as shown on the accompanying printout. These printouts were produced by the IBM APT-AC system. Similar messages will be produced by other APT processors. The print option was set for full diagnostic message printout.

Part A shows LX substituted for LY as the check surface in ISN 27 (internal sequence number 27). The error message in the portion of CL-printout below the part program segment shows that APT does not like the directional modifier GOFWD since it cannot intersect LX by proceeding in the direction it has been commanded. Of course, the diagnostic is misleading since it is not the directional modifier that is in error, but the APT system cannot correct the part programmer's logical errors. Contradictions such as

```
ISN 00019          FROM/SETPT              ## TOOL ALIGNMENT LOCATION           00002000
ISN 00020          INDIRV/1,0,0                                                00002100
ISN 00021          GO/PAST,L1              ## FOLLOWED BY RAMPING CUT           00002200
ISN 00022          TLRGT,GOFWD/L1,TANTO,C2 ## COUNTERCLOCKWISE CUTTING          00002300
ISN 00023          GOFWD/C2,TANTO,LX                                           00002400
ISN 00024          GOFWD/LX,TANTO,C4                                           00002500
ISN 00025          GOFWD/C4,TANTO,C3                        A                  00002600
ISN 00026          GOFWD/C3,TANTO,L2                                           00002700
ISN 00027          GOFWD/L2,PAST,[LX]                                          00002800
ISN 00028          GOLFT/LY,TANTO,C1                                           00002900
ISN 00029          GOFWD/C1,TANTO,L1                                           00003000
ISN 00030          TLLFT,GOBACK/L1,TO,LY   ## MOVE TOOL CLEAR OF PART          00003100
ISN 00031          GOTO/SETPT              ## AND RETURN TO SETPT              00003200
ISN 00032          FINI                                                        00003300

0026    GOTO/          C3
                1.60216509     0.71630441      0.0
                1.42644336     1.02239148      0.0
                1.19303106     1.28712932      0.0
                1.05454669     1.39657842      0.0
0027    **ERROR    2701    ISN      27  LABEL          SEQ NO 00002800   COND        16
0027    ERRONEOUS DIRECTIONAL MODIFIER IS SPECIFIED. FINAL STOPPING POINT IS IN OPPOSITE
        DIRECTION OF CUTTER MOTION.
0029
0029 SURFACE           C1                           CIRCLE    DS(IMP-TO)
                1.50000000     0.0             0.0
                0.0            0.0             1.00000000      1.50000000
```

```
ISN 00019          FROM/SETPT              ## TOOL ALIGNMENT LOCATION           00002000
ISN 00020          INDIRV/1,0,0                                                00002100
ISN 00021          GO/PAST,L1              ## FOLLOWED BY RAMPING CUT           00002200
ISN 00022          TLRGT,GOFWD/L1,TANTO,C2 ## COUNTERCLOCKWISE CUTTING          00002300
ISN 00023          GOFWD/C2,TANTO,LX                                           00002400
ISN 00024          GOFWD/LX,TANTO,C4                                           00002500
ISN 00025          GOFWD/C4,TANTO,C3                                           00002600
ISN 00026          GOFWD/C3,TANTO,L2                        B                  00002700
ISN 00027          GOFWD/L2,PAST,LY                                            00002800
ISN 00028          GOLFT/[LX],TANTO,C1                                         00002900
ISN 00029          GOFWD/C1,TANTO,L1                                           00003000
ISN 00030          TLLFT,GOBACK/L1,TO,LY   ## MOVE TOOL CLEAR OF PART          00003100
ISN 00031          GOTO/SETPT              ## AND RETURN TO SETPT              00003200
ISN 00032          FINI                                                        00003300

0026    GOTO/          C3
                1.60216509     0.71630441      0.0
                1.42644336     1.02239148      0.0
                1.19303106     1.28712932      0.0
                1.05454669     1.39657842      0.0
0027    GOTO/          L2
               -0.25000000     2.38163259      0.0
0028    **ERROR    2209    ISN      28  LABEL          SEQ NO 00002900   COND        16
0028    CUTTER IS OUT OF TOLERANCE OF DRIVE SURFACE AT START OF MOTION SEQUENCE.
0029 SURFACE           C1                           CIRCLE    DS(IMP-TO)
                1.50000000     0.0             0.0
                0.0            0.0             1.00000000      1.50000000
```

Figure 2.25   **Examples of diagnostics produced by errors in motion commands.**

this diagnostic implies should cause the part programmer to become suspicious and to look elsewhere for a remedy.

Part B shows LX substituted for LY as the drive surface in ISN 28. The error message says that the tool, when at the end of the motion command in ISN 27 (which is correct), is not within tolerance of the new drive surface LX. The drive surface of one

```
ISN 00019      FROM/SETPT                    ** TOOL ALIGNMENT LOCATION        00002000
ISN 00020      INDIRV/1,0,0                                                    00002100
ISN 00021      GO/PAST,L1                    ** FOLLOWED BY RAMPING CUT        00002200
ISN 00022      TLRGT,GOFWD/L1,TANTO,C2       ** COUNTERCLOCKWISE CUTTING       00002300
ISN 00023      GOFWD/C2,TANTO,LX                                              00002400
ISN 00024      GOFWD/LX,TANTO,C4                                              00002500
ISN 00025      GOFWD/C4,TANTO,C3                             C                00002600
ISN 00026      GOFWD/C3,TANTO,L2                                              00002700
ISN 00027      GOFWD/L2,PAST,LY                                              00002800
ISN 00028      GOLFT/LY,TANTO,C1                                              00002900
ISN 00029      GOFWD/C1,TANTO,L1                                              00003000
ISN 00030      TLRGT,GOBACK/L1,TO,LY         ** MOVE TOOL CLEAR OF PART        00003100
ISN 00031      GOTO/SETPT                    ** AND RETURN TO SETPT            00003200
ISN 00032      FINI                                                          00003300
```

```
ISN 00019      FROM/SETPT                    ** TOOL ALIGNMENT LOCATION        00002000
ISN 00020      INDIRV/1,0,0                                                    00002100
ISN 00021      GO/PAST,L1                    ** FOLLOWED BY RAMPING CUT        00002200
ISN 00022      TLRGT,GOFWD/L1,TANTO,C2       ** COUNTERCLOCKWISE CUTTING       00002300
ISN 00023      GOFWD/C2,TANTO,LX                                              00002400
ISN 00024      GOFWD/LX,TANTO,C4                                              00002500
ISN 00025      GOFWD/C4,TANTO,C3                             D                00002600
ISN 00026      GOFWD/C3,TANTO,L2                                              00002700
ISN 00027      GOFWD/L2,PAST,LY                                              00002800
ISN 00028      GOLFT/LY,TANTO,C1                                              00002900
ISN 00029      GOFWD/C1,TANTO,L1                                              00003000
ISN 00030    ⊗ GOBACK/L1,TO,LY              ** MOVE TOOL CLEAR OF PART         00003100
ISN 00031      GOTO/SETPT                    ** AND RETURN TO SETPT            00003200
ISN 00032      FINI                                                          00003300
```

```
0029  SURFACE          C1                           CIRCLE    DS(IMP-TO)
                 1.50000000      0.0          0.0
                 0.0             0.0          1.00000000    1.50000000
0029     GOTO/          C1
                -0.24569296    -0.18050229    0.0
                -0.17182790    -0.53386934    0.0
                -0.02722321    -0.86464690    0.0
                 0.18200248    -1.15883888    0.0
                 0.44699627    -1.40399721    0.0
                 0.75654554    -1.58974856    0.0
                 1.09755241    -1.70823328    0.0
                 1.45558794    -1.75443796    0.0
                 1.81550266    -1.72640756    0.0
                 1.99000000    -1.68000000    0.0
0030       **ERROR    -2210    ISN      30  LABEL          SEQ NO 00003100   COND        4
0030       WARNING...CUTTER IS ON THE WRONG SIDE OF DRIVE SURFACE. THE PROCESSING CONTINUES.
0030     GOTO/          L1
                 0.25000000    -2.18750000    0.0
0031     GOTO/          SETPT
                -1.00000000    -2.00000000    1.50000000
0032 ***** FINI *****
....END OF SECTION 3....
```

Figure 2.25 (Continued)

command does not have to be the same as the check surface of the previous command, but the tool must be within tolerance of the new drive surface in order to continue with contouring commands. This diagnostic is quite explicit, so there should be no problem in identifying the error. The severity of the error for parts A and B (condition code 16) means that the APT processor cannot continue with Section 2 processing.

The errors in parts C and D produce the diagnostic shown below the part program segments. The wrong tool position modifier was given in part C (it should have been TLLFT) while none was given in part D. The severity of the error (condition code 4) was not serious, so processing continued to completion. In this case, correct results were obtained (compare with Figure 2.4), but in general this is not the case. Warning messages should be disposed of as stated in Section 2.7.2. The error must be identified, the correct tool position modifier inserted, and the part program run again.

PROBLEMS

2.1. Figure P2.1 shows a tool path along symbolically labeled surfaces whose geometric definitions are as follows:

SETPT = POINT/ − 1, − 1		100
LX = LINE/XAXIS		110
LY = LINE/YAXIS		120
L1 = LINE/6.5,.75,3,3		130
L2 = LINE/ − 2.25,0,0,3		140
C1 = CIRCLE/4.5,1,1.5		150
C2 = CIRCLE/5,3.25,2		160
C3 = CIRCLE/2.25,2,2.5		170
C4 = CIRCLE/XLARGE,LY,YLARGE,OUT,C3,RADIUS,1.25		180

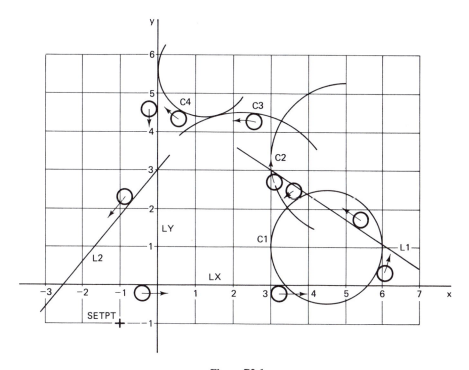

Figure P2.1

Assign an appropriate part number and, using an inside tolerance of 0.01 in. and an outside tolerance of 0.015 in., write the APT motion commands to take a 0.375-in.-diameter flat-end cutter from SETPT in a counterclockwise direction around the part as shown. Keep the tool endpoint to the right, left, or on the surface as marked and return to SETPT upon completion of the machining operation. The part surface is to be, by default, the $z = 0$ plane. Use appropriate startup and ending procedures to reduce the amount of cutting motion overlap and to avoid cutting into the part. Verify that the part program is correct by checking the cut vectors produced against the part geometry.

2.2. Repeat Problem 2.1 but take the tool in a clockwise direction around the part.

2.3. Repeat Problem 2.1 but for the tool path shown in Figure P2.3.

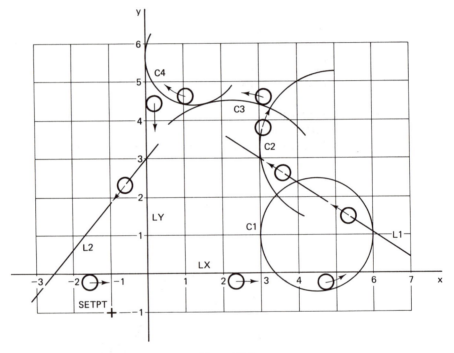

Figure P2.3

2.4. Repeat Problem 2.3 but take the tool in a clockwise direction around the part.

2.5. The following geometric definitions apply to the symbolically labeled surfaces shown in Figure P2.5.

SETPT = POINT/4, − 1		100
ORIG = POINT/0,0		110
LX = LINE/XAXIS		120
LY = LINE/YAXIS		130
ELP = ELLIPS/CENTER,ORIG,3,2,0		140
HYP = GCONIC/0,1,0,0,0, − 1		150
C1 = CIRCLE/ − 3.75,0,2		160

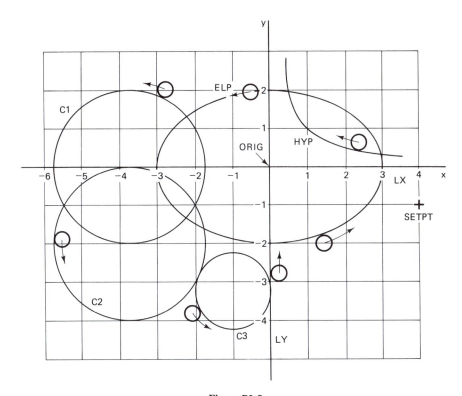

Figure P2.5

| C2 | = CIRCLE/ − 3.75, − 2,2 | 170 |
| C3 | = CIRCLE/XSMALL,LY,YSMALL,OUT,C2,RADIUS,1 | 180 |

Follow the instructions for Problem 2.1 except as noted: Use an inside tolerance of 0.0 in., an outside tolerance of 0.01 in., and a 0.5-in. diameter flat-end cutter.

2.6. Repeat Problem 2.5 but take the tool in a clockwise direction around the part.

2.7. Repeat Problem 2.5 but for the tool path shown in Figure P2.7.

2.8. Repeat Problem 2.7 but take the tool in a clockwise direction around the part.

2.9. The following geometric definitions apply to the symbolically labeled surfaces shown in Figure P2.9.

SETPT	= POINT/3,1	100
LX	= LINE/XAXIS	110
LY	= LINE/YAXIS	120
C1	= CIRCLE/1.5, − 1.5,1.5	130
C2	= CIRCLE/ − 1.25,1.25,1.25	140
C3	= CIRCLE/ − 2, − 2,2	150
L1	= LINE/RIGHT,TANTO,C1,RIGHT,TANTO,C2	160
L2	= LINE/LEFT,TANTO,C1,LEFT,TANTO,C2	170
L3	= LINE/ − 2, − 2,1.5, − 1.5	180

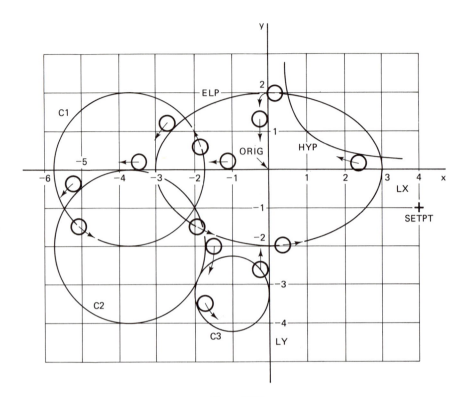

Figure P2.7

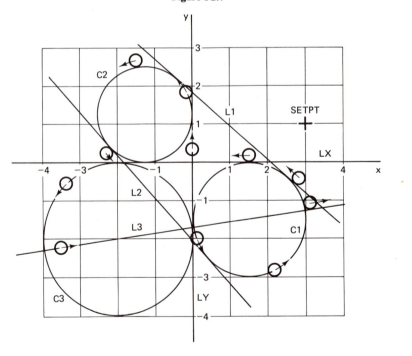

Figure P2.9

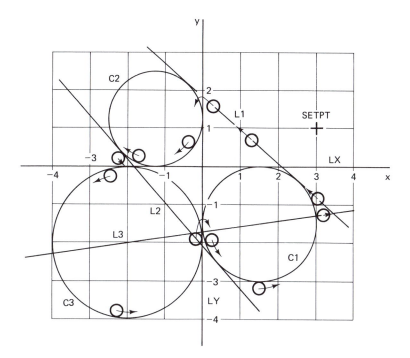

Figure P2.11

Follow the instructions for Problem 2.1 except as noted: Use an inside tolerance of 0.01 in., an outside tolerance of 0.0 in., and a 0.25-in.-diameter flat-end cutter.

2.10. Repeat problem 2.9 but take the tool in a clockwise direction around the part.

2.11. Repeat Problem 2.9 but for the tool path shown in Figure P2.11.

2.12. Repeat Problem 2.11 but take the tool in a clockwise direction around the part.

CHAPTER 3

Points, Lines, and Circles

The part programmer must describe the part geometry prior to writing tool motion commands for part production. This is done with geometric definition statements, each of which include, as part of their construction, a symbolic name for the geometric entity being described. These symbolic names are referred to in motion commands. Some definition formats by which points, lines, and circles may be described in APT are presented in this chapter. Definition statements for other geometric entities are described later. The material of this chapter, together with that of Chapter 2, allows a part programmer to prepare a complete simple APT part program.

3.1 GENERAL FORMAT OF A GEOMETRIC DEFINITION

Simple geometric surfaces are used constantly in the problems to which APT applies. The simplest include points, lines, and circles in the xy-plane. Although lines and circles are not really surfaces, this interpretation arises because the APT system processes them as if they were cylindrical surfaces, that is, the curve is the directrix and a line parallel with the z-axis is the generatrix. They are also considered to be infinite in the $\pm z$-direction. This interpretation allows their use as drive and check surfaces.

The geometric definitions follow a simple consistent format. In general, we have

Statement label) symbolic name = major section/minor section

The purpose of the statement label when used with geometric definitions is discussed in Chapter 4. The major section is a surface type, always a single reserved word which, for this chapter, is POINT, LINE, or CIRCLE. The minor section is the surface type definition, which may be as simple as a collection of numeric literals only or as complex as a combination of various modifiers, reserved words, other surface references, numeric literals, scalar variables, and so on. The minor section may also contain a combination of nested definitions as described in Section 3.5. The equals sign indicates that an identity relationship exists between the name and the geometric definition. It does not denote equality in an arithmetic sense. The geometric definition can now be represented as follows:

Symbolic name = surface type/surface type definition

3.1.1 Symbolic Names

A symbolic name, or symbol, is a part programmer-supplied name used to uniquely identify a particular entity. It consists of from one to six alphanumeric characters, at least one of which must be alphabetic. Except when used as subscripted names as described in Chapter 5, no special characters, as defined in Section 2.6.1, may be included in the name nor may the name be identical to any APT reserved word. A list of reserved words, also known as vocabulary words, appears in the Appendix. The symbolic name may be referenced in other statements, such as geometric definitions or motion commands, but must have been previously defined in the same part program.

Examples:

Valid:	C51	
	ALPHA	
	23J6	
	M13P08	
	7966A	
	CURVEA	
Invalid:	N3.44	Illegal character
	4733	No alphabetic character
	INCREMENT	Too many characters
	CIRCLE	Reserved word
	P-33	Illegal character

The list of reserved words is large and subject to change. To avoid naming conflicts, it is recommended that the symbol name include at least one numeric character.

3.1.2 Constants

Constants (numbers) are often required to express coordinate values, distances, angles, radii, and so on. They may be represented in either integer or real-number form since APT does not differentiate between the two number forms and treats all constants as real floating-point quantities; that is, they are represented internally in signed coefficient and exponent form. Allowed formats for constants are as described below.

Signed Magnitude Only
[±]*coeff*

where *coeff* is the real or integer constant of the form

$$n$$
$$n.$$
$$.n$$
$$n.m$$

The numbers *n* and *m* represent strings of decimal digits for the integral part and fractional part, respectively. No commas are permitted in the strings of digits.

Constants may include a + or − sign. When the constant is negative, the minus sign is required. If no sign is present, the constant is assumed to be positive. Leading zeros before the decimal point for numbers less than 1 are optional and trailing zeros following the last nonzero digit to the right of the decimal point are also optional. Constants may also be replaced by scalar variables as described in Chapter 4.

Examples:

Valid:	16	
	8.3	
	−2.1884	
	0.00001	
	−.0107	
	+12	
	−54.0	
Invalid:	35,000	Comma not allowed

Exponential Form

[±]*coeff* E[±]*exp*

where *coeff* = real constant of the form

$$n.$$
$$.n$$
$$n.m$$

exp = *n* (exponent, integer only)

All comments following the signed magnitude description also apply to the exponential form. The exponential form represents a constant (the coefficient) multiplied by 10 raised to an integer constant (the exponent) where the 10 is replaced by the letter E and the multiplication is implied. The following illustrates the conversion.

$$-1,125,000 = -11.25 \times 10^5 = -11.25E5$$
$$0.00000627 = 6.27 \times 10^{-6} = 6.27E-6$$

Examples:

Valid:	13.5E − 6	=	0.0000135
	.573E5	=	57,300
	480.E − 7	=	0.000048
	3.55E05	=	355,000
	− .975E + 06	=	− 975,000
	− 340.E − 08	=	− 0.0000034
	+ 28.E + 05	=	2,800,000
Invalid:	1425E − 2		No decimal point in coefficient
	− .0068E3.		Decimal point in exponent

The exponential form is recommended only for numbers greater than 100,000 or less than 0.00001 since it is difficult to construct (mathematical conversion is necessary and varied characters are used) and it leads to errors in reading (mathematical conversion is necessary and character combination is confusing). It is used to advantage when large numbers contain many following zeros before the decimal point and when very small numbers contain many leading zeros after the decimal point.

3.1.3 Angles

Numeric values for angles must be expressed in the form of constants in degrees. Minutes and seconds must be converted to fractional parts of a degree. Angles in radians are not permitted. When viewed in the plane of measurement, angles are positive when measured in a counterclockwise direction, negative when measured in a clockwise direction. Geometric definitions specifying an angular from-to relationship as given in Sections 3.2, 3.3, and 3.4 establish a measurement reference from which the sign must be determined as stated above.

3.2 POINT DEFINITIONS

A point is uniquely identified in APT by its position in the three-dimensional rectangular coordinate system. Should its definition not explicitly provide a z-coordinate, its projection onto the xy-plane is assumed unless the point is used in conjunction with a ZSURF statement as described in Section 9.2.3. A selection of point definitions is given in Table 3.1. Points defined with vectors and three-dimensional entities are described in Section 12.6.1.

3.3 LINE DEFINITIONS

A line is defined as the intersection of two planes. The APT system considers the line to be infinitely long and treats it as a plane surface perpendicular to the xy-plane. A selection of line definitions is given in Table 3.2. Lines defined with three-dimensional entities are described in Section 12.6.2.

TABLE 3.1 POINT DEFINITIONS

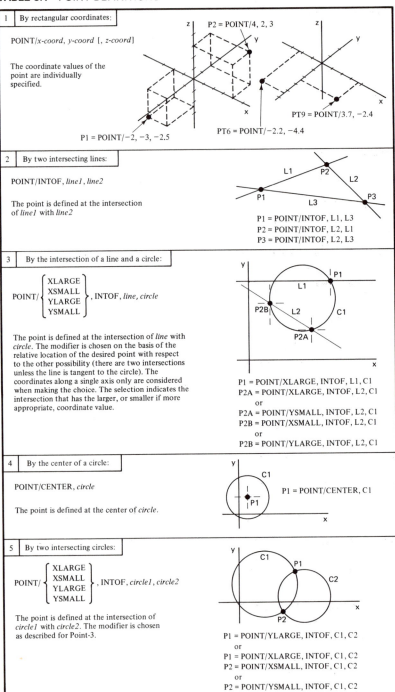

1	By rectangular coordinates:

POINT/*x-coord, y-coord* [, *z-coord*]

The coordinate values of the point are individually specified.

P2 = POINT/4, 2, 3
PT9 = POINT/3.7, −2.4
PT6 = POINT/−2.2, −4.4
P1 = POINT/−2, −3, −2.5

2	By two intersecting lines:

POINT/INTOF, *line1*, *line2*

The point is defined at the intersection of *line1* with *line2*

P1 = POINT/INTOF, L1, L3
P2 = POINT/INTOF, L2, L1
P3 = POINT/INTOF, L2, L3

3	By the intersection of a line and a circle:

POINT/ { XLARGE / XSMALL / YLARGE / YSMALL } , INTOF, *line*, *circle*

The point is defined at the intersection of *line* with *circle*. The modifier is chosen on the basis of the relative location of the desired point with respect to the other possibility (there are two intersections unless the line is tangent to the circle). The coordinates along a single axis only are considered when making the choice. The selection indicates the intersection that has the larger, or smaller if more appropriate, coordinate value.

P1 = POINT/XLARGE, INTOF, L1, C1
P2A = POINT/XLARGE, INTOF, L2, C1
or
P2A = POINT/YSMALL, INTOF, L2, C1
P2B = POINT/XSMALL, INTOF, L2, C1
or
P2B = POINT/YLARGE, INTOF, L2, C1

4	By the center of a circle:

POINT/CENTER, *circle*

The point is defined at the center of *circle*.

P1 = POINT/CENTER, C1

5	By two intersecting circles:

POINT/ { XLARGE / XSMALL / YLARGE / YSMALL } , INTOF, *circle1*, *circle2*

The point is defined at the intersection of *circle1* with *circle2*. The modifier is chosen as described for Point-3.

P1 = POINT/YLARGE, INTOF, C1, C2
or
P1 = POINT/XLARGE, INTOF, C1, C2
P2 = POINT/XSMALL, INTOF, C1, C2
or
P2 = POINT/YSMALL, INTOF, C1, C2

64

TABLE 3.1 (CONTINUED)

6	By the intersection of a circle and the angle of a radial line:

POINT/*circle*, ATANGL, θ

The point is defined at the intersection of the radial line with *circle*. The radial line intersects *circle* at one point only. The angle θ is determined by measuring <u>from</u> *circle*'s own positive x-axis <u>to</u> the radial line.

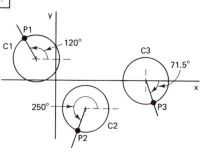

P1 = POINT/C1, ATANGL, 120
P2 = POINT/C2, ATANGL, 250
P3 = POINT/C3, ATANGL, −71.5
    or
P1 = POINT/C1, ATANGL, −240
P2 = POINT/C2, ATANGL, −110
P3 = POINT/C3, ATANGL, 288.5

7	On a circle with respect to another point on the circle:

POINT/*point*, DELTA, $\left\{ \begin{matrix} \text{CLW} \\ \text{CCLW} \end{matrix} \right\}$, ON, *circle*, ATANGL, θ

The point defined is on *circle* and is measured an angular distance θ from *point*, a reference point on *circle*. Angle θ is measured <u>from</u> the reference point <u>to</u> the desired point. Modifier CLW is used when θ is measured in a clockwise direction CCLW is used when measured in a counterclockwise direction.

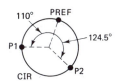

P1 = POINT/PREF, DELTA, CCLW, ON, CIR, ATANGL, 110
P2 = POINT/PREF, DELTA, CLW, ON, CIR, ATANGL, 124.5

8	By polar coordinates:

POINT/RTHETA, $\left\{ \begin{matrix} \text{XYPLAN} \\ \text{YZPLAN} \\ \text{ZXPLAN} \end{matrix} \right\}$, *radius*, θ

The point is defined by the polar coordinates *radius* and θ in a plane specified by two of the three principal axes. The plane is selected by choosing the appropriate modifier, whereby the first two characters of the modifier specify the principal axes of that plane. Angle θ is measured <u>from</u> the positive x-axis when XYPLAN and ZXPLAN are specified, and <u>from</u> the positive y-axis when YZPLAN is specified, <u>to</u> *radius*.

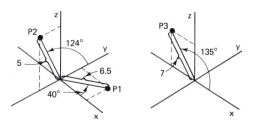

P1 = POINT/RTHETA, XYPLAN, 6.5, 40
P2 = POINT/RTHETA, YZPLAN, 5, 124
P3 = POINT/RTHETA, ZXPLAN, 7, 135

9	By a reference point, radius, and angle:

POINT/*point*, RADIUS, *radius*, ATANGL, θ

The point is defined by the polar coordinates *radius* and θ measured with respect to *point*, the origin of a displaced polar coordinate system. The x-axis of this displaced polar coordinate system is parallel to the x-axis of the principal coordinate system. Angle θ is measured <u>from</u> the x-axis of the displaced polar coordinate system <u>to</u> *radius*.

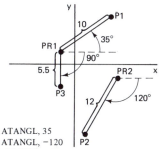

P1 = POINT/PR1, RADIUS, 10, ATANGL, 35
P2 = POINT/PR2, RADIUS, 12, ATANGL, −120
P3 = POINT/PR1, RADIUS, 5.5, ATANGL, −90

65

TABLE 3.1 (CONTINUED)

| 10 | By the intersection of a line and a conic: |

$$\text{POINT/}\left\{\begin{array}{c}\text{XLARGE}\\\text{XSMALL}\\\text{YLARGE}\\\text{YSMALL}\end{array}\right\}, \text{INTOF, } line, conic$$

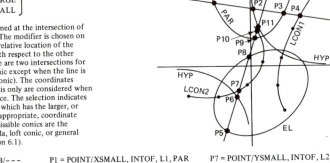

The point is defined at the intersection of *line* with *conic*. The modifier is chosen on the basis of the relative location of the desired point with respect to the other possibility (there are two intersections for each type of conic except when the line is tangent to the conic). The coordinates along a single axis only are considered when making the choice. The selection indicates the intersection which has the larger, or smaller if more appropriate, coordinate value. The permissible conics are the ellipse, hyperbola, loft conic, or general conic (see section 6.1).

HYP = HYPERB/– – –
PAR = GCONIC/– – –
EL = ELLIPS/– – –
LCON1 = LCONIC/– – –
LCON2 = LCONIC/– – –

P1 = POINT/XSMALL, INTOF, L1, PAR
P2 = POINT/XSMALL, INTOF, L1, HYP
P3 = POINT/XLARGE, INTOF, L1, PAR
P4 = POINT/YLARGE, INTOF, L1, LCON1
P5 = POINT/YSMALL, INTOF, L2, EL
P6 = POINT/XSMALL, INTOF, L2, HYP

P7 = POINT/YSMALL, INTOF, L2, LCON2
P8 = POINT/YLARGE, INTOF, L2, EL
P9 = POINT/YSMALL, INTOF, L2, PAR
P10 = POINT/YLARGE, INTOF, L2, LCON2
P11 = POINT/XLARGE, INTOF, L2, HYP

| 11 | By the intersection of a circle and a conic: |

POINT/INTOF, *circle, conic, point1, point2*

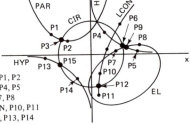

The point is defined at the intersection of *circle* with *conic* between the given points *point1* and *point2* which must lie on the conic. The permissible conics are the ellipse, hyperbola, loft conic, or general conic (see section 6.1).

HYP = HYPERB/– – –
PAR = GCONIC/– – –
EL = ELLIPS/– – –
LCON = LCONIC/– – –

P3 = POINT/INTOF, CIR, PAR, P1, P2
P6 = POINT/INTOF, CIR, HYP, P4, P5
P9 = POINT/INTOF, CIR, EL, P7, P8
P12 = POINT/INTOF, CIR, LCON, P10, P11
P15 = POINT/INTOF, CIR, HYP, P13, P14

| 12 | By the intersection of a line and a tabulated cylinder: |

POINT/INTOF, *line, tabcyl, point*

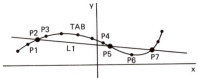

The point is defined at the intersection of *line* with *tabcyl*. Since a line may intersect a tabulated cylinder more than once, the desired intersection is identified by *point*, one of the points used to define the tabulated cylinder and which appears in the tabulated cylinder definition just prior to the point defined by the line crossing.

P2 = POINT/INTOF, L1, TAB, P1
P5 = POINT/INTOF, L1, TAB, P3
P5 = POINT/INTOF, L1, TAB, P4
P7 = POINT/INTOF, L1, TAB, P6

| 13 | As the nth point of a pattern: |

POINT/*patern, n*

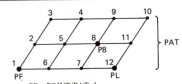

The point defined is the *n*th point of *patern*. The points of a pattern are numbered sequentially in the order of their definition or calculation. The first point is number 1.

PF = POINT/PAT, 1
P8 = POINT/PAT, 8
PL = POINT/PAT, (NUMF(PAT))
    or
PL = POINT/PAT, 12

TABLE 3.2 LINE DEFINITIONS

1	Through two points:

LINE/*x1, y1, z1, x2, y2, z2*

LINE/*x1, y1, x2, y2*

LINE/*point1, point2*

The line is defined as passing through the two
symbolically named points *point1* and *point2*.
The coordinates of the points may be given in
lieu of symbolic names.

L1 = LINE/3, −5, −2, 5, 3, 4

or

L1 = LINE/3, −5, 5, 3   (in xy-plane)

or

L1 = LINE/P1, P2

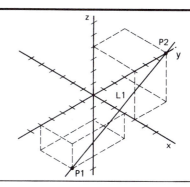

2	As the *x*- or *y*-axis or parallel to, but offset from, the *x*- or *y*-axis:

$$\text{LINE}/ \left\{ \begin{matrix} \text{XAXIS} \\ \text{YAXIS} \end{matrix} \right\} [, \textit{offset}]$$

The line is defined as the x-axis or y-axis by omitting
offset. It is parallel to but displaced from the
x-axis or y-axis by including *offset*, a positive or
negative value that represents its intercept with
the specified axis. The modifier, XAXIS or
YAXIS, is chosen according to the principal axis
involved.

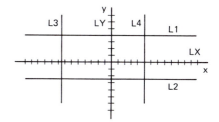

LX = LINE/XAXIS

L1 = LINE/XAXIS, 4

L2 = LINE/XAXIS, −2.5

LY = LINE/YAXIS

L3 = LINE/YAXIS, −7.5

L4 = LINE/YAXIS, 5

3	By a slope or angle and a coordinate axis intercept:

$$\text{LINE}/ \left\{ \begin{matrix} \text{SLOPE, } \textit{slope} \\ \text{ATANGL, } \theta \end{matrix} \right\}, \text{INTERC,} \left\{ \begin{matrix} \text{XAXIS} \\ \text{YAXIS} \end{matrix} \right\}, \textit{intercept}$$

The line is defined at an angle to the x-axis or y-axis.
The angle is specified by *slope* (the tangent of the
angle) or by θ, the angle directly, which is always
measured from the x-axis to the line being defined.
The line intersects the x-axis or y-axis at the value
intercept with the modifier, XAXIS or YAXIS,
chosen according to the principal axis involved.

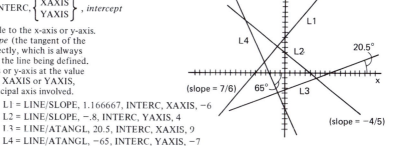

L1 = LINE/SLOPE, 1.166667, INTERC, XAXIS, −6

L2 = LINE/SLOPE, −.8, INTERC, YAXIS, 4

L3 = LINE/ATANGL, 20.5, INTERC, XAXIS, 9

L4 = LINE/ATANGL, −65, INTERC, YAXIS, −7

4	Through a point and parallel or perpendicular to a line:

$$\text{LINE}/\textit{point}, \left\{ \begin{matrix} \text{PARLEL} \\ \text{PERPTO} \end{matrix} \right\}, \textit{line}$$

The line is defined as passing through *point* and is
either parallel or perpendicular to *line* depending
on the choice of modifier, PARLEL or PERPTO,
respectively.

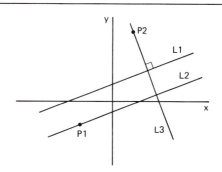

L2 = LINE/P1, PARLEL, L1

L3 = LINE/P2, PERPTO, L1

TABLE 3.2 (CONTINUED)

5	Through a point at an angle to the *x*- or *y*-axis:

$$\text{LINE}/point, \text{ATANGL}, \theta, \begin{Bmatrix} \text{XAXIS} \\ \text{YAXIS} \end{Bmatrix}$$

The line is defined as passing through *point* at an angle θ measured <u>from</u> the x- or y-axis, as determined by the modifier XAXIS or YAXIS, respectively, <u>to</u> the line being defined.

L1 = LINE/P1, ATANGL, −50, YAXIS

or

L1 = LINE/P1, ATANGL, 40

L2 = LINE/P2, ATANGL, 147, XAXIS

or

L2 = LINE/P2, ATANGL, 57, YAXIS

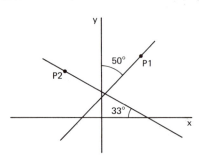

6	Through a point at an angle to a *line*:

LINE/*point*, ATANGL, θ, *line*

The line is defined as passing through *point* at an angle θ measured <u>from</u> *line* (the given line) <u>to</u> the line being defined.

L2 = LINE/P1, ATANGL, 58, L1

or

L2 = LINE/P1, ATANGL, −122, L1

L3 = LINE/P2, ATANGL, −120, L2

or

L3 = LINE/P2, ATANGL, 60, L2

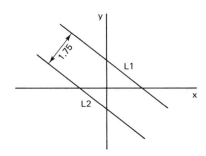

7	Parallel to and offset from a line:

$$\text{LINE}/\text{PARLEL}, line, \begin{bmatrix} \text{XLARGE} \\ \text{XSMALL} \\ \text{YLARGE} \\ \text{YSMALL} \end{bmatrix}, offset$$

The line is defined as parallel to *line* and displaced from it by the value *offset*. The modifier is chosen by observing the axis intercepts as both lines cross either the x- or y-axis and selecting that modifier which represents the relative intercept value of the desired line with respect to the given line. The offset distance is measured perpendicular to the lines.

L2 = LINE/PARALEL, L1, YSMALL, 1.75

or

L2 = LINE/PARALEL, L1, XSMALL, 1.75

8	Through a point with a slope relative to the *x*- or *y*-axis:

$$\text{LINE}/point, \text{SLOPE}, slope, \begin{Bmatrix} \text{XAXIS} \\ \text{YAXIS} \end{Bmatrix}$$

The line is defined as passing through *point* at an angle to the x- or y-axis. The angle is specified by *slope* (the tangent of the angle), which is measured relative to the axis selected by the modifier, XAXIS or YAXIS. The tangent is obtained from the angle measured <u>from</u> the principal axis <u>to</u> the line being defined.

L1 = LINE/P1, SLOPE, −1.3, XAXIS   (slope = −6.5/5)

or

L1 = LINE/P1, SLOPE, .76923, YAXIS   (slope = 5/6.5)

L2 = LINE/P2, SLOPE, .4, XAXIS   (slope = 3/7.5)

or

L2 = LINE/P2, SLOPE, −2.5, YAXIS   (slope = −7.5/3)

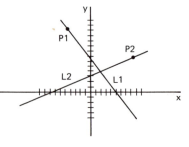

TABLE 3.2 (CONTINUED)

9	Through a point with a slope relative to a line:

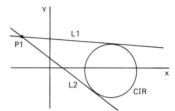

LINE/*point*, SLOPE, *slope*, *line*

The line is defined as passing through *point* at
an angle to *line*. The angle is specified by *slope*
(the tangent of the angle), which is measured
from the given line to the line being defined.

L2 = LINE/P1, SLOPE, −2.2, L1   (slope = −11/5 or
                                           slope = 11/−5)

L3 = LINE/P1, SLOPE, 1, L1   (slope = 5/5 or
                                           slope = −5/−5)

10	Through a point and tangent to a circle:

LINE/*point*, $\left\{ \begin{array}{c} \text{RIGHT} \\ \text{LEFT} \end{array} \right\}$, TANTO, *circle*

The line is defined as passing through *point* and
tangent to *circle*. The modifier is selected by
viewing along the line from the point to the circle
then basing the decision on which side of the
circle the line passes, to the left or to the right.
The point must not lie on the circle.

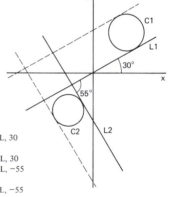

L1 = LINE/P1, LEFT, TANTO, CIR
L2 = LINE/P1, RIGHT, TANTO, CIR

11	Tangent to a circle at an angle to the *x*-axis:

LINE/ $\left\{ \begin{array}{c} \text{XLARGE} \\ \text{XSMALL} \\ \text{YLARGE} \\ \text{YSMALL} \end{array} \right\}$, TANTO, *circle*, ATANGL, θ

The line is defined tangent to *circle* and at an
angle θ to the x-axis. Angle θ is measured
from the x-axis to the line being defined. There
are two possible lines. The modifier is chosen
by observing the axis intercepts as both lines
cross either the x- or y-axis and selecting that
modifier which represents the relative intercept
value of the desired line with respect to the
other possibility.

L1 = LINE/YSMALL, TANTO, C1, ATANGL, 30
or
L1 = LINE/XLARGE, TANTO, C1, ATANGL, 30
L2 = LINE/YLARGE, TANTO, C2, ATANGL, −55
or
L2 = LINE/XSMALL, TANTO, C2, ATANGL, −55

12	Tangent to a circle and at an angle to a line:

LINE/ATANGL, θ, *line*, TANTO, *circle*, $\left\{ \begin{array}{c} \text{XLARGE} \\ \text{YLARGE} \\ \text{XSMALL} \\ \text{YSMALL} \end{array} \right\}$

The line is defined tangent to *circle* and at an
angle θ to *line*. Angle θ is measured from the
given line to the line being defined. There are
two possible lines. The modifier is chosen by
observing the axis intercepts as both lines cross
either the x- or y-axis and selecting that
modifier which represents the relative intercept
value of the desired line with respect to the
other possibility.

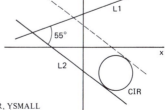

L2 = LINE/ATANGL, −55, L1, TANTO, CIR, YSMALL
or
L2 = LINE/ATANGL, −55, L1, TANTO, CIR, XSMALL

TABLE 3.2 (CONTINUED)

13	Tangent to two circles:

$$\text{LINE/} \left\{ \begin{matrix} \text{RIGHT} \\ \text{LEFT} \end{matrix} \right\} , \text{TANTO}, \textit{circle1}, \left\{ \begin{matrix} \text{RIGHT} \\ \text{LEFT} \end{matrix} \right\} , \text{TANTO}, \textit{circle2}$$

The line is defined as tangent to *circle1* and tangent to *circle2*. The modifiers RIGHT and LEFT are selected by viewing along the line in the direction from *circle1* to *circle2*, then basing the decisions on which side of each circle the line passes.

L1 = LINE/LEFT, TANTO, C1, LEFT, TANTO, C2
L2 = LINE/LEFT, TANTO, C1, RIGHT, TANTO, C2
L3 = LINE/RIGHT, TANTO, C1, LEFT, TANTO, C2
L4 = LINE/RIGHT, TANTO, C1, RIGHT, TANTO, C2

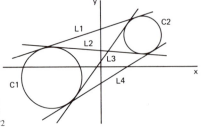

14	Through a point and tangent to a conic:

$$\text{LINE/}\textit{point}, \left\{ \begin{matrix} \text{RIGHT} \\ \text{LEFT} \end{matrix} \right\} , \text{TANTO}, \textit{conic}$$

The line is defined as passing through *point* and tangent to *conic*. The modifier is selected by viewing along the line in the direction from *point* to *conic*, then basing the decision on which side of the conic the line passes. The permissible conics are the ellipse, hyperbola, loft conic, or general conic (see Section 6.1).

L1 = LINE/P1, LEFT, TANTO, PAR
L2 = LINE/P1, RIGHT, TANTO, PAR
L3 = LINE/P1, LEFT, TANTO, HYP
L4 = LINE/P1, RIGHT, TANTO, HYP
L5 = LINE/P2, RIGHT, TANTO, EL
L6 = LINE/P2, LEFT, TANTO, EL
L7 = LINE/P2, RIGHT, TANTO, LCON

HYP = HYPERB/- - -
PAR = GCONIC/- - -
EL = ELLIPS/- - -
LCON = LCONIC/- - -

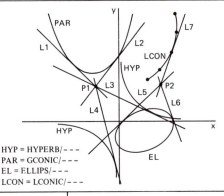

15	Through a point and perpendicular to a conic between two points on the conic:

LINE/*point1*, PERPTO, *conic*, *point2*, *point3*

The line is defined as passing through *point1* and perpendicular to *conic* between points *point2* and *point3*, which are on the conic. The permissible conics are the ellipse, hyperbola, loft conic, or general conic (see Section 6.1).

L1 = LINE/P1, PERPTO, HYP, P2, P3
L2 = LINE/P1, PERPTO, HYP, P4, P5
L3 = LINE/P1, PERPTO, LCON, P6, P7
L4 = LINE/P8, PERPTO, PAR, P9, P10
L5 = LINE/P8, PERPTO, EL, P11, P12
L6 = LINE/P8, PERPTO, EL, P12, P13

HYP = HYPERB/- - -
PAR = GCONIC/- - -
EL = ELLIPS/- - -
LCON = LCONIC/- - -

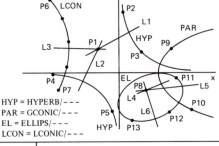

16	Through a point and tangent or perpendicular to a tabulated cylinder:

$$\text{LINE/}\textit{point1}, \left\{ \begin{matrix} \text{TANTO} \\ \text{PERPTO} \end{matrix} \right\} , \textit{tabcyl}, \textit{point2}$$

The line is defined as passing through *point1* and is tangent (or perpendicular) to *tabcyl*. Since there may be more than one point of tangency (or perpendicularity), the desired intersection is approximately identified by *point2*, which should be one of the points used to define the tabulated cylinder. This point is used as a "near" point for selecting the desired portion of the tabulated cylinder from among other possible portions.

L1 = LINE/P1, TANTO, TAB, P2
L2 = LINE/P1, TANTO, TAB, P3
L3 = LINE/P4, PERPTO, TAB, P2
L4 = LINE/P4, PERPTO, TAB, P5

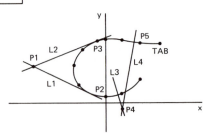

3.4 CIRCLE DEFINITIONS

A circle is defined as the locus of points which are equidistant (the radius) from a fixed point (the center). The APT system considers the circle as a circular cylindrical surface perpendicular to the xy-plane and of infinite extent in the $\pm z$-direction. A selection of circle definitions is given in Table 3.3.

3.5 NESTED DEFINITIONS

The geometric definitions discussed thus far have been structured as principal statements, that is, they contained a symbolic name and major and minor sections that conformed with their defining format. This required the preparation of one statement per definition. Because symbolic names appearing in a definition must have been previously defined, they too must have been structured as principal statements.

A feature in APT that applies to the structuring of statements using symbolic names is that known as **nesting.** Nesting means substituting principal definitions for symbolic names, that is, the definitions are embedded, or nested, within the statement that uses it. Nesting is effected by enclosing the definition within parentheses and inserting it within another statement in place of a symbolic name. Nesting may be performed in geometric definitions, in motion commands, or in modal commands. The APT system permits multiple levels of nesting in any one statement.

Geometric definitions may be nested with or without their symbolic name attached, reference then being made to named or unnamed nested definitions, respectively. Named nested definitions may subsequently be referenced by their symbolic name.

A maximum limit is placed on the number of unnamed geometric definitions in a part program. Whenever an unnamed geometric definition is used in a loop (Chapter 4) or in a macro (Chapter 11) it is counted toward this limit each time the loop or macro is executed. The following examples illustrate various constructions of nested definitions.

   (a) CA = CIRCLE/CENTER,(POINT/4,5),(POINT/6,3)

This example shows the direct substitution of two unnamed definitions for symbolic names. Except for defining the symbols P1 and P2, the statement is equivalent to the following:

    P1 = POINT/4,5
    P2 = POINT/6,3
    CA = CIRCLE/CENTER,P1,P2

   (b) C1 = CIRCLE/CENTER,(PTA = POINT/3,4),RADIUS,1.5

This example uses a named nested definition. Since the point is named, it may subsequently be referenced in the part program by its name, PTA.

TABLE 3.3 CIRCLE DEFINITIONS

1	By its center and its radius:

CIRCLE/*x-coord*, *y-coord*, *z-coord*, *radius*
CIRCLE/*x-coord*, *y-coord*, *radius*
CIRCLE/CENTER, *point*, RADIUS, *radius*

The circle is defined with its center at *point* or at
the coordinates *x-coord*, *y-coord*, and *z-coord*
and by its radius having a value of *radius*.

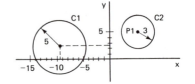

C1 = CIRCLE/−9.6, 2.3. 0, 5
or
C1 = CIRCLE/−9.6, 2.3, 5
C2 = CIRCLE/CENTER, P1, RADIUS, 3

2	By its center and a point on the circumference:

CIRCLE/CENTER, *point1*, *point2*

The circle is defined with its center at *point1*
and as passing through *point2*.

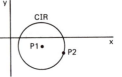

CIR = CIRCLE/CENTER, P1, P2

3	By three points on the circumference:

CIRCLE/*point1*, *point2*, *point3*

The circle is defined such that it passes through
point1, *point2*, and *point3*.

CIR = CIRCLE/P1, P2, P3

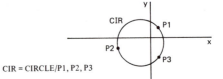

4	By its radius and two points on the circumference:

CIRCLE/ $\left\{ \begin{array}{l} \text{XLARGE} \\ \text{XSMALL} \\ \text{YLARGE} \\ \text{YSMALL} \end{array} \right\}$, *point1*, *point2*, RADIUS, *radius*

The circle is defined such that it passes through
point1 and *point2* with a radius value of *radius*.
Two possible circles satisfy the definition. The
modifier is chosen on the basis of the relative
location of the desired circle to the other
possibility. The choice is made by comparing
the x- or y-coordinate value of the centers of
both circles and selecting that modifier which
represents the relative location of the desired
circle to the other possibility.

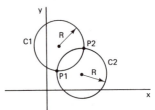

C1 = CIRCLE/YLARGE, P1, P2, RADIUS, R
or
C1 = CIRCLE/XSMALL, P1, P2, RADIUS, R
C2 = CIRCLE/YSMALL, P1, P2, RADIUS, R
or
C2 = CIRCLE/XLARGE, P1, P2, RADIUS, R

5	By its center and a tangent line:

CIRCLE/CENTER, *point*, TANTO, *line*

The circle is defined tangent to *line*
with its center at *point*.

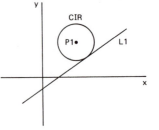

CIR = CIRCLE/CENTER, P1, TANTO, L1

TABLE 3.3 (CONTINUED)

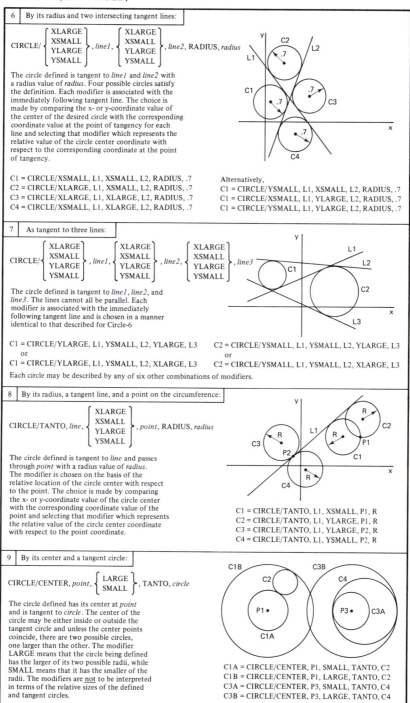

6 | By its radius and two intersecting tangent lines:

$$CIRCLE/ \begin{Bmatrix} XLARGE \\ XSMALL \\ YLARGE \\ YSMALL \end{Bmatrix}, line1, \begin{Bmatrix} XLARGE \\ XSMALL \\ YLARGE \\ YSMALL \end{Bmatrix}, line2, RADIUS, radius$$

The circle defined is tangent to *line1* and *line2* with a radius value of *radius*. Four possible circles satisfy the definition. Each modifier is associated with the immediately following tangent line. The choice is made by comparing the x- or y-coordinate value of the center of the desired circle with the corresponding coordinate value at the point of tangency for each line and selecting that modifier which represents the relative value of the circle center coordinate with respect to the corresponding coordinate at the point of tangency.

C1 = CIRCLE/XSMALL, L1, XSMALL, L2, RADIUS, .7
C2 = CIRCLE/XLARGE, L1, XSMALL, L2, RADIUS, .7
C3 = CIRCLE/XLARGE, L1, XLARGE, L2, RADIUS, .7
C4 = CIRCLE/XSMALL, L1, XLARGE, L2, RADIUS, .7

Alternatively,
C1 = CIRCLE/YSMALL, L1, XSMALL, L2, RADIUS, .7
C1 = CIRCLE/XSMALL, L1, YLARGE, L2, RADIUS, .7
C1 = CIRCLE/YSMALL, L1, YLARGE, L2, RADIUS, .7

---

7 | As tangent to three lines:

$$CIRCLE/ \begin{Bmatrix} XLARGE \\ XSMALL \\ YLARGE \\ YSMALL \end{Bmatrix}, line1, \begin{Bmatrix} XLARGE \\ XSMALL \\ YLARGE \\ YSMALL \end{Bmatrix}, line2, \begin{Bmatrix} XLARGE \\ XSMALL \\ YLARGE \\ YSMALL \end{Bmatrix}, line3$$

The circle defined is tangent to *line1*, *line2*, and *line3*. The lines cannot all be parallel. Each modifier is associated with the immediately following tangent line and is chosen in a manner identical to that described for Circle-6.

C1 = CIRCLE/YLARGE, L1, YSMALL, L2, YLARGE, L3
  or
C1 = CIRCLE/YLARGE, L1, YSMALL, L2, XLARGE, L3

C2 = CIRCLE/YSMALL, L1, YSMALL, L2, YLARGE, L3
  or
C2 = CIRCLE/YSMALL, L1, YSMALL, L2, XLARGE, L3

Each circle may be described by any of six other combinations of modifiers.

---

8 | By its radius, a tangent line, and a point on the circumference:

$$CIRCLE/TANTO, line, \begin{Bmatrix} XLARGE \\ XSMALL \\ YLARGE \\ YSMALL \end{Bmatrix}, point, RADIUS, radius$$

The circle defined is tangent to *line* and passes through *point* with a radius value of *radius*. The modifier is chosen on the basis of the relative center with respect to the point. The choice is made by comparing the x- or y-coordinate value of the circle center with the corresponding coordinate value of the point and selecting that modifier which represents the relative value of the circle center coordinate with respect to the point coordinate.

C1 = CIRCLE/TANTO, L1, XSMALL, P1, R
C2 = CIRCLE/TANTO, L1, YLARGE, P1, R
C3 = CIRCLE/TANTO, L1, YLARGE, P2, R
C4 = CIRCLE/TANTO, L1, YSMALL, P2, R

---

9 | By its center and a tangent circle:

$$CIRCLE/CENTER, point, \begin{Bmatrix} LARGE \\ SMALL \end{Bmatrix}, TANTO, circle$$

The circle defined has its center at *point* and is tangent to *circle*. The center of the circle may be either inside or outside the tangent circle and unless the center points coincide, there are two possible circles, one larger than the other. The modifier LARGE means that the circle being defined has the larger of its two possible radii, while SMALL means that it has the smaller of the radii. The modifiers are <u>not</u> to be interpreted in terms of the relative sizes of the defined and tangent circles.

C1A = CIRCLE/CENTER, P1, SMALL, TANTO, C2
C1B = CIRCLE/CENTER, P1, LARGE, TANTO, C2
C3A = CIRCLE/CENTER, P3, SMALL, TANTO, C4
C3B = CIRCLE/CENTER, P3, LARGE, TANTO, C4

TABLE 3.3 (CONTINUED)

| 10 | By its radius, a point on the circumference, and tangent to a circle: |

$$CIRCLE/ \begin{Bmatrix} XLARGE \\ XSMALL \\ YLARGE \\ YSMALL \end{Bmatrix}, \begin{Bmatrix} RIGHT \\ LEFT \end{Bmatrix}, TANTO, circle, THRU, point, RADIUS, radius$$

The circle defined is tangent to *circle* and passes through *point* with a radius value of *radius*. RIGHT and LEFT are positional modifiers that specify the location of the center of the defined circle relative to a line connecting the point with the center of the given circle. The sense of direction is established by viewing along the line from the point to the given circle center. The other modifier is chosen on the basis of the location of the desired tangent point relative to the other possibility, if one exists. The choice is made by comparing the corresponding x- or y-coordinate values of the tangent points and selecting that modifier which conforms to the desired circle location. If only one possibility exists, the choice of the first modifier is immaterial.

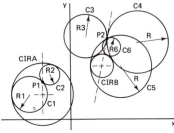

CIRA and CIRB are given circles.

C1 = CIRCLE/XSMALL, RIGHT, TANTO, CIRA, THRU, P1, RADIUS, R1 } First modifier immaterial
C2 = CIRCLE/XSMALL, LEFT, TANTO, CIRA, THRU, P1, RADIUS, R2
C3 = CIRCLE/YLARGE, RIGHT, TANTO, CIRB, THRU, P2, RADIUS, R3
C4 = CIRCLE/XLARGE, LEFT, TANTO, CIRB, THRU, P2, RADIUS, R } Same radius
C5 = CIRCLE/XSMALL, LEFT, TANTO, CIRB, THRU, P2, RADIUS, R
C6 = CIRCLE/XSMALL, LEFT, TANTO, CIRB, THRU, P2, RADIUS, R6 } First modifier immaterial

| 11 | By its radius, a tangent line, and a tangent circle: |

$$CIRCLE/ \begin{Bmatrix} XLARGE \\ XSMALL \\ YLARGE \\ YSMALL \end{Bmatrix}, line, \begin{Bmatrix} XLARGE \\ XSMALL \\ YLARGE \\ YSMALL \end{Bmatrix}, \begin{Bmatrix} IN \\ OUT \end{Bmatrix}, circle, RADIUS, radius$$

The circle defined is tangent to both *line* and *circle* and has a radius value of *radius*. The third modifier is applied first. IN is used if one circle is contained within the other, OUT if neither circle is within the other.

The first modifier describes the location of the center of the desired circle with respect to the tangent point of the line. The choice is made by comparing the x- or y-coordinate value of the center of the desired circle with the corresponding coordinate value of the point of tangency and selecting that modifier which represents the relative value of the circle center coordinate with respect to the tangency point coordinate.

The second modifier is applied last. It describes the location of the center of the desired circle with respect to the other remaining possible location. The choice is made by comparing the x- or y-coordinate value of the center of the desired circle with the corresponding coordinate value for the other location and selecting that modifier which represents the relative value of the desired circle center coordinate with respect to the other possibility.

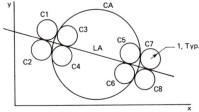

C1 = CIRCLE/YLARGE, LA, XSMALL, OUT, CA, RADIUS, 1
C2 = CIRCLE/YSMALL, LA, XSMALL, OUT, CA, RADIUS, 1
C3 = CIRCLE/YLARGE, LA, XSMALL, IN, CA, RADIUS, 1
C4 = CIRCLE/YSMALL, LA, XSMALL, IN, CA, RADIUS, 1
C5 = CIRCLE/YLARGE, LA, XLARGE, IN, CA, RADIUS, 1
C6 = CIRCLE/YSMALL, LA, XLARGE, IN, CA, RADIUS, 1
C7 = CIRCLE/YLARGE, LA, XLARGE, OUT, CA, RADIUS, 1
C8 = CIRCLE/YSMALL, LA, XLARGE, OUT, CA, RADIUS, 1

C1 = CIRCLE/YLARGE, LB, XSMALL, OUT, CB, RADIUS, 1
C2 = CIRCLE/YLARGE, LB, XSMALL, IN, CB, RADIUS, 1
C3 = CIRCLE/YLARGE, LB, XLARGE, IN, CB, RADIUS, 1
C4 = CIRCLE/YLARGE, LB, XLARGE, OUT, CB, RADIUS, 1

TABLE 3.3 (CONTINUED)

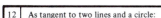

12	As tangent to two lines and a circle:

$$\text{CIRCLE/} \begin{Bmatrix} \text{XLARGE} \\ \text{XSMALL} \\ \text{YLARGE} \\ \text{YSMALL} \end{Bmatrix}, \text{TANTO}, line1, \begin{Bmatrix} \text{XLARGE} \\ \text{XSMALL} \\ \text{YLARGE} \\ \text{YSMALL} \end{Bmatrix}, \text{TANTO}, line2, \begin{Bmatrix} \text{XLARGE} \\ \text{XSMALL} \\ \text{YLARGE} \\ \text{YSMALL} \end{Bmatrix}, \begin{Bmatrix} \text{IN} \\ \text{OUT} \end{Bmatrix} \text{TANTO}, circle$$

The circle defined is tangent to *line1*, *line2*, and *circle*. Modifier IN is used if one circle is contained within the other, OUT if neither circle is within the other.

The first and second modifiers describe the location of the center of the desired circle with respect to the tangent point of the line associated with that modifier. The choices are made as described for Circle-11.

The third modifier is applied last. It describes the location of the center of the desired circle with respect to the other remaining possible location. The choice is made as described for Circle-11.

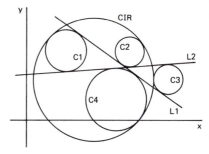

C1 = CIRCLE/YSMALL, TANTO, L1, YLARGE, TANTO, L2, XLARGE, IN, CIR
C2 = CIRCLE/YLARGE, TANTO, L1, YLARGE, TANTO, L2, XSMALL, IN, CIR
C3 = CIRCLE/YLARGE, TANTO, L1, YSMALL, TANTO, L2, XLARGE, OUT, CIR
C4 = CIRCLE/YSMALL, TANTO, L1, YSMALL, TANTO, L2, XSMALL, OUT, CIR

13	By its radius and two tangent circles:

$$\text{CIRCLE/} \begin{Bmatrix} \text{XLARGE} \\ \text{XSMALL} \\ \text{YLARGE} \\ \text{YSMALL} \end{Bmatrix}, \begin{Bmatrix} \text{IN} \\ \text{OUT} \end{Bmatrix}, circle1, \begin{Bmatrix} \text{IN} \\ \text{OUT} \end{Bmatrix}, circle2, \text{RADIUS}, radius$$

The circle defined is tangent to *circle1* and *circle2* and has a radius value of *radius*. Modifier IN is used if one circle is contained within the other, OUT if neither circle is within the other. The comparison is made between the circle being defined and the circle immediately following the modifier being chosen.

The first modifier is chosen after the IN-OUT modifiers have been assigned. It describes the location of the center of the desired circle with respect to the other possible location. The choice is made by comparing the x- or y-coordinate value of the center of the desired circle with the corresponding coordinate value for the other location and selecting that modifier which represents the relative value of the desired circle center coordinate with respect to the other possibility.

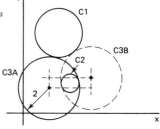

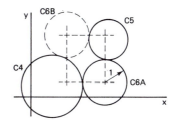

C3A = CIRCLE/XSMALL, OUT, C1, IN, C2, RADIUS, 2
    or
C3A = CIRCLE/YSMALL, OUT, C1, IN, C2, RADIUS, 2
C6A = CIRCLE/XLARGE, OUT, C4, OUT, C5, RADIUS, 1
    or
C6A = CIRCLE/YSMALL, OUT, C4, OUT, C5, RADIUS, 1
C9 = CIRCLE/YLARGE, IN, C7, IN, C8, RADIUS, .5

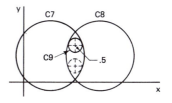

TABLE 3.3 (CONTINUED)

| 14 | By its radius, a tangent line, and a tangent tabulated cylinder: |

$$\text{CIRCLE/TANTO, } line, \left\{ \begin{array}{l} \text{XLARGE} \\ \text{XSMALL} \\ \text{YLARGE} \\ \text{YSMALL} \end{array} \right\}, tabcyl, \left\{ \begin{array}{l} \text{XLARGE} \\ \text{XSMALL} \\ \text{YLARGE} \\ \text{YSMALL} \end{array} \right\}, point, \text{RADIUS, } radius$$

The circle is defined as tangent to *line* and *tabcyl* with a radius value of *radius*. Since there may be more than one point of tangency, the desired location is approximately identified by *point*, one of the points used to define the tabulated cylinder and which appears in the tabulated cylinder definition just prior to the point of tangency. This point is used as a "near" point for selecting the desired portion of the tabulated cylinder from among other possible portions.

The modifiers describe the location of the center of the circle with respect to the tangent point of the line and tabcyl associated with that modifier. The choices are made as described for Circle-11.

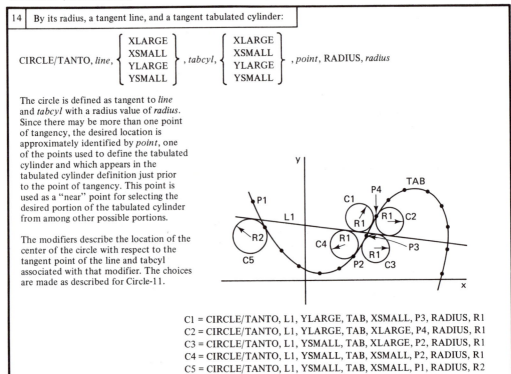

C1 = CIRCLE/TANTO, L1, YLARGE, TAB, XSMALL, P3, RADIUS, R1
C2 = CIRCLE/TANTO, L1, YLARGE, TAB, XLARGE, P4, RADIUS, R1
C3 = CIRCLE/TANTO, L1, YSMALL, TAB, XLARGE, P2, RADIUS, R1
C4 = CIRCLE/TANTO, L1, YSMALL, TAB, XSMALL, P2, RADIUS, R1
C5 = CIRCLE/TANTO, L1, YSMALL, TAB, XSMALL, P1, RADIUS, R2

(c) LAA = LINE/PTK,RIGHT,TANTO,(CIRCLE/CENTER,(POINT/−3.2,4.7), $
TANTO,L12)

This example uses two levels of nested unnamed definitions. It also shows selective use of nesting, that is, all symbolic names need not be replaced by nested definitions. The use of nested definitions often requires that a statement be continued on another line, hence the need for the continuation symbol $.

(d) CIR5 = CIRCLE/YSMALL,LINEA,XSMALL,OUT,(CIRCLE/TANTO, $
(LINEA=LINE/P1,P2),YSMALL,P1,RADIUS,2),RADIUS,1

This example shows a combination of named and unnamed nested definitions in one statement. Named nested definitions may be referenced within the same statement provided that the order of evaluation is observed so that the name is defined before it is referenced. Nests are evaluated first with evaluation proceeding from the innermost nest outward. For this example, the nest defining LINEA is evaluated first (since it is the innermost

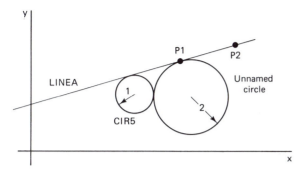

Figure 3.1   Geometry for example (d) of Section 3.5.

nest), followed by the unnamed circle nest, and finally by CIR5. Nests on the same level are evaluated from right to left, thus permitting a symbol to be defined in one nest and referenced subsequently in another that occurs to the left of the defining one. The following statement is incorrectly constructed since LINEA is defined in a nest that is evaluated after another which references it.

CIR5 = CIRCLE/YSMALL,(LINEA = LINE/P1,P2),XSMALL,OUT, $
          (CIRCLE/TANTO,LINEA,YSMALL,P1,RADIUS,2),RADIUS,1

The geometry of the correct statement is shown in Figure 3.1.

(e) L3 = LINE/LEFT,TANTO,(CIRCLE/CENTER,P1,TANTO, $
          (LINE/RIGHT,TANTO,(CIRCLE/TANTO,(LINE/ $
          (P1 = POINT/1,1,0),ATANGL, − 30),YSMALL,P1, $
          RADIUS,.5),LEFT,TANTO,C2)),LEFT,TANTO,C2

This example contains five levels of nests. P1 is defined within the fifth level (innermost) nest and referenced in the first- and third-level nests. This example also shows that the order of statement evaluation is modified by the presence of nests. That P1 can be referenced seemingly both before and after its definition is a direct consequence of this evaluation order. The geometry is shown in Figure 3.2. Here is a sequential listing of the definitions:

P1 = POINT/1,1,0
L1 = LINE/P1,ATANGL, − 30
C1 = CIRCLE/TANTO,L1,YSMALL,P1,RADIUS,.5
L2 = LINE/RIGHT,TANTO,C1,LEFT,TANTO,C2
C3 = CIRCLE/CENTER,P1,TANTO,L2
L3 = LINE/LEFT,TANTO,C3,LEFT,TANTO,C2

(f) FROM/(SETPT = POINT/ − 1, − 1,1)
INDIRP/(POINT/1,0,0)
GO/TO,(LI = LINE/P1,RIGHT,TANTO,(C1 = CIRCLE/CENTER,P2, $
    RADIUS,1.5))

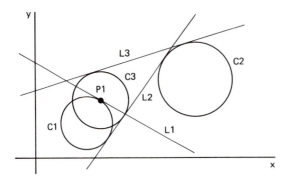

Figure 3.2   Geometry for example (e) of Section 3.5.

This example shows the use of nested definitions in motion commands.

There are pros and cons relative to the use of nesting. The following recommendations pertain to its use.

1. Nesting is to be avoided when used simply to reduce the number of separate statements. The examples above show that statement clarity is reduced with nesting and that complexity can lead to incorrect geometry description. Cleverness in constructing a complex statement is no consolation when difficulty is encountered in debugging the program.

2. Geometric entities cannot be redefined and reassigned to the same symbol with any of the geometric definitions presented thus far. There are legitimate needs for redefining the geometry, such as described in Chapter 4. The alternatives include use of the CANON definition as described in Section 3.7 and unnamed nested redefinition. When required, nesting should be used with discretion and to a depth consistent with statement clarity.

3.6 CONSTRUCTING SIMPLE GEOMETRIC STATEMENTS

The principal purpose of the geometric definitions is to describe the shape of the part for tool motion command reference. Also, they may be used for intermediate tool positioning to establish a tool-to-part relationship for subsequent moves. The discussion of this section is limited to their construction for the principal purpose. The other purpose will be discussed in conjunction with parts of more complex geometry and where additional APT features are introduced. One should realize that the shape of the part may not be visualized from geometric definitions alone; the motion commands must also be considered.

The APT part programming effort culminates in the computation of a set of cut vectors which, when converted to a form suitable for a given machine, will result in production of the desired part. Among the factors that contribute to the success of this effort is the care taken to develop the geometric definitions and motion commands. It is recommended that the part programmer follow an organized procedure while preparing geometric definitions, such as the following:

1. Determine from the drawing or sketch, or by some other means, the important or critical part features, such as dimensions, surfaces, tolerances, and so on.
2. Choose an appropriate three-dimensional rectangular coordinate system consistent with the considerations of step 1.
3. Symbolically label obvious part surfaces (curves) and points for reference in geometric definitions and motion commands.
4. Write the geometric definitions necessary to describe the part.
5. Establish and define the initial and final tool locations.

Let us now prepare geometric definitions for the part of Figure 3.3. It consists of flat stock with a slot and two holes. First, assume the following critical for tolerance specification:

1. The diameters of the two holes
2. The flats of the slots and their parallel alignment

Next, observe that the following are also critical:

1. The spacing between the two holes
2. The angular separation of the upper hole from the slot
3. The flats of the slot with respect to their spacing and alignment parallel with and symmetrical about a line connecting their centerline with the center of the lower hole

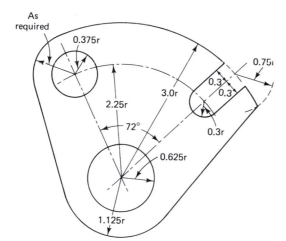

Figure 3.3   Example part for geometric statement construction.

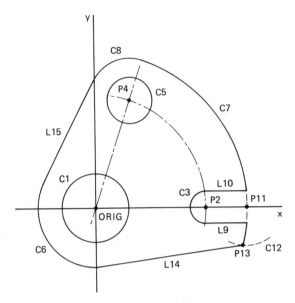

Figure 3.4   Coordinate system placement and symbolic labeling of geometric entities.

The coordinate system will now be chosen. First, we can place the part such that its bottom surface is on the *xy*-plane (actually on rest buttons). This places the positive *z*-axis above the part, as viewed in the drawing. Next, to maintain angular alignment of the upper hole with the slot, we select the coordinate system origin at the center of the lower hole. If the part is now rotated such that the centerline of the slot is on the *x*-axis, the flats will be parallel with the positive *x*-axis and the upper hole will be inclined as shown in Figure 3.4, where all drive surfaces have been labeled and the symbolically labeled part placed on the *xy*-plane. Other labels will be added to aid geometric statement construction. The drawing of Figure 3.3 incorporates the usual drafting assumptions of parallel or perpendicular straight lines unless otherwise specified and has been prepared with APT in mind. Note that the radius of the upper left part periphery is unspecified, as are the lengths of certain straight lines and dimensions of other intersecting curves.

We concentrate on the important considerations mentioned above, remembering that the APT system sequentially processes the statements and that a symbol must be defined before it is referenced. The following is a suitable set of geometric definitions.

ORIG =	POINT/0,0	100
C1 =	CIRCLE/CENTER,ORIG,RADIUS,.625	110
P2 =	POINT/RTHETA,XYPLAN,2.25,0	120
C3 =	CIRCLE/CENTER,P2,RADIUS,.3	130
P4 =	POINT/RTHETA,XYPLAN,2.25,72	140
C5 =	CIRCLE/CENTER,P4,RADIUS,.375	150
C6 =	CIRCLE/CENTER,ORIG,RADIUS,1.125	160
C7 =	CIRCLE/CENTER,ORIG,RADIUS,3	170
C8 =	CIRCLE/CENTER,P4,SMALL,TANTO,C7	180

L9	= LINE/YSMALL,TANTO,C3,ATANGL,0	190
L10	= LINE/YLARGE,TANTO,C3,ATANGL,0	200
P11	= POINT/3,0	210
C12	= CIRCLE/CENTER,P11,RADIUS,.75	220
P13	= POINT/YSMALL,INTOF,C7,C12	230
L14	= LINE/P13,LEFT,TANTO,C6	240
L15	= LINE/LEFT,TANTO,C6,LEFT,TANTO,C8	250

The angular separation of the upper hole and the slot were defined in lines 100 through 150. This required locating hole C3 at point P2 and hole C5 at point P4. Circles C6, C7, and C8 were then defined followed by the flats of the slot, L9 and L10. Finally, geometric construction was required to locate the intersection of L14 with C7. This required that P11, C12, and P13 be defined.

The method used to define L9 and L10 parallels the flats of the slot with the slot centerline and ties the slot width to the radius specification of C3, which is centered on the slot centerline. The critical aspects of the problem have been taken care of.

The definitions given above are not unique. For example, we could have defined circle C3 by the coordinates of its center (2.25, 0) and by its radius. Similarly, we could have located P11 by specifying its coordinates and we could have used this technique in several other instances. However, except for P2 and P4, each given dimension of the problem appears in only one definition. Thus, if this part were to differ from another only in the size of one or more dimensions, only a minimum number of defining statements would require change. Those definitions that do not include dimensions are dependent on symbols defined in other statements.

Advantages sometimes result from the selection of certain defining formats over others. At other times we may be able merely to simplify the statements by choosing from equivalent defining formats. For this problem we could have defined L9 and L10 by using the line format for locating a line parallel with another but offset a given distance. Of course this means that we must include dimensions in two additional statements and the part program would not be as flexible with respect to dimension changes.

3.7 THE CANONICAL FORM

We have seen a variety of formats by which points, lines, and circles may be defined. One of the functions of the APT processor is to transform each instance of an entity in a part program into a compact internal representation of data storage for ease in computer processing. This internal representation is known as the **canonical form**. Various entities, including matrices, vectors, and all geometric surfaces, are stored in their canonical form, which is usually related to an analytic representation of the surface. Surface in this context includes matrices, vectors, and so on.

The format of the canonical form varies from surface to surface and includes information for processor use in addition to the analytic representation of the surface. The former is not required by the part programmer and will not be discussed further. The implication of the canonical form is that each instance of a given surface is reduced to the same internal representation, thus its defining identity is lost. For example, every point can be represented by its three coordinate values in the Cartesian coordinate system regardless of how it was defined. Similarly, each line can be represented by the coefficients of its equation regardless of how it was defined. By knowing the canonical form, it is possible to define and to redefine an instance of a surface with the CANON statement.

3.7.1 The CANON Statement

The general form of the CANON statement conforms with that of the general form described in Section 3.1. More specifically, we have

$$symbol1 = surface/[symbol2,]CANON, list$$

where *symbol1* is the name of the instance of the surface being defined, *surface* is the type of entity, *symbol2* is the name of a previously defined entity of type *surface*, and *list* is a list of canonical parameters corresponding to *surface*. The definition of the parameters and their left-to-right ordering as they appear in *list* for POINT, LINE, and CIRCLE surfaces is given in Table 3.4.

By omitting *symbol2*, since it is optional, we are able to define a new symbol by specifying the parameters of the canonical form for *surface*. When *symbol1* is the same as *symbol2*, we are able to redefine a symbol. When *symbol1* differs from *symbol2*, we are able to define a new symbol as equivalent in many respects to another. *Symbol2* may be a nested definition.

A printout of the canonical form parameters for a surface often appears on the CL-printout. Such is the case for circles in Example 2.1, where the seven parameters are

TABLE 3.4 POINT, LINE, AND CIRCLE CANONICAL DEFINITIONS

Surface	Parameter		Definition
Point	1	x	Cartesian coordinates of the point in the given reference system
	2	y	
	3	z	
Line	1	A	Coefficients of the plane equation $Ax + By + Cz - D = 0$,
	2	B	where (A, B, C) are for a normalized vector (see Chapter 8);
	3	$C = 0$	in APT, a line is a plane normal to the xy-plane; therefore, C
	4	D	$= 0$
Circle	1	x	Cartesian coordinates of the center point
	2	y	
	3	z	
	4	$A = 0$	Coefficients for the vector of the circle axis; in APT a circle is
	5	$B = 0$	a right circular cylinder whose axis is normal to the xy-plane;
	6	$C = 1$	therefore, $(A, B, C) = (0, 0, 1)$
	7	r	Radius value

printed in two lines just prior to the cut vectors resulting from motion statements along a circle (see Figure 2.4).

3.7.2 Defining a New Symbol: Method 1

For this method, we use the following statement format:

$symbol$ = $surface$/CANON,p_1,p_2, . . . ,p_n

Symbol is the name of the new surface. Each parameter p_i must be supplied as required for the canonical definition of *surface*. The parameters are scalar quantities supplied as constants, variables, or expressions (see Section 4.1.1).

Examples:

P1 = POINT/CANON, − 2,4,3

The point is at x = − 2, y = 4, z = 3. This definition is the same as Point-1 in Table 3.1.

L1 = LINE/CANON,0,1,0,5

The line is parallel with the *x*-axis and crosses the *y*-axis at y = 5.

L2 = LINE/CANON, − .5,.866,0,0

The line passes through the origin at an angle of + 30° to the *x*-axis.

C1 = CIRCLE/CANON,2, − 3,0,0,0,1,6

The circle is located at the point (2, − 3, 0) and has a radius of 6.

3.7.3 Defining a New Symbol: Method 2

For this method, we use the following statement format:

$symbol1$ = $surface$/$symbol2$,CANON,$[p_1]$,$[p_2]$, . . . ,$[p_n]$

Symbol1 is the name of the new surface, *symbol2* is the name of a previously defined surface of type *surface*, and parameter p_i is supplied as required for the canonical definition of *surface* when its value for *symbol1* differs from that for *symbol2*. See Section 3.7.2 for parameter specification.

For this method, a copy of the canonical definition of *symbol2* provides a starting point from which to define *symbol1*. The definition of *symbol2* is unchanged. The parameter values of *symbol2* are used to define *symbol1* except for those p_i supplied in this definition. The replacements for the p_i are made in positions i in the canonical form for *symbol1*.

The list of parameters is position sensitive. The comma associated with a parameter must not be omitted when the parameter is omitted so that position i may be determined. That is, successive commas are needed when successive parameters are omitted. The exception to this rule is that trailing commas need not follow the last parameter being supplied.

Examples:

P1 = POINT/P5,CANON,,,4

P1 is a new point. Presumably, its coordinate value $z = 4$ differs from that of P5.

L1 = LINE/L3,CANON,.707,.707

L1 is a new line. Its slope is $+135°$ (or $-45°$) and it is the same distance from the origin as is L3. Trailing commas are not required.

C1 = CIRCLE/C4,CANON,,,,,,6

C1 is a new circle. It is concentric with C4 but has a radius value of 6.

3.7.4 Redefining a Symbol

The CANON statement provides a means by which the symbol for a surface may be redefined. Although a given APT system may have exceptions, attempts to redefine a symbol by other means produce the multiple definition error diagnostic. The format for this statement is

symbol = *surface/symbol*,CANON,$[p_1],[p_2], \ldots ,[p_n]$

where *symbol* is the name of a previously defined surface of type *surface* being redefined and parameter p_i is supplied as required for the canonical definition of *surface* according to redefinition needs. See Section 3.7.2 for parameter specification.

The list of parameters is position sensitive. Only those parameters supplied in this statement are used to replace parameters in the corresponding positions of the symbol already defined. The comma associated with a parameter must not be omitted when the parameter is omitted, so that position i may be determined. That is, successive commas are needed when successive parameters are omitted. The exception to this rule is that trailing commas need not follow the last parameter being supplied.

Examples:

P1 = POINT/P1,CANON,,-4

P1 is redefined for a new coordinate value. Presumably, its coordinate value $y = -4$ differs from its previous value.

L1 = LINE/L1,CANON,,,,7

L1 is redefined to be parallel with its former self but now to be 7 dimension units from the origin.

C1 = CIRCLE/C1,CANON,3,4,5

C1 is now redefined to be located at the point (3, 4, 5). Its radius is unchanged.

Once a symbol has been defined, its surface type may never be altered. However, its parameter values may be changed as often as desired by redefining as described above. Redefinition is often required within a loop (Chapter 4) or within a macro (Chapter 11).

Instead of using the format above to redefine a symbol, and also to avoid the multiple definition error diagnostic, the APT processor mode can be changed to allow redefinition with the CANON/ON statement. While in this mode, all subsequent geometric definitions will not be subject to multiple definition error checking. The CANON/OFF statement will return the mode to the error-checking condition. In the following example, P1 is defined initially in line 100. It is legitimately redefined in lines 110 and 130. However, its redefinition in line 150 will produce an error diagnostic.

P1 = POINT/1,1,1	100
P1 = POINT/P1,CANON,,−1	110
CANON/ON	120
P1 = POINT/2,2,2	130
CANON/OFF	140
P1 = POINT/3,3,3	150

Use of the CANON/ON statement is not recommended because it bypasses multiple definition error checking.

3.7.5 Precautions for Using the CANON Statement

Since the CANON statement allows for redefining a symbol without causing a multiple definition error diagnostic, the potential exists for difficulty in debugging a part program. Specifics in verifying a part program are given in Section 3.8. Until then, these comments should be carefully considered when using CANON statements:

1. Placement of values in incorrect parameter positions can occur inadvertently. This error can occur when several consecutive commas precede the desired parameter position but when an incorrect number of commas was entered. The APT system has no way of knowing that an error was made, thus no diagnostic is issued. To help avoid such errors, it is necessary to become thoroughly familiar with the canonical form for the type of surface for which the CANON statement is to be used, to carefully sight check each CANON statement appearing in the part program listing, and to reconsider the need for it by possibly restructuring the program.

2. Incorrect *symbol1* and/or *symbol2* entries may arise from carelessness in labeling geometric figures, from incorrect keyboard entry of characters, and/or from inadvertently interchanging symbols in the CANON statement. As long as the symbols are legitimate, the APT system cannot know that an error was made. Carefully preparing the part program as mentioned above is suggested.

3.8 PROGRAM DEBUGGING: EMPHASIS ON GEOMETRIC DEFINITIONS

The material in this chapter, together with that of Chapter 2, allows us to write complete APT part programs, simple though they may be. But we should not forget the discussion of Section 2.7 on debugging. We now supplement that discussion by addressing the problem as it applies to geometric definitions.

3.8.1 Preliminaries

The basic philosophy of making the part program understandable is continued. However, it becomes more important as the part programmer assumes responsibility for writing the entire program. Although the part programmer may be ingenious in preparing the program statements, good or bad they are displayed for all to see. We will suggest ways for making the program statements ''good.''

The part programmer must select symbolic names for the geometric entities. They should be chosen according to a naming convention that allows them to be readily associated with a particular geometric entity. For example, the names of all points could begin with the same prefix but differ in suffix only, which could be numerically sequenced. All lines and circles could be named in a similar manner. Always annotate the drawing with the symbol names.

Select ''natural'' geometric definitions; those that conform with the geometry being described and that fit existing definition formats without making additional computations. For example, if tangent conditions exist, use the formats that exploit tangent conditions. If dimensions are in terms of polar coordinates, select those formats when available. Do not hide geometric entities or remove them from their physical relationship to other entities of the part by limiting the selection of formats.

Limit the use of unnamed nested definitions. They are useful as single occurrences in geometrical constructions supporting the development of definitions for the principal geometry and for the applications described in Chapter 4.

3.8.2 Debugging

The following suggestions are extensions to those presented in Section 2.7.2.

Compilation-Time Errors

1. Be sure that symbols are defined before they are referenced. This implies an ordering of the statements with possible impact on the naming convention.
2. Secondary errors often arise from one definition error where the symbol, now rejected by the APT system, is not defined for use in subsequent definitions. This problem is analogous to that of statement ordering.
3. Impossible mathematical situations may be encountered. There may be a negative

radius, nonintersecting lines in a point definition, three collinear points in a circle definition, or the wrong symbol may have been referenced. The latter condition often occurs because of multiply defined symbols which, when corrected, did not have subsequent references changed to the new symbol.

4. Reserved words may have been used for symbol names. Such words must be avoided for this purpose. This is easily done by following a suitable naming convention.

Execution-Time Errors

1. Ambiguous modifiers may have been selected (e.g., XLARGE when XSMALL was required). The geometry is now incorrect, but not inconsistent for definition purposes, so errors relating to improper check surfaces or the tool not within tolerance of drive surfaces may be produced.

2. Do not bypass program safeguards during execution time. Use the CANON statement with care. Avoid it if possible to retain geometric definition error reporting capability.

3.8.3 Examples of Errors in Geometric Definitions

Figure 3.5 shows the geometric definitions for Figure 3.4 with errors deliberately inserted to force APT Section 1 diagnostics. These printouts were produced by the IBM APT-AC system. Similar messages will be produced by other APT processors. The print option was set for full diagnostic message printout.

In part A, RTHETA was misspelled in ISN 9. The error message follows the statement in error and is quite clear. The severity of the error (condition code 4) was not serious so processing continued, but Section 2 processing will be inhibited. This error

```
ISN 00005 ORIG    = POINT/0,0                              00000500
ISN 00006 C1      = CIRCLE/CENTER,ORIG,RADIUS,.625         00000600
ISN 00007 P2      = POINT/RTHETA,XYPLAN,2.25,0             00000610
ISN 00008 C3      = CIRCLE/CENTER,P2,RADIUS,.3             00000700
ISN 00009 P4      = POINT/RTHTA,XYPLAN,2.25,72             00000800

** ERROR 01082. COND. CODE 04.. ISN 00009.. LABEL  (NONE).. SEQ. NO. 00000800 *
****** SPECIFIED STATEMENT REFERENCES UNDEFINED VARIABLE SYMBOL... RTHTA ******

ISN 00010 C5      = CIRCLE/CENTER,P4,RADIUS,.375           00000900
ISN 00011 C6      = CIRCLE/CENTER,ORIG,RADIUS,1.125        00001000
ISN 00012 C7      = CIRCLE/CENTER,ORIG,RADIUS,3            00001100
ISN 00013 C8      = CIRCLE/CENTER,P4,SMALL,TANTO,C7        00001200
ISN 00014 L9      = LINE/YSMALL,TANTO,C3,ATANGL,0          00001400
ISN 00015 L10     = LINE/YLARGE,TANTO,C3,ATANGL,0          00001500
ISN 00016 P11     = POINT/3,0                              00001501
ISN 00017 C12     = CIRCLE/CENTER,P11,RADIUS,.75           00001502
ISN 00018 P13     = POINT/YSMALL,INTOF,C7,C12              00001510
ISN 00019 L14     = LINE/P13,LEFT,TANTO,C6                 00001600
ISN 00020 L15     = LINE/LEFT,TANTO,C6,LEFT,TANTO,C8       00001700
```

Ⓐ

Figure 3.5   Examples of diagnostics produced by errors in geometric definitions.

```
ISN 00005 ORIG      = POINT/0,0                                        00000500
ISN 00006 C1        = CIRCLE/CENTER,ORIG,RADIUS,.625                    00000600
ISN 00007 P2        = POINT/RTHETA,XYPLAN,2.25,0                        00000610
ISN 00008 C3        = CIRCLE/CENTER,P2,RADIUS,.3                        00000700
ISN 00009 P4        = POINT/RTHETA,XYPLAN,2.25,72                       00000800
ISN 00010 C5        = CIRCLE/CENTER,P4,RADIUS,.375                      00000900
ISN 00011 C6        = CIRCLE/CENTER,ORIG,RADIUS,1.125                   00001000
ISN 00012 C7        = CIRCLE/CENTER,ORIG,RADIUS,3                       00001100
ISN 00013 C8        = CIRCLE/CENTER,P4,SMALL,TANTO,C7                   00001200
ISN 00014 L9        = LINE/YSMALL,TANTO,C3,ATANGL,0                     00001400
ISN 00015 L10       = LINE/YLARGE,TANTO,C3,ATANGL,0                     00001500
ISN 00016 P11       = POINT/3,0                                         00001501
ISN 00017 C12       = CIRCLE/CENTER,P11,RADIUS, 7.5                     00001502
ISN 00018 P13       = POINT/YSMALL,INTOF,C7,C12                         00001510
```
                                                                    (B)

```
** ERROR 01092. COND. CODE 04.. ISN 00018.. LABEL  (NONE).. SEQ. NO. 00001510 *
****** INVALID 'POINT' DEF... INTERSECTION OF TWO CIRCLES DOES NOT EXIST ******
```

```
ISN 00019 L14       = LINE/P13,LEFT,TANTO,C6                            00001600
ISN 00020 L15       = LINE/LEFT,TANTO,C6,LEFT,TANTO,C8                  00001700
```

```
ISN 00005 ORIG      = POINT/0,0                                        0000050L
ISN 00006 C1        = CIRCLE/CENTER,ORIG,RADIUS,.625                    00000600
ISN 00007 P2        = POINT/RTHETA,XYPLAN,2.25,0                        00000610
ISN 00008 C3        = CIRCLE/CENTER,P2,RADIUS,.3                        00000700
ISN 00009 P4        = POINT/RTHETA,XYPLAN,2.25,72                       00000800
ISN 00010 C5        = CIRCLE/CENTER,P4,RADIUS,.375                      00000900
ISN 00011 C6        = CIRCLE/CENTER,ORIG,RADIUS,1.125                   00001000
ISN 00012 C7        = CIRCLE/CENTER,ORIG,RADIUS,3                       00001100
ISN 00013 C8        = CIRCLE/CENTER,P4,SMALL,TANTO,C7                   00001200
ISN 00014 L9        = LINE/YSMALL,TANTO,C3,ATANGL,0                     00001400
ISN 00015 L10       = LINE/YLARGE,TANTO,C3,ATANGL,0                     00001500
ISN 00016 P11       = POINT/3,0                                         00001501
ISN 00017 C12       = CIRCLE/CENTER,P11,RADIUS,.75                      00001502
ISN 00018 P13       = POINT/YSMALL,INTOF,C7,C12                         00001510
ISN 00019 L14       = LINE/P13, RIGHT ,TANTO,C6                         00001600
ISN 00020 L15       = LINE/LEFT,TANTO,C6,LEFT,TANTO,C8                  00001700
```
                                                                    (C)

```
ISN 00005 ORIG      = POINT/0,0                                        00000500
ISN 00006 C1        = CIRCLE/CENTER,ORIG,RADIUS,.625                    00000600
ISN 00007 P2        = POINT/RTHETA,XYPLAN,2.25,0                        00000610
ISN 00008 CS        = CIRCLE/CENTER,P2,RADIUS,.3                        00000700
```

```
** ERROR 01030. COND. CODE 08.. ISN 00008.. LABEL  (NONE).. SEQ. NO. 00000700 *
**** ILLEGAL ELEMENT LEFT OF AN '=' SIGN. VARIABLE SYMBOL ONLY PERMISSIBLE. ***
```

```
ISN 00009 P4        = POINT/RTHETA,XYPLAN,2.25,72                       00000800
ISN 00010 C5        = CIRCLE/CENTER,P4,RADIUS,.375                      00000900
ISN 00011 C6        = CIRCLE/CENTER,ORIG,RADIUS,1.125                   00001000
ISN 00012 C7        = CIRCLE/CENTER,ORIG,RADIUS,3                       00001100
ISN 00013 C8        = CIRCLE/CENTER,P4,SMALL,TANTO,C7                   00001200
ISN 00014 L9        = LINE/YSMALL,TANTO,C3,ATANGL,0                     00001400
```
                                                                    (D)

```
** ERROR 01082. COND. CODE 04.. ISN 00014.. LABEL  (NONE).. SEQ. NO. 00001400 *
******* SPECIFIED STATEMENT REFERENCES UNDEFINED VARIABLE SYMBOL... C3 ********
```

```
ISN 00015 L10       = LINE/YLARGE,TANTO,C3,ATANGL,0                     00001500
```

```
** ERROR 01082. COND. CODE 04.. ISN 00015.. LABEL  (NONE).. SEQ. NO. 00001500 *
****** SPECIFIED STATEMENT REFERENCES UNDEFINED VARIABLE SYMBOL... C3 ********
```

```
ISN 00016 P11       = POINT/3,0                                         00001501
ISN 00017 C12       = CIRCLE/CENTER,P11,RADIUS,.75                      00001502
ISN 00018 P13       = POINT/YSMALL,INTOF,C7,C12                         00001510
ISN 00019 L14       = LINE/P13,LEFT,TANTO,C6                            00001600
ISN 00020 L15       = LINE/LEFT,TANTO,C6,LEFT,TANTO,C8                  00001700
```

Figure 3.5 (Continued)

did not generate secondary errors by the APT processor. Since P4 is now undefined, it would not be surprising if some APT processors reported secondary errors for C5, C8, and L15.

In part B the radius of C12 was given as 7.5, rather than .75. The APT system could not know of this logical error. However, it could not define P13 since now there is no intersection of C7 with C12.

The error in part C (RIGHT was specified rather than LEFT) does not cause a diagnostic and processing continues. This error will have to be found during part program verification.

The primary error in part D in ISN 8 (C3 is misspelled as CS) generated two secondary errors (ISNs 14 and 15). The primary error (CS is a reserved word) is a serious one (condition code 8). The secondary errors are not really errors at all and these messages will disappear when ISN 8 is corrected.

PROBLEMS

3.1 Figure P3.1 shows a collection of points, lines, and circles that define the periphery of a part shown by the heavy outline. Write the APT geometric definitions to define the part and the motion commands to take a ½-in.-diameter tool counterclockwise outside the part outline shown. Begin by establishing a coordinate system, defining a set point, and with the scaled grid, define points P1, P2, and P3 (with coordinates estimated to the nearest ¼-inch). Thereafter, use APT geometric definition formats and no more information from the figure than the values of the radii given, the angle given, and the fact that various tangency conditions exist between the lines and circles as shown in the figure.

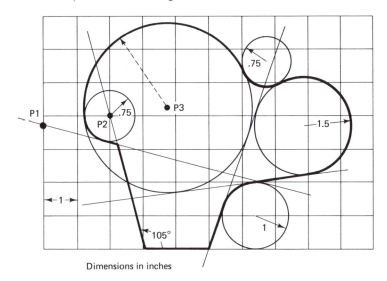

Dimensions in inches

Figure P3.1

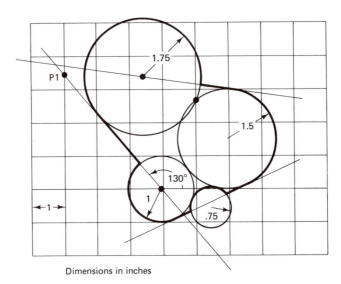

Dimensions in inches

Figure P3.2

3.2 The part outline of Figure P3.2 is determined by the various geometric figures shown. Write the APT geometric definitions to define the part and the motion commands to take a 0.375-in.-diameter tool counterclockwise outside the part outline shown. Begin by establishing a coordinate system, defining a set point, and defining the point P1 (with coordinates estimated to the nearest ¼-inch). Thereafter, use APT geometric definition formats and no more information from the figure than the values of the radii given, the angle given, and the fact that various tangency and intersection conditions exist between the lines and circles as shown in the figure.

3.3 The heavy outline shown in Figure P3.3 is the part periphery determined by the geometric figures shown. Write the APT geometric definitions to define the part and the motion commands to take a 0.25-in.-diameter tool counterclockwise outside the part outline shown. Begin by establishing a coordinate system in accordance with the location of the part with respect to the scaled grid, define a set point, and locate the 2-inch radius circle (with coordinates estimated to the nearest ¼-inch). Thereafter, use APT geometric definition formats and no more information from the figure than the values of the radii given and the fact that various intersection, perpendicularity, tangency, and parallelism conditions exist between the lines and circles as shown in the figure.

3.4 Figure P3.4 shows a part outline as determined by the points, lines, and circles shown. Write an APT part program that includes the geometric definitions and motion commands to take a 0.25-in.-diameter tool clockwise around the part. Begin by establishing a coordinate system, defining a set point, and with the scaled grid, define the points P1 and P2 (with coordinates estimated to the nearest ¼-inch). Thereafter, use APT geometric definition formats and no more information from the figure than the values of the radii given and the fact that various intersection, tangency, and parallelism conditions exist between the lines and circles as shown in the figure.

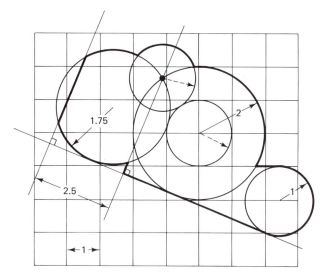

Dimensions in inches

Figure P3.3

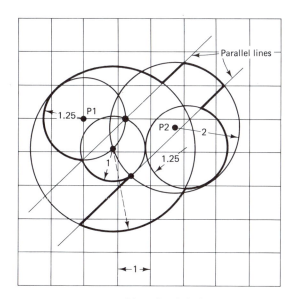

Dimensions in inches

Figure P3.4

CHAPTER 4

Computation, Decisions, and Looping

The material presented thus far restricts the part programmer to determining all constants, geometric definitions, and motion commands completely for a part prior to the time the part program is prepared and executed. Thus all computation and decision making must be performed by the programmer. Because APT uses the computer's arithmetic capability to compute cut vectors, this capability is also made available to the part programmer for calculations to support geometric definitions and tool motion commands. The manner in which the arithmetic feature can be used is discussed in this chapter.

We also include a discussion of the decision-making and looping capability of APT. These features are often combined with the arithmetic feature and together they greatly extend the power of the APT language. These features free the part programmer from attention to arithmetic details. Full attention can then be directed to the logic of the part program. These features allow the development of general-purpose part programs that otherwise would be restricted to a part with a single application.

4.1 COMPUTATION

The computation of cut vectors is aided by the geometric definitions feature since the part programmer can symbolically define the geometry and direct the tool according to the part profile, the cut vectors being automatically computed by the APT system. Sometimes, the geometry is such that the part programmer must compute some coordinate values, angles, radii, matrix elements, and other dimensions and values. For these purposes, the part programmer may use the computational features of the APT language.

4.1.1 Arithmetic Assignment Statement

Computation in APT is formalized with the arithmetic assignment statement, which in its most general form, is as follows:

[Statement label)]   scalar variable = arithmetic expression

The purpose of the statement label is discussed in Section 4.2.2. The format of this statement differs from those for geometric definitions and motion commands in that there are no major or minor sections.

The arithmetic expression represents a mathematical functional relationship between operands (constants, scalar variables, arithmetic functions, or other arithmetic expressions), either individually or in combination. This relationship is indicated by combining the operands with arithmetic operators to form the expression. When the arithmetic assignment statement is executed, the value of the arithmetic expression is computed and assigned as the value of the variable appearing on the left-hand side of the equals sign. One can also interpret the statement to mean that the value of the variable is to be replaced by the value of the arithmetic expression. The equals sign as used here does not mean equality in the usual mathematical sense but signifies an assignment or replacement operation.

Constants

Constants (numeric literals) are represented in arithmetic assignment statements as described in Section 3.1.2.

Scalar Variables

Scalar variables (also referred to simply as variables) are interpreted in APT in the same sense as used in mathematics, that is, they represent quantities that are referred to symbolically. The value of a variable may remain constant for an entire part program or it may assume varying values as the computational nature of the part program dictates. With few exceptions, scalar variables may be used anywhere in the part program where numeric literals are permitted. Variables are denoted by symbols constructed according to the rules for naming geometric entities, that is, they consist of from one to six alphanumeric characters, at least one of which must be alphabetic. See Section 3.1.1 for details on preparing names. Actually, the symbolic name designates a computer memory location in which the current value for that variable may be found. Since all symbolic names used in APT are constructed according to the same set of rules, the APT system determines the type of entity represented by the symbol (e.g., scalar variable, geometric definition, statement label, or other) from the context of its first appearance in the part program. Symbols defined as one type of entity cannot be redefined as another type of entity within the same part program.

Arithmetic Operators

The following special characters are used as binary (requiring two operands) or unary operators (requiring a single operand) in arithmetic expressions.

Symbol	Binary operation	Unary operation
**	Exponentiation	
/	Division	
*	Multiplication	
+	Addition	
−	Subtraction	Negation

Arithmetic Expressions

An arithmetic expression may consist of a single simple operand such as a constant or scalar variable, or it may consist of a collection of operands separated by arithmetic operators. An expression may never contain adjacent operators, except to denote exponentiation, nor adjacent operands.

Single constants or scalar variables are expressions that may appear by themselves as the expression in an arithmetic assignment statement. Examples are the following:

    W = 4.78
    P1X = XVAL
    A = − A

In the first statement, the value $+4.78$ is assigned as the value of the variable W. When the second statement is executed, the current value of XVAL is assigned as the value of the variable P1X. The third statement indicates that the negative of the current value of A is to become the new value of A.

Constants and simple scalar variables may be combined with arithmetic operators to form more complex expressions as follows:

    N = N + 1
    RA = A1 + .119 − CNN/36
    AX7 = 3.3*X**2 + 1.06*X − 4

In the first statement, the current value of N is augmented by 1. In the second statement, the value of CNN will be divided by 36 and this result will be subtracted from the sum of 0.119 and the value of A1. The final value will then be assigned to the variable RA. If CNN = 72 and A1 = -6.3, then RA will be assigned the value -8.181. Note that integer and real numbers may be mixed in an expression. The third statement is in the form of a quadratic equation as follows:

$$y = ax^2 + bx + c$$

Also note in the third statement that multiplication is never implied but always requires use of the multiplication symbol.

The order of evaluation of an expression is governed by the hierarchy of arithmetic operators and functions and proceeds from left to right among operations of the same priority as modified by the presence of parentheses. The hierarchy is given in Table 4.1.

TABLE 4.1

Operator	Priority
Functions	1 (highest)
**	2
*,/	3
+,−	4 (lowest)

For the quadratic form shown above, the evaluation would proceed as follows (the W_i indicate intermediate evaluation results):

$$
\begin{aligned}
AX7 &= 3.3*X**2 + 1.06*X - 4 \\
&= 3.3*W_1 + 1.06*X - 4 \\
&= W_2 + 1.06*X - 4 \\
&= W_2 + W_3 - 4 \\
&= W_4 - 4 \\
&= W_5 \quad \text{(evaluation completed)}
\end{aligned}
$$

The order of evaluation may be modified with parentheses. Although parentheses are not operators, they override the precedence order and function as if they had a priority higher than any operator. Note the differences in the following.

Algebraic notation	Suitable APT statement
$w = \dfrac{xz}{y}$	W = X*Z/Y
$w = \dfrac{xz}{y}$	W = (X*Z)/Y
$w = \dfrac{x}{yz}$	W = X/(Y*Z)
$w = \dfrac{x}{yz}$	W = X/Y/Z

The second entry requires that X*Z be evaluated prior to the division. The third entry requires that Y*Z be evalated prior to the division.

Parentheses are required where shown in the following examples.

Algebraic notation	APT statement
$l = \dfrac{ij + k}{h}$	$L = (I*J + K)/H$
$b = (x + y)(w - v)$	$B = (X + Y)*(W - V)$
$k = \dfrac{d}{-f}$	$K = D/(-F)$
$r = a + \dfrac{b + d}{c}$	$R = A + (B + D)/C$
$r = a + \dfrac{b}{c + d}$	$R = A + B/(C + D)$
$s = \dfrac{t^{k+1}}{w}$	$S = T**(K + 1)/W$
$g = t^{k/w}$	$G = T**(K/W)$
$f = (a + b - c)^3$	$F = (A + B - C)**3$

Expressions may be nested within other expressions by enclosing them within parentheses. The evaluation of nested expressions begins with the innermost expression. Evaluation of an expression is always from left to right and according to the operator hierarchy rule unless modified by parentheses. Following are examples of nested expressions.

Algebraic notation	APT statement
$h = \dfrac{r}{g + k(x - y)}$	$H = R/(G + K*(X - Y))$
$m = \dfrac{p + \dfrac{q}{a - b}}{xy - z}$	$M = (P + Q/(A - B))/(X*Y - Z)$
$y = a + bx + cx^2 + dx^3 + ex^4$	$Y = A + X*(B + X*(C + X*(D + E*X)))$

The second entry will be evaluated as follows (the W_i indicate intermediate evaluation results).

$$
\begin{aligned}
M &= (P + Q/(A - B))/(X*Y - Z) \\
&= (P + Q/W_1)/(X*Y - Z) \\
&= (P + W_2)/(X*Y - Z) \\
&= W_3/(W_4 - Z) \\
&= W_3/W_5 \\
&= W_6 \quad \text{(evaluation completed)}
\end{aligned}
$$

Redundant sets of parentheses (those not used for altering operator precedence) may be used to effect program clarity and understanding.

4.1.2 Arithmetic Functions

The requirement for computing certain elementary mathematical functions occurs so frequently in NC machining that various functions have been provided in APT for the convenience of the part programmer. These functions may be required for computing coordinate values, angles, or other geometric quantities. Without their availability, it would be necessary for the part programmer to develop an algorithm for each function required, consistent, of course, with convergence and accuracy requirements sufficient to satisfy the problem. Table 4.2 contains a list of functions representative of those found in APT systems. Most of the functions are used in examples in this book.

The general form of a function is

name (*list*)

where *name* is the symbolic name of the function and *list* is a list of one or more arguments required by the function. When used, *name* acts as a scalar variable that is assigned a scalar value upon evaluation of the function.

Arguments of the arithmetic and trigonometric functions are arithmetic expressions which may be numeric literals, scalar variables, or complicated expressions containing arithmetic functions and nested expressions. The arguments of the geometric functions are geometric symbols. The arguments are always enclosed within a pair of parentheses immediately following the function symbol. Arguments that are angles and function values that are angles are in degrees as described in Section 3.1.3.

The functions are used by writing their assigned name, with associated argument list, in the proper location of an expression just as any scalar variable is used in an expression. The evaluation of expressions containing functions is subject to the hierarchy rules of Table 4.1, where it is seen that functions are assigned a priority higher than that of arithmetic operators. The result from evaluating a function is the substitution of a scalar value in the position in the expression where the function appears. The evaluation of the expression is then completed using this scalar value. Of course, the arguments of the function must be computed before a value is returned for the function.

The following examples use functions in the construction of arithmetic statements.

Algebraic notation	APT statement
$\alpha = \arctan 63.49$	ALPHA = ATANF (63.49)
$r = \sqrt{x^2 + y^2}$	R = SQRTF(X*X + Y*Y)
$area = \dfrac{ab}{2} \sin\theta$	AREA = .5*A*B*SINF(THETA)
$a = \sqrt{b^2 + c^2 - 2bc\cos\beta}$	A = SQRTF(B*B + C*C − 2*B*C*COSF(BETA))
$w = t - k\sin\left(\beta + \arctan\dfrac{y}{x}\right)$	W = T − K*SINF(BETA + ATANF(Y/X))
$t = \dfrac{\log\sqrt{a^2 - x^2}}{x^2}$	T = LOG10F(SQRTF(A*A − X*X))/(X*X)
$w = e^{at}\left(\dfrac{1}{t} + \ln t\right)$	W = EXPF(A*T)*(1/T + LOGF(T))
$d = 7.5a^{-|x^2 - 3.6|}\tan^2\theta$	D = 7.5*A**(−ABSF(X*X − 3.6))*TANF(THETA)**2

TABLE 4.2 FUNCTIONS

Name	Algebraic notation or geometric interpretation	Symbol	Comments
Arithmetic			
Absolute value	$\lvert x \rvert$	ABSF(x)	Function value is positive x.
Square root	$\sqrt{x}$	SQRTF(x)	x must be nonnegative.
Natural logarithm	$\cdot \ln x$	LOGF(x)	Base e.
Common logarithm	$\log x$	LOG10F(x)	Base 10.
Exponential	e^x	EXPF(x)	e is base of natural logarithm.
Algebraic minimum	$\min \{x_1, x_2, \ldots, x_n\}$	MIN1F($x_1, x_2, \ldots, x_n$)	Function value is algebraically smallest value from argument list.
Algebraic maximum	$\max \{x_1, x_2, \ldots, x_n\}$	MAX1F($x_1, x_2, \ldots, x_n$)	Function value is algebraically largest value from argument list.
Integer	Int(x)	INTGF(x)	Truncates fractional value.
Trigonometric			

Name		Symbol	Comments
Sine	$\sin \alpha$	SINF(α)	α in degrees.
Cosine	$\cos \alpha$	COSF(α)	α in degrees.
Tangent	$\tan \alpha$	TANF(α)	α in degrees.
Arctangent	$\arctan x$	ATANF(x)	Function value is angle in degrees normalized to first or fourth quadrant; $x = a/b$.
Arctangent	$\arctan a/b$	ATAN2F(a, b)	Function value is angle in degrees in correct quadrant; a and b must not both be zero; first and second quadrant angles are measured CCW, third and fourth quadrants CW.
Geometric			
Angle		ANGLF(cir, pt)	Function value is angle α in degrees measured CCW; cir is symbol for a circle; pt is symbol for a point.
Distance		DISTF(g_1, g_2)	Function value is distance between two parallel planes, two parallel lines, or two points. g_1 and g_2 are symbols for points, lines, or planes.
Number of points in a PATERN		NUMF($patt$)	Function value is the number of points in $patt$, a symbol for a pattern. See example in Point-13 in Section 3.2.

4.1.3 Examples Using Computation

The first example below shows arithmetic statements for computing quantities required in geometric definitions. The second example shows how scalar variables may be used in a part program so that changes may be readily made without changing other statements that depend on the same values.

Example 4.1: Eliminating Hand Computation of Geometric Elements

Figure 4.1 shows the geometry of a part bounded by LX, L2, C1, and L1. Perhaps some features of the part are dimensioned in an unorthodox manner, but for illustration purposes we choose to depart from accepted drafting practice. In this drawing we have a mixture of decimal and fractional numbers, an angle expressed in degrees and minutes and a radius expressed as a fraction of a line length D. The length D will be computed by applying the law of cosines. Recall that geometric definitions using these quantities require numeric values in decimal form. We will satisfy this requirement by substituting in their place scalar variables whose values will be computed from the information given.

That portion of the part program that includes the necessary computations and geometric definitions is as follows:

LX	= LINE/XAXIS	100
P1X	= −(1 + 23/64)	110
ALPHA	= 37 + 41/60	120
L1	= LINE/(POINT/P1X,0),ATANGL,ALPHA	130
P2X	= SQRTF(7.77)	140
L2	= LINE/(POINT/P2X,0),ATANGL,(2*ALPHA)	150
D1	= −P1X/COSF(ALPHA)	160
D2	= −P1X + P2X	170
R	= 2/3*SQRTF(D1*D1 + D2*D2 − 2*D1*D2*COSF(ALPHA))	180
C1	= CIRCLE/CENTER,(POINT/INTOF,L1,L2),RADIUS,R	190

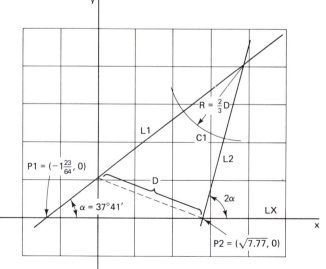

Figure 4.1   Geometry for Example 4.1.

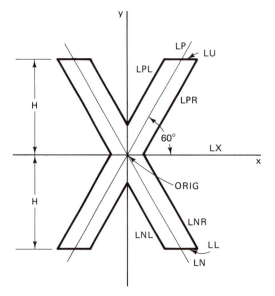

Figure 4.2 Geometry for Example 4.2.

These statements are a straightforward application of APT computing features. Note that scalar variables have been defined prior to their reference. When this part program is executed, required computations will be made and values assigned to the scalar variables. These values will then be substituted in the geometric definitions as shown. Intermediate values are computed as needed. For example, D1 and D2 are two sides of the triangle for which D is computed. This part program is flexible in the sense that should angle α be changed, it will be necessary to replace only statement 120 to effect the change. This is an advantage over hand calculation since it avoids recalculating all scalar variables dependent on α.

Example 4.2: The Letter X

Figure 4.2 shows the geometry for engraving the letter X with a flat-end engraving tool. Of course, with the tool we do not expect to get the sharp corners shown. The height of the letter, 2H, is to be 10 times the tool diameter D and the width of the arms of the X is to be 1.5 times the tool diameter. All scaling is to be performed within the part program such that, when the tool diameter is specified, the size of the letter will have been determined. When positioned at SETPT, the tool is to be directly over the center of the X. A part program that performs the engraving for a letter 10 in. high is as follows (tool diameter = 1 in.):

SETPT	= POINT/0,0,2	100
ORIG	= POINT/0,0,0	110
LX	= LINE/XAXIS	120
THETA	= 60	130
LP	= LINE/ORIG,ATANGL,THETA	140
LN	= LINE/ORIG,ATANGL, − THETA	150
D	= 1	160

H	= 5*D	170
LU	= LINE/PARLEL,LX,YLARGE,H	180
LL	= LINE/PARLEL,LX,YSMALL,H	190
W	= .75*D	200
LPR	= LINE/PARLEL,LP,YSMALL,W	210
LPL	= LINE/PARLEL,LP,YLARGE,W	220
LNR	= LINE/PARLEL,LN,YLARGE,W	230
LNL	= LINE/PARLEL,LN,YSMALL,W	240
	CUTTER/D	250
	FROM/SETPT	260
	GO/TO,LPR	270
	TLLFT,GOLFT/LPR,TO,LU	280
	GOLFT/LU,TO,LPL	290
	GOLFT/LPL,TO,LL	300
	GOLFT/LL,TO,LPR	310
	GOLFT/LPR,TO,LNR	320
	GOLFT/LNR,TO,LU	330
	GOLFT/LU,TO,LNL	340
	GOLFT/LNL,TO,LL	350
	GOLFT/LL,TO,LNR	360
	GOLFT/LNR,TO,LPL	370
	GOTO/SETPT	380

4.2. DECISION MAKING AND TRANSFER OF CONTROL

Thus far the APT system has been used to compute cut vectors according to a fixed sequence of instructions requiring a fixed set of parameters. However, if the part exhibits quasi-repetitious geometry, if tool motion is dependent on the dynamic behavior of the machining operation, if iterative computations are required to define geometric parameters and surfaces, if a choice of tool paths exists because of the relation of tool size to part geometry, or if other similar factors exist, it may be desirable or necessary to incorporate within the part program the decision logic for dynamically controlling part program execution.

The APT language provides for conditional (IF) and unconditional (JUMPTO) transfer of program control. Statements using IF and JUMPTO are combined with other statements in a manner such that decision making and the iterative execution of part program segments is achieved. A special program structure is required for this purpose.

4.2.1 Bounding Transfer of Control Statements

The group of statements among which transfer of control occurs must be bounded by the LOOPST-LOOPND statement pair. This part of the program segment is structured as follows:

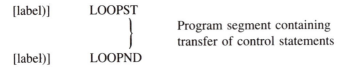

LOOPST and LOOPND must each appear in a statement by themselves. They may also be labeled. These statements serve no purpose other than to delineate those statements among which transfer of control takes place. They are not executable statements.

Transfer of control may not occur from outside the program segment bounded by LOOPST-LOOPND into the segment. Conversely, transfer of control may not occur from within a segment bounded by LOOPST-LOOPND to a statement not within the same segment. Consequently, the segment bounded by LOOPST-LOOPND can be entered and exited only once during program execution.

4.2.2 The Statement Label

Statements to which transfer of control occurs are identified by a statement label. The label is an identification tag that is used for reference purposes only. It is formed of from one to six alphanumeric characters. Often, numeric-only labels are not permitted. No special characters are allowed in constructing the label. When used to identify a statement, the label must begin in column 1 and must be terminated by a right parenthesis. When referenced in other statements, the parenthesis is omitted. The parenthesis does not contribute to the character count of the label.

Examples:

Valid:	A13)	
	2L)	
	LABEL1)	
Invalid:	END)	Reserved word
	A/4)	Special character
	FINISHED)	Too many characters
	1.33)	Special character

Identical labels in separate LOOPST-LOOPND bounded program segments are often permitted. However, this practice is discouraged because it leads to error-prone coding, confusion in reading part programs, and difficult debugging procedures.

4.2.3 Transfer of Control Statements

Under normal conditions the APT system executes statements sequentially as they appear in the part program. This sequential order may be altered through conditional or unconditional transfer of program control. IF and JUMPTO statements are used for this purpose. In effect, a transfer or jump to another statement in the part program is predetermined when the part program is written or is effected through conditional choices during execution of the part program.

The unconditional transfer statement format is

[*label*)]   JUMPTO/*ref-label*

where *ref-label* is another statement label within the same LOOPST-LOOPND pair bounding this JUMPTO statement. When executed, this statement causes part program control to be transferred to the statement containing *ref-label* as its statement label. This transfer may be to a statement preceding or following this JUMPTO as the logic of the part program dictates.

The conditional JUMPTO transfer statement format is

[*label*)]   JUMPTO/(*label-list*),*index*

where *label-list* is a list of statement labels separated by commas and within the same LOOPST-LOOPND pair bounding this JUMPTO statement, and *index* is a scalar variable, or nested computing expression, with an integer value from 1 to *n*, where *n* is the number of labels in *label-list*. This statement is also referred to as a computed JUMPTO. When executed, this statement causes part program control to be transferred to the statement containing one of the labels in *label-list* as its statement label. Selection of this label is conditioned on the value of *index*, which determines the label by its numerical position in *label-list*. The value of *index* cannot be zero or negative nor can it exceed the number of labels in *label-list*. When this statement is used, *index* is generally constrained to assume an integer value through assignment; otherwise, it will be truncated to become an integer for this statement. The following is a suitable flowchart symbol for this statement.

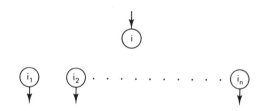

Examples:

JUMPTO/(ID3,ID1,ID2,ID4),IND
JUMPTO/(LAB1,LAB2,LAB3),(J/2+1)

In the first statement, IND must have a value of 1, 2, 3, or 4. In the second, J must have a value of 0, 1, 2, 3, or 4. When J is 0 or 1 control is transferred to the statement with the label LAB1. When J is 2 or 3 control is transferred to the statement with the label LAB2.

The arithmetic IF transfer statement format is

[*label*)]   IF(*arith-exp*) s_1,s_2,s_3

where the s_i are other statement labels within the same LOOPST-LOOPND pair bounding this IF statement and *arith-exp* is any arithmetic expression formed as described in Section

4.1. *Arith-exp* must be enclosed within parentheses and the s_i must be separated by commas. When this statement is executed, the value of *arith-exp* is computed and transfer of control occurs as follows:

1. If *arith-exp* < 0, control transfers to the statement whose label is s_1.
2. If *arith-exp* = 0, control transfers to the statement whose label is s_2.
3. If *arith-exp* > 0, control transfers to the statement whose label is s_3.

This statement is named from the need to evaluate an arithmetic expression. It serves as a three-way branch, which functions as a two-way branch whenever any two of the s_i are identical. It must always be followed by a labeled statement; otherwise, there is no way for the following statement to be executed. The following is a suitable flowchart symbol for this statement. The exit conditions are shown for illustration purposes. They are placed near the exit lines leading to the part program logic to which program control is transferred when the condition is met. Annotating the flowchart this way expands the program documentation and is recommended.

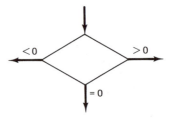

Examples:

IF (K − 5) A1, A3, A2
IF (B**2 − 4*A*C) NEG,ZER,POS

In the first statement, if K is less than 5, control is transferred to the statement whose label is A1. If K = 5, control is transferred to A3, and if K is greater than 5, control is transferred to A2. In the second statement B**2 − 4*A*C is evaluated first, then transfer of control is determined in a similar manner.

The TRANTO statement is also an unconditional transfer of control statement but for use with multiple check surface motion statements. It is described in Section 4.4.

4.2.4 Examples Illustrating Transfer of Control

The following examples show the application of JUMPTO and IF statements. In these examples the flow of control is always forward. Backward jumps are illustrated in connection with looping in Section 4.3.

Example 4.3: Using the Arctangent Function

This example shows the computation of the proper angle with the ATANF arctangent function provided that the opposite and adjacent sides of a right triangle are known as signed quantities.

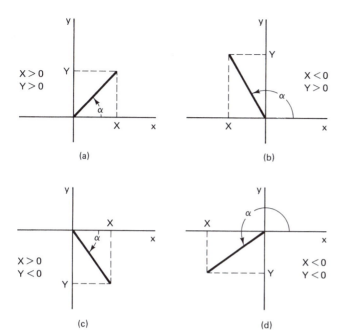

Figure 4.3   Determining the quadrant of an angle given the opposite and adjacent sides of a right triangle as signed quantities.

What is computed here should really be done with the ATAN2F function. However, not all APT systems include the latter function.

The ATANF function returns an angle that is normalized to the first or fourth quadrant only. If the opposite and adjacent sides of a triangle are known, as shown in Figure 4.3, it is possible to place the angle in the proper quadrant with the proper logic in the part program. A flowchart for this logic is given in Figure 4.4, where the desired angle α is computed

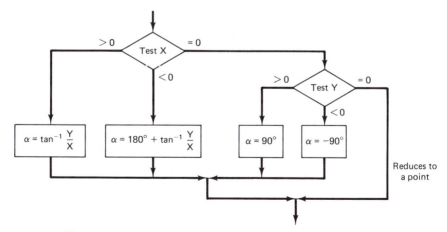

Figure 4.4   Flowchart for placing the angle in the correct quadrant.

when the opposite side Y and adjacent side X are known as signed quantities. It is possible that both X and Y are zero, in which case the geometry reduces to a point and a different segment of the part program must be entered as indicated. Note that infinite values for tan ($\pm 90°$) are avoided. Since branching is required, a LOOPST-LOOPND pair is required to bound the IF and JUMPTO statements and the statements with labels. The part program segment for this logic is as follows:

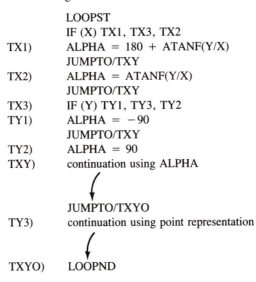

```
          LOOPST
          IF (X) TX1, TX3, TX2
TX1)      ALPHA = 180 + ATANF(Y/X)
          JUMPTO/TXY
TX2)      ALPHA = ATANF(Y/X)
          JUMPTO/TXY
TX3)      IF (Y) TY1, TY3, TY2
TY1)      ALPHA = -90
          JUMPTO/TXY
TY2)      ALPHA = 90
TXY)      continuation using ALPHA


          JUMPTO/TXYO
TY3)      continuation using point representation


TXYO)     LOOPND
```

Example 4.4: Calculating the Index for the Computed JUMPTO

This example illustrates how a portion of a part program can be made universal and it shows variations on structuring a program for the computed JUMPTO.

Figure 4.5 shows geometry from which four different cutter paths may be derived if we assume that the lines are either parallel with the x-axis or inclined to it at a $\pm 45°$ angle. Clearly, the variables of the problem are the angles α and β. First, we observe that we can offset L1 from the x-axis by 1 unit or 3 units by the calculation

$$\text{offset} = 2 - \alpha/45$$

If $\alpha = -45°$, the offset is 3 units. If $\alpha = +45°$, the offset is 1 unit. Although there are four unique paths, commonality of path segments and the uniqueness of the tangent of 45° suggests that an index for a computed JUMPTO be calculated so that the motion statements can be predetermined but selected during program execution as determined by the values for α and β. A suitable formula for this purpose is

$$\text{index} = 1 + (1 + \tan \alpha) + \tfrac{1}{2}(1 + \tan \beta)$$

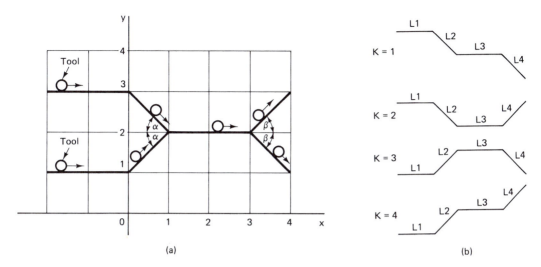

(a)   (b)

Figure 4.5   (a) Basic geometry; (b) possible tool paths.

For example, for α and β both at $+45°$, the index computes to

$$\text{index} = 1 + (1 + 1) + \tfrac{1}{2}(1 + 1) = 4$$

Three other values for the index can be computed for the various combinations of positive and negative signs for α and β.

The logic for two program structures appropriate for this problem is given in Figure 4.6. Assuming that the tool has already been placed on L1, a part program following the logic of Figure 4.6(a) follows. The symbols of the part program are marked in Figure 4.5(b).

P1	= POINT/1,2	100
P2	= POINT/3,2	110
ALFA =	individually selected as $+45°$ or $-45°$	120
BETA =		130
L1	= LINE/XAXIS,(2 − ALFA/45)	140
L2	= LINE/P1,ATANGL,ALFA	150
L3	= LINE/P1,P2	160
L4	= LINE/P2,ATANGL,BETA	170
K	= 1.5 + (1 + TANF(ALFA)) + .5*(1 + TANF(BETA))	180

$\left(\begin{array}{l}\text{motion statements that position the}\\\text{tool on L1}\end{array}\right)$

INDIRV/1,0,0   300
LOOPST   310

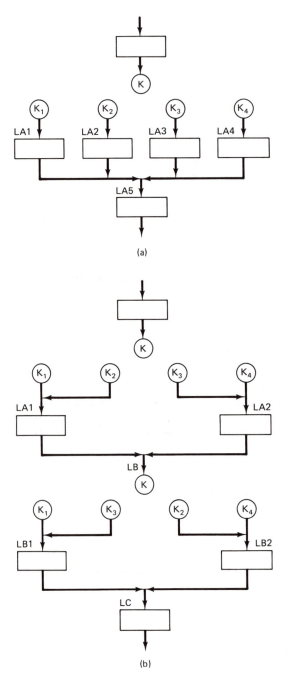

(a)

(b)

Figure 4.6   (a) Flowchart with a single computed JUMPTO; (b) flowchart with two computed JUMPTOs.

```
       JUMPTO/(LA1,LA2,LA3,LA4),K                        320
LA1)  GOFWD/L1,PAST,L2                                    330
       GOFWD/L2,TO,L3                                     340
       GOFWD/L3,PAST,L4                                   350
       JUMPTO/LA5                                         360
LA2)  GOFWD/L1,PAST,L2                                    370
       GOFWD/L2,TO,L3                                     380
       GOFWD/L3,TO,L4                                     390
       JUMPTO/LA5                                         400
LA3)  GOFWD/L1,TO,L2                                      410
       GOFWD/L2,PAST,L3                                   420
       GOFWD/L3,PAST,L4                                   430
       JUMPTO/LA5                                         440
LA4)  GOFWD/L1,TO,L2                                      450
       GOFWD/L2,PAST,L3                                   460
       GOFWD/L3,TO,L4                                     470
LA5)  GOFWD/L4, . . .                                     480
       LOOPND                                             490
```

The value 0.5 is added to K in line 180 to ensure that when truncated, K will have the desired value. It is assumed that the check surface for the statement labeled LA5 is the same for all four cases. Observe that repetition of motion statements occurs among the four program segments. This is recognized in the flowchart of Figure 4.6(b), for which the statements between the LOOPST-LOOPND pair follow.

```
       LOOPST                                             310
       JUMPTO/(LA1,LA1,LA2,LA2),K                         320
LA1)  GOFWD/L1,PAST,L2                                    330
       GOFWD/L2,TO,L3                                     340
       JUMPTO/LB                                          350
LA2)  GOFWD/L1,TO,L2                                      360
       GOFWD/L2,PAST,L3                                   370
LB)   JUMPTO/(LB1,LB2,LB1,LB2),K                          380
LB1)  GOFWD/L3,PAST,L4                                    390
       JUMPTO/LC                                          400
LB2)  GOFWD/L3,TO,L4                                      410
LC)   GOFWD/L4, . . .                                     420
       LOOPND                                             430
```

Although this program segment contains six fewer statements than the other one, its program logic is more difficult to follow. Thus it should not be used for this problem. This is not to imply that similar logic would not be appropriate for other problems.

Alternative solutions for this problem are given by the flowcharts of Figures 4.7 and 4.8, where the arithmetic IF replaces the computed JUMPTO in most cases. The statements within the labeled blocks of Figure 4.8 would be those for the program for Figure 4.6(a). Instead of computing K, the angles are tested directly. None of these flowcharts is preferred over that of Figure 4.6(a) because of the complexity introduced by the arithmetic IF statement.

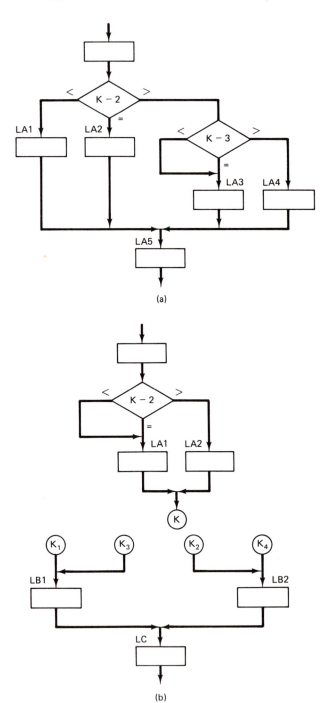

(a)

(b)

Figure 4.7   (a) Alternative flowchart for Figure 4.6(a); (b) alternative flowchart for Figure 4.6(b).

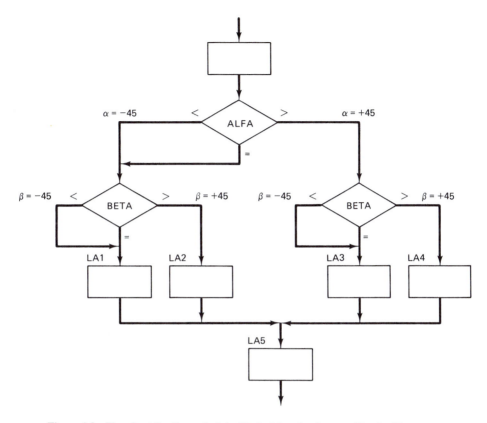

Figure 4.8   Flowchart for Example 4.4 with decisions implemented by the IF statement.

4.3 LOOPING

Let us write the part program segment to compute the natural logarithm (base e) of a number x with the following series representation. Normally, this value would be computed with the LOGF function.

$$\ln x = \frac{x - 1}{x} + \frac{(x - 1)^2}{2x^2} + \frac{(x - 1)^3}{3x^3} + \cdots \qquad x \geq \frac{1}{2}$$

The number of terms of the series that must be computed is governed by accuracy considerations. This number cannot be infinite, of course, but may be predetermined and fixed, or variable with the number determined by convergence criteria within the part program. The following part program segment will use a fixed number of terms n to

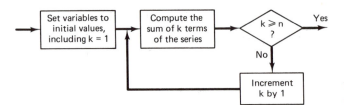

Figure 4.9   Flowchart for the logarithm example.

compute the natural logarithm of a number x, whose value must be equal to or greater than ½.

	LOOPST	100
	N = n	110
	LOGN = 0	120
	T1 = (X − 1)/X	130
	K = 1	140
LA1)	LOGN = LOGN + T1**K/K	150
	IF (N − K) LA3, LA3, LA2	160
LA2)	K = K + 1	170
	JUMPTO/LA1	180
LA3)	LOOPND	190

From the flowchart for the part program segment (Figure 4.9) we note the presence of a program loop. A loop is a sequence of statements provided with a feedback path for the purpose of repeatedly executing these statements. Certain testing criteria must be met for departure from a loop.

Initialization statements are required to ensure that certain scalar variables have the values needed for proper operation of the loop and to compute the proper value for the logarithm. Among these are lines 110 and 140 for the variables N and K, which control the number of loop iterations, and line 120 for LOGN, which holds partial sum values until the final value is computed. The computation of each term of the series and the partial sum is formed in line 150. This is followed by the IF statement, used as a two-way branch, resulting in transfer to LA3 if the required number of iterations have been performed, or to LA2, where the value of K is augmented by 1. The loop is reentered at LA1.

For this example the initialization statements could have been placed before LOOPST since they are not within the range of any of the loop transfer statements. They were included within the LOOPST-LOOPND pair to emphasize the fact that initialization is an important element of loop structure. This example follows the loop structure diagram of Figure 4.10(a). Figure 4.10 also shows other loop structures. Although initialization usually serves as entry into the loop, the remaining parts of the loop may be permuted

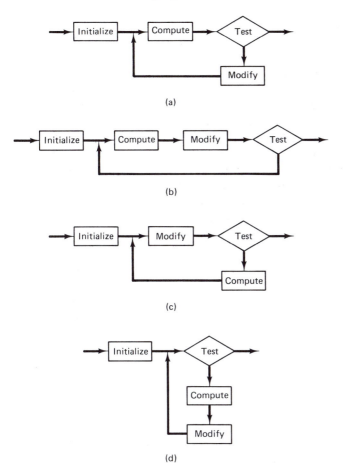

(a)

(b)

(c)

(d)   **Figure 4.10   Some loop structures.**

in any order. The following part program segment computes the logarithm according to
the flowchart of Figure 4.10(c). K must be initialized differently.

	LOOPST	100
	N = n	110
	LOGN = 0	120
	T1 = (X − 1)/X	130
	K = 0	140
LA1)	K = K + 1	150
	IF (N − K) LA3, LA3, LA2	160
LA2)	LOGN = LOGN + T1**K/K	170
	JUMPTO/LA1	180
LA3)	LOOPND	190

Care must be taken when constructing the initialization and testing parts of a loop to avoid performing the loop $n + 1$, $n - 1$, ∞, or 0 times when it really should be performed n times.

The part program segments above were written so that the loop was executed a predetermined and fixed number of times n. We will now rewrite the program so that the number of times the loop is executed is determined by convergence criteria within the loop. For this example we will compute the natural logarithm of $x = 5$.

From the series representation, for x greater than 1 each term will be positive. Therefore, as each term is computed, the sum will increase, finally reaching the value for the logarithm when an infinite number of terms are computed. Of course, we cannot compute an infinite number of terms. Instead, when the difference between one partial sum and its immediately preceding value is less than a small number, say 0.01, we will declare the computation complete and exit the loop. The following part program segment computes an approximation to the logarithm in this manner.

```
        X = 5                                        100
        K = 1                                        110
        T1 = (X − 1)/X                               120
        LOGB = 0                                     130
        LOOPST                                       140
LA1)    LOGA = LOGB                                  150
        LOGB = LOGB + T1**K/K                        160
        K = K + 1                                    170
        IF (.01 − ABSF(LOGA − LOGB)) LA1, LA2, LA2   180
LA2)    LOOPND                                       190
```

This example follows the loop structure diagram of Figure 4.10(b). Two consecutive partial sums for the logarithm are computed, LOGA and LOGB. The absolute value of their difference is subtracted from .01. If the result is negative, we branch back to LA1, where we save the most recent partial sum LOGB in LOGA and compute a new partial sum in LOGB. This process continues until the convergence criteria is met and the loop exited. The exact form of the convergence test will, in general, depend on the nature of the computation. Series representations that have terms with alternating positive and negative signs may require a different convergence test.

Slowly converging or oscillating values or too accurate a convergence test may cause the loop to be executed a very large number of times, possibly infinite. To prevent this possibility from occurring, it is wise to employ a loop limit test as described in Section 4.6.4.

Example 4.5: A Logarithmically Calibrated Dial

Figure 4.11 shows the geometry for engraving a circular dial with logarithmically spaced calibration marks. The engraving tool diameter is zero. The tool will be positioned on point PX, the intersection of line LX with circle COUT, then directed clockwise to engrave a long or short tick mark, each beginning from COUT and extending to CLONG or CSHORT,

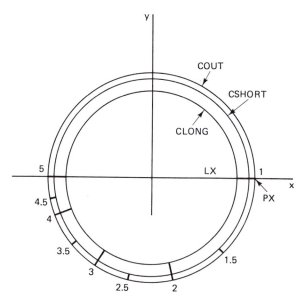

Figure 4.11 Logarithmically engraved dial of Example 4.5.

respectively. The maximum number of whole graduations (long tick marks) and the total angular distribution θ are given. The natural logarithm function is used in the computation.

The flowchart of Figure 4.12 applies to the following part program segment that directs the tool for engraving the dial.

```
        CUTTER/0                                                  100
SETPT   = POINT/0,0,3                                             110
        THETA = 180                                               120
        M = 5                                                     130
        R1 = 2                                                    140
ORIG    = POINT/0,0                                               150
COUT    = CIRCLE/CENTER,ORIG, RADIUS,R1                           160
CSHORT= CIRCLE/CENTER,ORIG,RADIUS,(.95*R1)                        170
CLONG = CIRCLE/CENTER,ORIG,RADIUS,(.875*R1)                       180
LX      = LINE/XAXIS                                              190
PX      = POINT/XLARGE,INTOF,LX,COUT                              200
        K = 1                                                     210
        J = 1                                                     220
        N = 1                                                     230
        NM = LOGF(M)                                              240
        JEND = 2*(M − 1)                                          250
        FROM/SETPT                                                260
        INDIRP/PX                                                 270
        GO/ON,COUT                                                280
        TLON,GOBACK/LX,ON,CLONG                                   290
        GOBACK/LX,ON,COUT                                         300
```

LA1)	LOOPST	310
	N = N + .5	320
	J = J + 1	330
	K = −K	340
	ANG = −LOGF(N)/NM*THETA − 90	350
	LX = LINE/LX,CANON,COSF(ANG),SINF(ANG)	360
	GORGT/COUT,ON,LX	370
	IF (K) LA2, LA2, LA3	380
LA2)	GORGT/LX,ON,CSHORT	390
	JUMPTO/LA4	400
LA3)	GORGT/LX,ON,CLONG	410
LA4)	GOBACK/LX,ON,COUT	420
	IF (JEND − J) LA5, LA1, LA1	430
LA5)	LOOPND	440

The distribution of tick marks is over an angle of 180°, as given by θ (line 120), along the outer circle COUT with a radius of 2 in. (line 160). CSHORT and CLONG are proportionally related to COUT by their radius specification (lines 170 and 180). Calibration marks are made in a clockwise direction beginning at 0° (conforming with the value ln 1) and continuing to the maximum angular distribution specified, θ = 180° (conforming with the value ln 5, where M = 5 is given in line 130). This will provide for M = 5 long tick marks, including the end marks, for whole values of N and for 4 short tick marks when N contains fractional values. N is incremented by 0.5 for each mark in line 320. Scalar variable M specifies the maximum value of the calibration scale while scalar variable N is used to compute the distance along COUT at which a calibration mark is to be engraved. The angular placement of the tick marks is determined by the ratio −θ ln N/ln 5 (line 350). Thus ln 5 must be computed for subsequent use. This is done in line 240.

The geometric variable for the part program segment above is line LX, along which the tick mark is engraved. Thus LX must be redefined for each pass through the loop as given in line 360 with the CANON form of line definition. The components of the normal vector must be computed from the angle at which the tick mark is to be engraved. This is done in line 350, which is the expression above with 90° subtracted from its value to account for the angle at which the normal lies. An alternative way of writing the part program is to use unnamed nested line definitions as given below.

The loop is initialized and the long tick mark at 0° is engraved prior to entering the loop, lines 210 through 300. Thereafter, loop modification takes place and the tool is commanded along COUT to the location of the next tick mark, lines 320 through 370. A test is made in line 380 to determine whether a long or short tick mark should be engraved. The variable K is used for this purpose by alternating its value between +1 and −1 (line 340). After engraving the tick mark, loop testing for completion is done in line 430. J is used as a counter to compare against the final value, JEND, for loop exit purposes. Since J assumes only whole-number values, its use is preferred over variables that assume fractional values to avoid problems with inexact number representation.

The alternative way of defining the lines for the tick marks requires that statements within the loop be modified as follows. All else remains unchanged.

LA1)	LOOPST	310
	N = N + .5	320
	J = J + 1	330
	K = −K	340

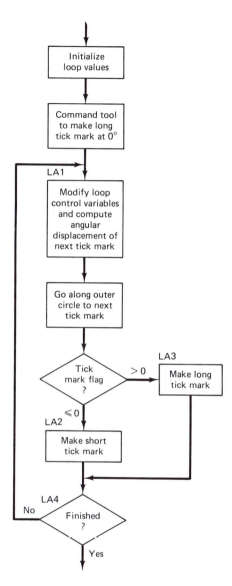

Figure 4.12 Flowchart for computing logarithmic calibrations and engraving dial.

	ANG = −LOGF(N)/NM*THETA	350
	GORGT/COUT,ON,(LINE/ORIG,ATANGL,ANG)	360
	IF (K) LA2, LA2, LA3	370
LA2)	GORGT/(LINE/ORIG,ATANGL,ANG),ON,CSHORT	380
	JUMPTO/LA4	390
LA3)	GORGT/(LINE/ORIG,ATANGL,ANG),ON,CLONG	400
LA4)	GOBACK/(LINE/ORIG,ATANGL,ANG),ON,COUT	410
	IF (JEND − J) LA5, LA1, LA1	420
LA5)	LOOPND	430

4.4 MULTIPLE CHECK SURFACES AND TRANTO

At times the check surface for a motion command is not known a priori because of factors that arise while machining the part. Among these factors are dimensional and geometrical changes in the part or tool because tool paths and geometrical entities are computed and defined dynamically or because of tool diameter differences caused by tool changes. We might incorporate logic within the part program that decides which of alternative check surfaces should be chosen. This logic may be based on computations made for this purpose or it may be incorporated within the motion command itself. For the latter case, we must use the multiple check surfaces motion command. Its format is as follows:

$$
\begin{bmatrix} \text{TLLFT,} \\ \text{TLON,} \\ \text{TLRGT,} \end{bmatrix}
\begin{Bmatrix} \text{GOFWD} \\ \text{GOBACK} \\ \text{GORGT} \\ \text{GOLFT} \end{Bmatrix}
/dsurf,
\begin{Bmatrix} \text{TO} \\ \text{ON} \\ \text{PAST} \\ \text{TANTO} \end{Bmatrix}
,[n, \text{INTOF,}] \qquad \$
$$

$$
csurf1, lab1,
\begin{Bmatrix} \text{TO} \\ \text{ON} \\ \text{PAST} \\ \text{TANTO} \end{Bmatrix}
,[n, \text{INTOF,}] csurf2, lab2
$$

This statement should be compared with that given in Section 2.3.2, where *dsurf* and *csurf1* are the drive and check surfaces given there. We are now specifying a second check surface *csurf2*. Associated with the immediately preceding check surfaces are the statement labels *lab1* and *lab2*.

     With this statement we are providing a choice of check surfaces. The choice is not arbitrary, however. It is dependent on conditions that arise while the cutter paths are being computed, that is, while the tool is in motion. Simultaneous with tool motion, the APT system is searching for the two check surfaces. As soon as one of them is reached, according to the modifiers, program control is transferred to the statement with the label that appears immediately after the check surface.

     An application for this statement is shown in Figure 4.13, where the tool may first

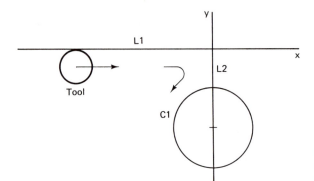

Figure 4.13 Geometry illustrating the need for a multiple check surface motion command.

reach either line L2 or circle C1. A multiple check surface statement for resolving the issue is as follows:

TLRGT,GOFWD/L1,TO,L2,CSL1,TO,C1,CSL2

If the tool reaches L2 first, control transfers to the statement labeled CSL1; otherwise, it will reach C1 first and control will transfer to the statement labeled CSL2. Presumably, tool motion will continue at each of those statements along the proper drive surface. Figure 4.14 is a flowchart illustrating operation of this statement. Regardless of which path was followed, sometimes the two paths must merge, as shown in the figure. This implies that some part program segment must be bypassed much as was done in Examples 4.3 and 4.4. There, bypassing was done with the unconditional JUMPTO statement.

With the multiple check surface motion command we have a special case whereby the JUMPTO cannot be used. Instead, the TRANTO statement is provided for this purpose. Its format is

TRANTO/*ref-label*

where *ref-label* is a statement label appearing on a statement placed such that transfer is forward only. This is a consequence of the manner in which the APT system processes the statements and retains identity of labeled statements. It is not necessary to bound the multiple check surface motion command, the TRANTO statement, and the labeled statements referenced by them within a LOOPST-LOOPND pair.

Following is a part program segment with a multiple check surface motion command for resolving the check surface decision of Figure 4.13.

TLRGT,GOFWD/L1,TO,L2,1CS,TO,C1,2CS
1CS) GORGT/L2,TO,C1
      GORGT/C1, . . .
      TRANTO/3CS
2CS) GOBACK/C1, . . .
3CS) continuation

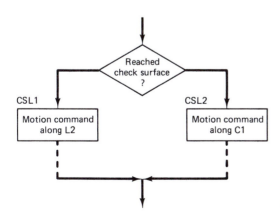

Figure 4.14 Equivalent logic of the multiple check surface motion command.

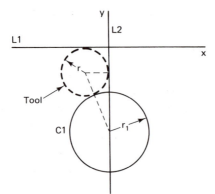

Figure 4.15 Geometry for computing the radius of a tool tangent to L1, L2, and C1.

A part program segment that does not use a multiple check surface motion command must rely on computation to resolve the geometry. For this purpose we will use the geometry shown in Figure 4.15 to determine that critical value of r, the radius of the tool, that will permit us to decide which check surface to select. The right-triangle relationship gives

$$(r_1 + r)^2 = r^2 + (y_1 - r)^2$$

where y_1, assumed to be known, is the magnitude of the y-coordinate of the center of circle C1. For this special case, this leads to the quadratic equation,

$$r^2 - 2(y_1 + r_1)r + (y_1{}^2 - r_1{}^2) = 0$$

for which the solution is

$$r = (y_1 + r_1) \pm \sqrt{(y_1 + r_1)^2 - (y_1^2 - r_1)^2}$$

The correct value for r is obtained by using the negative value of the radical. The other value satisfies the right triangle relationship but does not satisfy the geometry. The relevant segment of the part program becomes

```
        LOOPST
        R = Y1 + R1 − SQRTF((Y1+R1)**2 − (Y1*Y1−R1*R1))
        IF (D/2 − R) 1CS, 2CS, 2CS
  1CS)  TLRGT,GOFWD/L1,TO,L2
        GORGT/L2,TO,C1
        GORGT/C1, . . .
        JUMPTO/3CS
  2CS)  TLRGT,GOFWD/L1,TO,C1
        GOBACK/C1, . . .
  3CS)  LOOPND
```

D is the diameter of the tool. Although no loops are present, the LOOPST-LOOPND pair is still required because of the transfer of control statements.

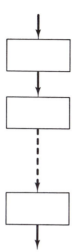

Figure 4.16 Flowchart showing forward program flow.

4.5 STRUCTURING THE PROGRAM FOR DECISION MAKING AND REPETITION

Thus far, the part program segments have included few decisions. Realistically, multiple decisions must be made and multiple loops must be written. This section includes a discussion of good and bad practices for incorporating these features in a part program. At all times we must remember that the program must exhibit the following characteristics.

1. It must be easily understood.
2. It must allow for a reduction of debugging iterations.
3. It must exhibit forward flow of statement execution.

4.5.1 Selection of Alternatives

In conjunction with the TRANTO discussion of Section 4.4 it was stated that transfer of control must be in a forward direction only. We now say that it is a recommended practice that part programs exhibit forward flow throughout the progam for ease in understanding the part program. Of course, for a loop it is necessary to transfer backward. This case is discussed in Section 4.5.2. At this time we will consider consequences of program flow using the IF statement and the computed JUMPTO.

In Figure 4.16 we show that normal program flow is forward. The program flowchart is simple and easily understood. When we select paths with the IF statement, we can still maintain forward flow with a linear placement of statements in the program as shown in Figure 4.17 even though the flowchart shows parallel paths. By emphasizing this fact with the dashed lines of Figure 4.18, we can assume that the logic within the dashed lines is really no more than one of the blocks of Figure 4.16. This is because LOOPST-

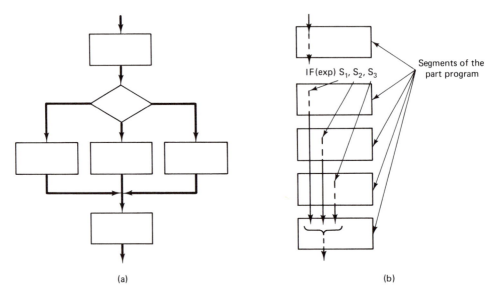

(a)   (b)

Figure 4.17 (a) Path selection with an IF statement; (b) forward program flow depicted with a linear placement of program statements.

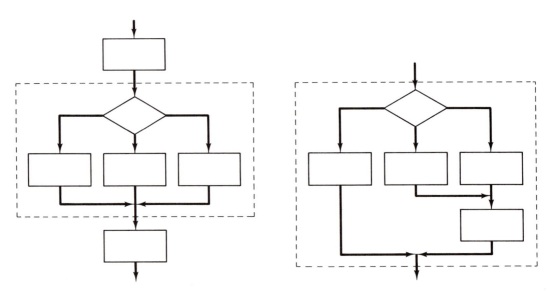

Figure 4.18 Maintaining forward flow with the single-entry, single-exit program structure within a LOOPST-LOOPND pair.

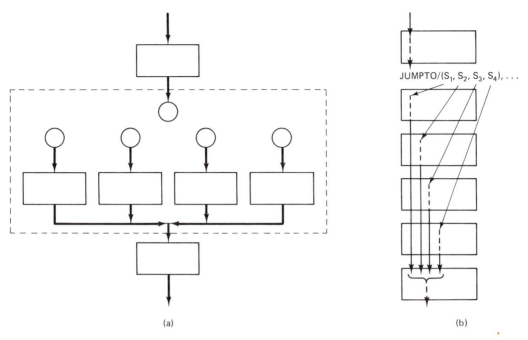

(a)   (b)

Figure 4.19   Interpreting forward flow with a computed JUMPTO statement: (a) flowchart segment; (b) linear placement of program statements.

LOOPND pairs have only one entry and only one exit (see Section 4.2.1). A variation in program flow is included in Figure 4.18 to maintain the forward flow concept. Interpretation of the computed JUMPTO in a similar manner is illustrated in Figure 4.19.

The TRANTO statement is flowcharted as a two-exit IF statement. Thus it is a special case of the logic shown in Figure 4.17. Multiple alternatives are merely an extension of the concepts already presented. An example is given in Figure 4.20, where several two-outcome decisions are shown. We will now discuss good programming practices in conjunction with the coding of decision trees such as shown in Figure 4.20.

Eventually, multiple paths resulting from a decision block should merge to continue forward program flow. Figure 4.20 is drawn to emphasize the single-entry single-exit structure of the decision logic. Merging of the paths from two-way or three-way decision blocks should proceed in an inside-out manner as drawn. For Figure 4.20, this occurs first for decision blocks 3 and 4, then block 2, and finally for block 1. The recommended linear placement of program statements begins by systematically following left paths far down the decision tree until merging of paths for the original decision block must occur. After retracing up the tree to the last decision block entered, forward path tracing should resume from that block by following the other branches in a left-to-right manner, always preferring left branches not yet followed. Retracement should continue until all path merging is realized. The order in which the statements are to be written for Figure 4.20 is shown in the program statement placement diagram given there. The arrows indicate

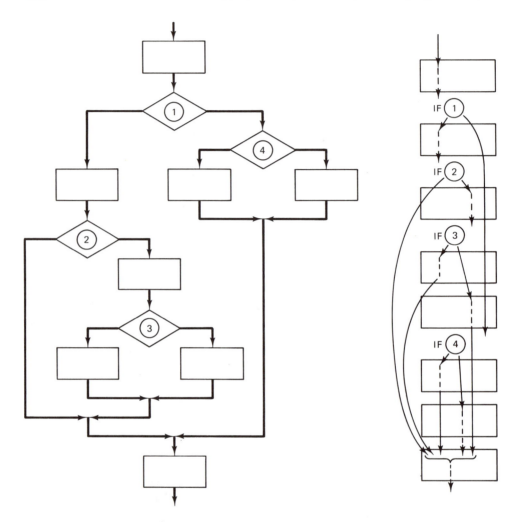

Figure 4.20   Skeleton flowchart with multiple alternatives and its linear placement of program statements.

the left-right branching and statement bypassing needed for correct program logic. Difficulties in programming transfer of control logic in APT are a consequence of the lack of structured programming features in the language.

Transfer of control statements should never contain the label of an unconditional JUMPTO statement. If so, implementing cascaded JUMPTOs in this manner complicates program understanding by disguising the logic and it leads to secondary errors caused by relabeling statements when changes are made. Furthermore, labels should be named by a positional naming scheme so that they may be readily located.

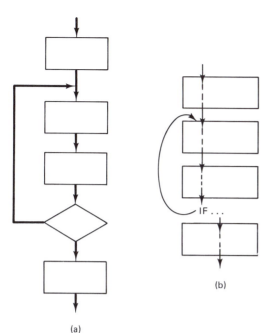

(a)

(b)

Figure 4.21   (a) Preferred loop structure; (b) its linear placement of program statements.

4.5.2 Repetition

As discussed in Section 4.3, repetition, or looping, is effected by transferring control backward in the part program. The need for loops was discussed there. However, one should use loops with discretion. Often, there is justification for using in-line code rather than loops.

The flowchart preferred for loops is shown in Figure 4.21 together with the resulting linear placement of program statements. Only one backward branch is required and no statements are bypassed. This diagram should be compared to that of Figure 4.22, where bypassing of statements results in a program more difficult to understand. Loop structures with unorthodox exit locations, such as that of Figure 4.23, cause awkward program organization and should be avoided.

Multiple loops are often required in a part program. Such loops may be cascaded (in series form) or nested as shown in Figure 4.24, where two cascaded loops are nested within another loop. Such loop structures require reinitialization of the inner loops for each pass through the range of the outer loop. The high-frequency loops are the inner ones (those most often executed), while the lowest-frequency loop is the outer one.

Preferred loop structures have been identified in Figures 4.21 and 4.24. Complex loop structures that reuse code for multiple loops, that often overlap, and that reenter a loop at other than its normal entry point are shown in Figure 4.25. These complicated loop structures should be avoided for reasons already given.

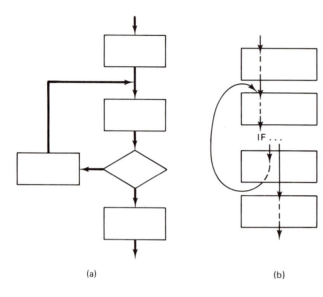

(a)

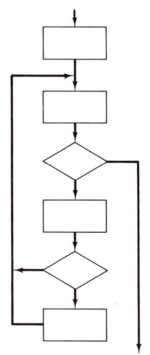

(b)

**Figure 4.22   (a) Alternative loop struc-
ture; (b) its linear placement of program
statements showing statement bypassing.**

**Figure 4.23   Loop structure with an
unorthodox exit.**

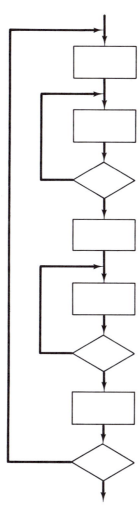

Figure 4.24 Skeleton diagram showing cascaded and nested loops.

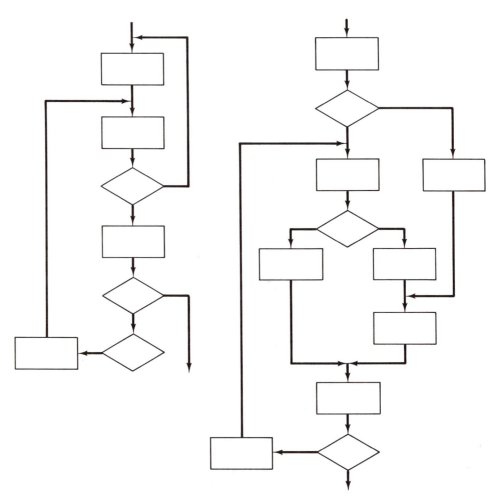

Figure 4.25   Two skeleton diagrams showing complicated loop structures that are to be avoided.

4.6 PROGRAM DEBUGGING: HANDLING COMPLEX
PROGRAM STRUCTURES

Learning features of the APT language as presented in this book is a reinforcing process. One must prepare motion commands (Chapter 2) with reference to geometric entities (Chapter 3) for each program that is written. As a result, the debugging procedures introduced thus far are also continually being reinforced. Now, confronted with new program structures, we must expand our arsenal of debugging techniques. We maintain the philosophy given in Section 1.8 that programmer time is expensive relative to computer hardware costs.

4.6.1 Preliminaries

Not only must we strive to make the program understandable, but we must adopt a strategy for handling program complexity. This is necessary because the program structures introduced in this chapter introduce complexity in the program. An attitude stressing organization must be adopted. For this purpose, we put forth the following guidelines.

1. Simplify the problem for program design purposes. Do not incorporate all bells and whistles in the first design of the program. Expand the design to include all features through several design iterations.

2. Not all programs need be written for general-purpose application. The utility of a program or program segment may be limited, thus justifying a modest straight-forward approach to program development.

3. Write and check out the part program in manageable increments. The design of the program generally suggests partitions that may be added as development progresses. This guideline is derived from Section 1.8, where it was stated that a limited number of things, typically seven or less, can be kept mentally organized in short term memory at one time.

General suggestions for preparing the program:

1. Adopt a meaningful labeling scheme, such as one based on grouping statements physically in selected parts of the program or by choosing a numerically ascending sequencing scheme. Choose labels that cannot be confused with symbol or variable names.

2. Avoid using the same labels within different LOOPST-LOOPND pairs.

3. Before writing the code, draw a skeleton flowchart of the part program segment being programmed. Use the preferred transfer of control and repetitive program structures of Section 4.5.

4. Avoid testing for equality for loop exit with the IF statement when the arithmetic expression produces values with fractional parts. Fractional values are often inexactly represented in a computer. Thus equality may be impossible to achieve and loop exit may not occur. This is one cause of infinite loop execution. Test for loop exit on an ''equal to or greater than'' or ''equal to or less than'' basis or compute values with whole numbers only.

4.6.2 Debugging

As much as possible must be learned from each run of the program, whether or not errors were encountered. This suggests using language features designed for debugging purposes. Some such features are included in the following suggestions.

Compilation-time Errors

Request cross-reference listings (XREF) for initial program runs. Be sure that the PRINT/OPTION,ON,XREF statement is included. Use this listing to identify statements with improper symbol and variable references. Also use this list to locate all occurrences of names being changed or deleted from the program.

Execution-time Errors

1. Check the labels printed during loop execution to verify proper loop operation. Use a loop limit counter to avoid infinite loops during program checkout.
2. Often the APT Section number (see Figure 1.3) in which the error was detected is printed out. The CL-printout may contain incompletely processed results. By noting the Section number, it may be necessary to interpret the results differently from those produced upon completion.
3. Request auxiliary printout for verification purposes. Use the PRINT/3,symbols-&-variables statement form. Strategically place the PRINT/ON and PRINT/OFF statements to eliminate or reduce unwanted printout.

4.6.3 The PRINT and XREF Statements

Statements providing printout for program debugging and verification purposes are important tools for the part programmer. PRINT and XREF are two such statements. These statements do not generate cut vectors and their output will not appear in the CL-file. Formats of these statements for program development as discussed thus far are as follows:

$$\text{PRINT/}\begin{cases} \text{OPTION,ON,XREF} \\ \text{ON} \\ \text{OFF} \\ \text{3,ALL} \\ \text{3,}\textit{list} \\ \text{2,}\textit{list} \end{cases}$$

$$\text{XREF/}\begin{cases} \text{ON} \\ \text{OFF} \end{cases}$$

List is a list of scalar variables and/or geometric entity symbols separated by commas. Combinations of these statements may coexist in a part program as dictated by printout needs.

The PRINT/ON statement causes the canonical form of each surface and scalar variable to be printed at the time it is generated. This is a modal command, staying in effect until turned off by the PRINT/OFF statement. This command should be used with discretion since considerable printout can occur, especially if loops are executed many times. More than one instance of this print pair may be used in the program.

The PRINT/3,ALL statement causes the canonical form of each surface and scalar variable defined thus far in the part program to be printed out at the time this statement is executed. Repeated printouts will occur if this statement is in a loop.

Should canonical form printout for all surface and scalar variables not be desired, selective printout may be requested with the PRINT/3,*list* statement. The variables and symbols selected for printout are to appear in *list*. An example of this statement appears in Figure 4.26, ISN 33, where printout for line LX is requested. The canonical form values for LX, as defined in Table 3.4, are printed out. A heading precedes the printout of values. For LX, canonical form values A and B are nonzero while C is always zero, and, since LX passes through the origin, D is zero. Printout occurs outside the loop statements listing. For this example, the PRINT statement was executed eight times. During checkout of this part program, it would have been useful to include ANG in the PRINT statement.

The PRINT/2,*list* statement functions like the PRINT/3,*list* statement except that the heading line is not printed. This statement should be used with care since the values are not otherwise identified. An example of its use appears in Section 4.6.4.

When available, a printout of the names of variables and symbols and their occurrences in the program by line number is useful to the part programmer. From this printout, called a cross-reference table, misspelled variables, not otherwise causing diagnostics, may be identified and deletions or substitution of variables or symbols is aided since all references to these names are included in the table. When the XREF/ON statement is encountered, data for this table are gathered until the XREF/OFF or FINI statements are reached. Printout of this table occurs at the end of compilation time when the PRINT/OPTION,ON,XREF statement is included in the part program. An example of a cross-reference printout for the IBM APT-AC system appears in Figure 4.27. This is caused to occur by the statements in lines with ISN numbers 3 and 4. From this table, for example, we see that variable J occurs in statements with ISN numbers 17, 28 (twice), and 39.

An aid to debugging programs with loops is the printout of statement labels for labeled statements being executed. Figure 4.28 is the first sheet of CL-printout for the part program of Figure 4.26. LA2-LA4 or LA3-LA4 label pairs are printed out depending on whether short or long tick marks, respectively, are being engraved. Only the first LA2-LA4 label pair appears on this sheet.

4.6.4 Avoiding Infinite Loops

We cautioned in Section 4.3 that infinite loops could occur with a convergence testing exit. A part program segment was given there that had the potential for infinitely looping

```
ISN 00001 PARTNO   LOGARITHMICALLY CALIBRATED DIAL.                    00000100
ISN 00002          CLPRNT                                              00000200
ISN 00003          XREF/ON                                             00000210
ISN 00004          PRINT/OPTION,ON,XREF                                00000220
ISN 00005          CUTTER/0                                            00000300
ISN 00006 SETPT    = POINT/0,0,3                                       00000400
ISN 00007          THETA = 180                                         00000500
ISN 00008          M = 5                                               00000510
ISN 00009          R1 = 2                                              00000600
ISN 00010 ORIG     = POINT/0,0                                         00000900
ISN 00011 COUT     = CIRCLE/CENTER,ORIG,RADIUS,R1                      00001000
ISN 00012 CSHORT   = CIRCLE/CENTER,ORIG,RADIUS,(.95*R1)                00001100
ISN 00013 CLONG    = CIRCLE/CENTER,ORIG,RADIUS,(.875*R1)               00001110
ISN 00014 LX       = LINE/XAXIS                                        00001200
ISN 00015 PX       = POINT/XLARGE,INTOF,LX,COUT                        00001400
ISN 00016          K = 1                                               00001500
ISN 00017          J = 1                                               00001600
ISN 00018          N = 1                                               00003100
ISN 00019          NM = LOGF(M)                                        00003210
ISN 00020          JEND = 2*(M - 1)                                    00003220
ISN 00021          FROM/SETPT                                          00003300
ISN 00022          INDIRP/PX                                           00003500
ISN 00023          GO/ON,COUT                                          00003510
ISN 00024          TLON,GOBACK/LX,ON,CLONG                             00003600
ISN 00025          GOBACK/LX,ON,COUT                                   00003800
ISN 00026 LA1)     LOOPST                                              00003900
ISN 00027          N = N + .5                                          00004000
ISN 00028          J = J + 1                                           00004100
ISN 00029          K = -K                                              00004200
ISN 00030          ANG = -LOGF(N)/NM*THETA - 90                        00004300
ISN 00031          LX = LINE/LX,CANON,COSF(ANG),SINF(ANG)              00004310
ISN 00032          PRINT/3,LX                                          00004320
ISN 00033          GORGT/COUT,ON,LX                                    00004400
ISN 00034          IF (K) LA2, LA2, LA3                                00004500
ISN 00035 LA2)     GORGT/LX,ON,CSHORT                                  00004600
ISN 00036          JUMPTO/LA4                                          00004700
ISN 00037 LA3)     GORGT/LX,ON,CLONG                                   00004800
ISN 00038 LA4)     GOBACK/LX,ON,COUT                                   00004900
ISN 00039          IF (JEND - J) LA5, LA1, LA1                         00004910
ISN 00040 LA5)     LOOPND                                              00005000
     LX              LINE         4 ITEMS     UNITS 1.00000000
   -.711380336    -.702807240    .000000000     .000000000

     LX              LINE         4 ITEMS     UNITS 1.00000000
   -.976378214    -.216068468    .000000000     .000000000

     LX              LINE         4 ITEMS     UNITS 1.00000000
   -.976378214     .216068468    .000000000     .000000000

     LX              LINE         4 ITEMS     UNITS 1.00000000
   -.839912537     .542721778    .000000000     .000000000

     LX              LINE         4 ITEMS     UNITS 1.00000000
   -.641324143     .767270058    .000000000     .000000000

     LX              LINE         4 ITEMS     UNITS 1.00000000
   -.421929090     .906628833    .000000000     .000000000

     LX              LINE         4 ITEMS     UNITS 1.00000000
   -.204215011     .978926059    .000000000     .000000000

     LX              LINE         4 ITEMS     UNITS 1.00000000
    .000000000    1.00000000     .000000000     .000000000

ISN 00041          GOTO/SETPT                                          00005500
ISN 00042          FINI                                                00005600
```

Figure 4.26   Illustration of a PRINT/3 printout.

```
                TOPOLOGICAL CROSS-REFERENCE LISTING (BY ISN) OF SYMBOLS GLOBAL TO PART PROGRAM

ANG       00030  00031  00031
CLONG     00013  00024  00037
COUT      00011  00015  00023  00025  00033  00038
CSHORT    00012  00035
J         00017  00028  00028  00039
JEND      00020  00039
K         00016  00029  00029  00034
LX        00014  00015  00024  00025  00031  00031  00032  00033  00035  00037  00038
M         00008  00019  00020
N         00018  00027  00027  00030
NM        00019  00030
ORIG      00010  00011  00012  00013
PX        00015  00022
R1        00009  00011  00012  00013
SETPT     00006  00021  00041
THETA     00007  00030

                                                  END OF CROSS-REFERENCE LISTING
```

Figure 4.27   Cross-reference listing for Figure 4.26.

given that the convergence criteria was too stringent. We now modify that program to place a limit on the number of times that the loop may be executed regardless of the convergence tendency.

Figure 4.29 is the part program with a loop limit test modification. The IF statement of ISN 11 performs a limit test on K. K, a whole number, is really used in the computation of each term of the series, but since it is incremented by 1 each time through the loop it is also a counter. Thus it serves a dual purpose. Its value is tested against the constant 16 in the IF statement. The loop will be exited whenever K is equal to or greater than 16, meaning that there were 15 passes through the loop since K was initialized to 1. This additional IF statement causes a slight modification in the preferred loop structure of Figure 4.10(b).

ISN 12 contains a PRINT/2 statement for the scalar variable LOGB, the value of the logarithm. This PRINT statement format was discussed in Section 4.6.3. Its printout lacks the heading line produced by the PRINT/3 statement, but in this case we know what variable is being printed out and do not need the heading. There were 11 passes through the loop, so the limit of 15 was not reached. Therefore, convergence occurred and the last two values computed were within 0.01 of each other (actually, 0.00780903, as a simple subtraction shows).

No motion commands were used in the part program, so Section 3 was not entered. No CL-printout was produced, so no label printouts can be checked. Only Section 1 calculations were made.

....SECTION 3....

```
ISN                                                              LABEL  REC M   CARD
0001 PARTNO/  LOGARITHMICALLY CALIBRATED DIAL.                          00002 00000100
0005 CUTTER/                                                            00004 00000300
0021 FROM/    0.0        SETPT      0.0        3.00000000               00006 00003300

0022                                                                    00007 00003500
0023 GOTO/    COUT                                                      00009 00003510
              2.00000000            0.0        0.0
0024 GOTO/    LX                                                        00011 00003600
              1.75000000            0.0        0.0
0025 GOTO/    LX                                                        00013 00003800
              2.00000000            0.0        0.0

0040 SURFACE             CIRCLE     DS(IMP-TO)                    LA5 00014M00005000
0033 SURFACE  COUT       0.0        0.0        0.0                      00016 00004400
                         0.0        1.00000000 2.00000000

0033 GOTO/    COUT                                                      00017 00004400
              1.99951668           -0.04396647 0.0
              1.99565151           -0.13181443 0.0
              1.98792865           -0.21940759 0.0
              1.97636302           -0.30657662 0.0
              1.96097698           -0.39315302 0.0
              1.94180027           -0.47896944 0.0
              1.91886996           -0.56385998 0.0
              1.89223038           -0.64766055 0.0
              1.86193302           -0.73020917 0.0
              1.82803645           -0.81134625 0.0
              1.79060619           -0.89091496 0.0
              1.74971460           -0.96876148 0.0
              1.70544073           -1.04473534 0.0
              1.65787014           -1.11868967 0.0
              1.60709482           -1.19048152 0.0
              1.55321289           -1.25997211 0.0
              1.49632853           -1.32702711 0.0
              1.43655168           -1.39151689 0.0
              1.40561448           -1.42276067 0.0

0035 GOTO/    LX                                                  LA2 00019 00004600
              1.33533376           -1.35162264 0.0
0038 GOTO/    LX                                                  LA4 00021 00004900
              1.40561448           -1.42276067 0.0

0033 SURFACE             CIRCLE     DS(IMP-TO)                          00023 00004400
0033 SURFACE  COUT       0.0        0.0        0.0
                         0.0        1.00000000 2.00000000

0033 GOTO/    COUT                                                      00024 00004400
              1.37456014           -1.45278505 0.0
              1.31054273           -1.51078713 0.0
              1.24408034           -1.56597066 0.0
              1.17529697           -1.61823269 0.0
              1.10432095           -1.66747571 0.0
              1.03128469           -1.71360786 0.0
              0.95632444           -1.75654307 0.0
              0.87958006           -1.79620124 0.0
```

Figure 4.28 CL-printout for Figure 4.26 showing label identification.

134

```
ISN 00001 PARTNO   LOOP WITH IF TOLERANCE TEST AND COUNT LIMIT TEST.
ISN 00002          CLPRNT
ISN 00003          X = 5
ISN 00004          K = 1
ISN 00005          T1 = (X - 1)/X
ISN 00006          LOGB = 0
ISN 00007          LOOPST
ISN 00008 LA1)     LOGA = LOGB
ISN 00009          LOGB = LOGB + T1**K/K
ISN 00010          K = K + 1
ISN 00011          IF (16 - K) LA3, LA3, LA2
ISN 00012 LA2)     PRINT/2,LOGB
ISN 00013          IF (.01 - ABSF(LOGA-LOGB)) LA1, LA3, LA3
ISN 00014 LA3)     LOOPND
   .800000000

   1.12000000

   1.29066666

   1.39306666

   1.45860266

   1.50229333

   1.53225264

   1.55322416

   1.56813724

   1.57887466

   1.58668369

ISN 00015          FINI
```

Figure 4.29   Part program that avoids infinite looping.

PROBLEMS

4.1. Write APT arithmetic assignment statements for the following.

(a) $x = a + b - 2c$

(b) $w = \dfrac{m(3n - 1)}{3n + 1}$

(c) $v = \dfrac{4\pi r^3}{3}$

(d) $y = \dfrac{x}{1 + \dfrac{2x}{3 + \dfrac{4x}{5 + 8x}}}$

(e) $h = \theta + \dfrac{\theta^2}{2!} - \dfrac{\theta^3}{3!} + \dfrac{\theta^4}{4!} - \dfrac{\theta^5}{5!}$

(f) $c = \sqrt[3]{a^2 + b^2} + \sqrt{w + 4ab}$

(g) $\tan x = \dfrac{\sin x}{\cos x}$

(h) $r = \cos\sqrt{(x_1 - x_2)(y_1 - y_2)}$

(i) $t = \sin^2 \dfrac{\theta}{2} \arctan \dfrac{x + 1}{y}$

(j) $y = \sqrt{\left(1 + \dfrac{1}{1 + \dfrac{t^2}{1 + t^3}}\right) \sin(\omega t - n\pi)}$

Write APT programs for Problems 4.2 through 4.6 to perform the computation indicated. Use more than one arithmetic assignment statement where necessary. Numerical values are given and are to be assigned to the appropriate variables in separate assignment statements and used in the computation by referencing the variable names. Use a PRINT statement to output the results.

4.2. Write a program to convert an angle from degrees, minutes, and seconds to radians (suggested value $= 37°20'44''$).

4.3. Write a program to compute the radius of a circle inscribed within a regular polygon of N sides, each side of length L, according to the formula, $r = L/2 \cot (180/N)$ (suggested values: $N = 6, L = 2$).

4.4. Write a program to compute the radius of a circle inscribed in a triangle whose sides are a, b, and c according to the following formula:

$$r = \frac{\sqrt{s(s - a)(s - b)(s - c)}}{s}$$

where $s = \frac{1}{2}(a + b + c)$ (suggested values: $a = 3, b = 4, c = 5$).

4.5. The distance between two points in the xy-plane is given as follows:

$$d = \sqrt{r_1^2 + r_2^2 - 2r_1 r_2 \cos(\theta_1 - \theta_2)}$$

where r_1 and r_2 are the distances of the points from the origin and θ_1 and θ_2 are their angles with respect to the positive x-axis. Compute the distance d for $r_1 = 10.5$, $\theta_1 = 35°$, $r_2 = 6.3$, and $\theta_2 = -28°$.

4.6. The solution to the equation $ax^2 + bx + c = 0$ is given by the quadratic formula as

$$x_1, x_2 = \frac{-b \pm \sqrt{b^2 - 4ac}}{2a}$$

where one of the roots, say x_1, is obtained with the plus sign and the other, x_2, with the minus sign. A particular choice for the coefficients a, b, and c will make the radicand positive (two unequal real roots), negative (complex conjugate roots), or zero (two equal real roots). Write a program to compute the radicand, test its value, and branch to the appropriate statements to compute the roots, except for the case of a negative radicand, in which case the value of the radicand shall be assigned to each root instead of carrying out the indicated computation. Run the program for each of the following sets of coefficients that test each path in the program.
(a) $a = -3, b = 12, c = 2.5$

(b) $a = 6$, $b = 6$, $c = 1.5$

(c) $a = -5.4$, $b = 4.5$, $c = -3.1$

4.7. Given the following expansion (valid for $y^2 < x^2$),

$$(x + y)^n = x^n + nx^{n-1}y + \frac{n(n-1)}{2!} x^{n-2}y^2$$

$$+ \frac{n(n-1)(n-2)}{3!} x^{n-3}y^3 + \cdots$$

write a program containing loops to compute the $n + 1$ terms for the value of $(x + y)^n$ according to the series expansion. Provide for an arbitrary integer value of n. Perform the series computation using the values $x = 2.5$, $y = 1.25$, and $n = 5$ and also compute the value of $(x + y)^n$ directly in an arithmetic assignment statement.

4.8. Figure P4.8 shows a part whose shape is determined by two radial lines and various circular edges. Assume that this part may be made with various values for r_1, r_2, r_3, α, and for the number n of complete semicircular cutouts on the larger circular edge (n may be zero) but that the corner cutouts of radius r_3 are always included. The figure shows $n = 2$ semicircular cutouts which are equally distributed angularly on r_1. Write an APT part program that treats the foregoing quantities as variables and is of general purpose use in the sense that when numerical values are assigned to these variables in arithmetic assignment statements, no further part program changes are required to produce the part. Run the program with the numerical values $r_1 = 8$ in., $r_2 = 3$ in., $r_3 = 1$ in., $\alpha = 120°$, and with $n = 3$. Use a ⅛-in.-diameter tool, select appropriate tolerances and a set point, and take the tool in a counterclockwise direction to produce the part.

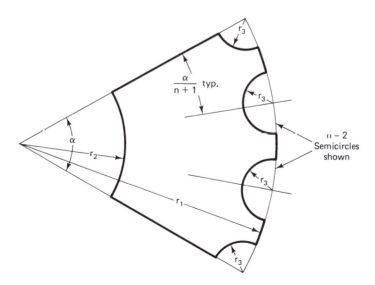

Figure P4.8

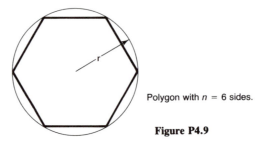

Polygon with $n = 6$ sides.

Figure P4.9

4.9. Figure P4.9 shows a polygon inscribed within a circle of radius r. A hexagon is shown, although any figure with three or more sides may be so drawn. Write an APT general-purpose part program to produce the polygon given the radius r and the number of sides n as program variables whose values are to be assigned in arithmetic assignment statements only. No other part program changes will be necessary to run the program for different values of r and n. Run the program with the values $r = 3$ in., $n = 7$, tool diameter $\frac{1}{2}$ in., and with appropriately assigned tolerances and set point definition. Take the tool in a counterclockwise direction around the part.

4.10. A rectangular-shaped part containing various semicircular cutouts is shown in Figure P4.10. Write an APT general-purpose part program to produce the part given the radii of the cutouts (all the same), the overall dimensions of the rectangle, and the number n of semicircles equally spaced on each long dimension (the figure shows two on each side). These values are to be assigned to the appropriate variables in arithmetic assignment statements. Should different values be required, no other part program changes shall be made to effect the change. Run the program with the values $d_1 = 4$ in., $d_2 = 10$ in., $r = 0.75$ in., $n = 4$, tool diameter $= \frac{1}{2}$ in., and with appropriately selected tolerances and set point definition. Take the tool in a counterclockwise direction around the part.

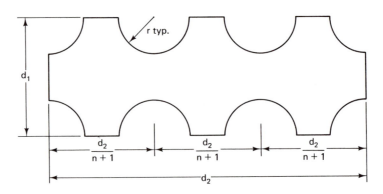

n = number of semicircular cutouts on each long dimension side ($n = 2$ shown)

Figure P4.10

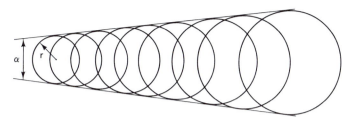

n = number of circles (n = 9 shown)
R = ratio of adjacent circle diameters

Figure P4.11

4.11. An APT program could be written to direct the pen of a numerically controlled drafting machine to produce the drawing of Figure P4.11 if the tool diameter is reduced to zero in the part program. Write such a part program for the drawing such that the angle α (always greater than zero), smallest circle radius r, number of tangent circles n, and the ratio of the circle diameters for adjacent circles R, are treated as variable quantities and assigned in arithmetic assignment statements. Note that the tangent lines are part of the drawing. Run the program with the values $\alpha = 15°$, $r = 1$ in., $R = 1.1$, $n = 6$, and with appropriately selected tolerances and set point definition.

4.12. Figure P4.12 shows a part whose boundary is composed of circular segments and radial lines. The exact geometry depends on the initial angle α, the angular increment β, the initial radius r, the radius increment dr, and the number of circles m. The figure is drawn with $\alpha = -10°$, $\beta = 15°$, $r = 3$ in., $dr = 1$ in., and $m = 6$. Prepare a general-purpose APT part program that uses looping techniques in conjunction with unnamed geometric definitions within motion commands to properly machine the part. The part program should be written with α, β, r, dr, and m as variables whose values must be entered in arithmetic assignment statements for a given part geometry. Assume that α is limited to the range $\pm 60°$, that is, the part is completely contained within two quadrants as shown. Select suitable tolerances, define a tool set point, and use the numeric values given above to take a $\frac{1}{2}$-in.-diameter tool from set point counterclockwise around the part and back to set point.

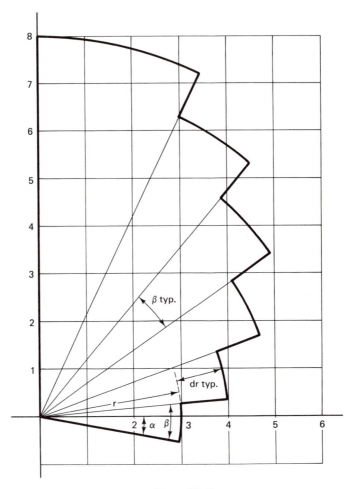

Figure P4.12

CHAPTER 5

Subscripted Variables

In APT we may store information in the form of an array, also known as a table, which is a particular data structure in which the array elements are referred to by an index (its subscript) that accompanies the array name. An array may contain as its elements either scalar values or geometric definitions. The significance of the array concept is its notational convenience, which allows part programs to be written so that computation and looping often facilitate the preparation of a part program and contribute to enhancement of the logical features of the part program. The manner that arrays are defined and used through the vehicle of subscripted variables is described in this chapter.

5.1. ARRAYS

Mathematical formulations are often simplified through the symbolism of subscripted variables. Subscripting is a notational concept implying order. For example, we may identify a point in three-dimensional space through its coordinate values as (x_1, y_1, z_1), or (x_2, y_2, z_2), or in general as (x_i, y_i, z_i), where the subscript i assumes integer values. This means that the triplet of scalar variables having the same subscript value is associated with the same point, and that point is now uniquely identified. Subscripts are also used in equations. For example, the recursive relationship,

$$y_n = ay_{n-1} + bx_n$$

is interpreted as meaning that the nth value of y is obtained by multiplying the $(n - 1)$st value of y by a and adding this to the product of b and the nth value of x. If we required the eighth value of y, we must use the seventh value of y and the eighth value of x in the computation. We may also use the subscript notation for labeling curves. The symbols $C_1, C_2, \ldots, C_n$ could denote a family of circles, each circle individually defined.

The subscript notation is usually applied to suggest a family relationship among its members, that is, the members share common characteristics. However, this is not a requirement for the use of subscripts. When used in computer languages, subscripting implies a particular data structure. In APT, this structure is a one-dimensional array, or table. A one-dimensional array can be visualized as a single column of entries, or elements. Each row of the array contains only one entry or element. The use of double-subscript notation, such as $x_{i,j}$ (not available in APT), implies a two-dimensional array for which the first subscript refers to the rows and the second to the columns. Three-dimensional arrays (also not available in APT) have three subscripts, which refer to rows, columns, and layers.

5.2 RESERVING STORAGE FOR ARRAYS

The number of elements that an array is expected to accommodate may not be evident from the manner in which the part program is written, especially when the elements are defined within loops. To ensure that a sufficient number of computer memory locations are available to contain all elements of an array, we must specify the maximum size (number of elements) of the array. We do this with the RESERV statement. Its format is

$$\text{RESERV}/a_1,n_1,a_2,n_2, \ldots ,a_k,n_k$$

where a_i = array name (scalar variable or geometric definition symbol)

   n_i = maximum number of elements for a_i (n_i must be a constant)

This is a specification statement (declaration) and is nonexecutable. It must appear in the part program before the array names defined by it are referenced but it cannot appear within a loop or macro (see Chapter 11). Here are examples of the RESERV statement:

   RESERV/CIR,15

   RESERV/X,33,Y,33,LIN,50,PT,63,ALPHA,180

5.3 REFERENCING ARRAY ELEMENTS

Once defined in a RESERV statement, an array name must be subscripted wherever referenced. The general format for referencing an array element is

   name(exp)

where *name*, the array name, may be a scalar variable or the symbolic name of a geometric entity and *exp*, the subscript, is an arithmetic expression enclosed within parentheses as shown. The subscript may be any arithmetic expression but is usually a constant or a scalar variable. It is treated as an integer value, that is, noninteger values are truncated to integers for subscripting purposes.

Subscripted name examples:

Valid:	CIR(3)	
	X(J)	
	PT(L + 1)	
	VAL(3*I + J − 1)	
	LIN(2 + K)	
	A(J(K) + 1)	
Invalid:	A(0)	Subscript may not be zero
	BJ(−4)	Subscript may not be negative

There is a distinction between CIR(3) and CIR3. The former is the third element of the array named CIR, while the latter is a nonsubscripted name.

Each element of the array is referenced uniquely through its subscript value. Subscript values are assigned sequentially and correspond to the row number of the array (see Figure 5.1). The first element is always referenced by subscript number 1, the last element by the value given in the RESERV statement. Intermediate elements are referenced by the value of the subscript, often called a pointer since it "points" to an element without ambiguity. Reference to an element outside the range given above causes an error message to be printed. Here are examples of subscript variable referencing:

L1   = LINE/A(1),A(2),A(3),A(4)

P5   = POINT/X(1),Y(1),Z(1)

C3   = CIRCLE/CENTER,P(6),TANTO,L(4)

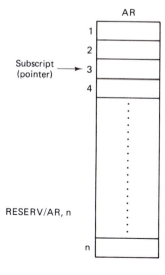

Figure 5.1   Representation of an array as a linear sequence of memory locations.

The part program segment below uses a scalar variable as a subscript in computing the angle for Example 4.5. Other examples are found in Section 5.5.

	RESERV/ANG,20	100
	THETA = 180	110
	M = 5	120
	J = 1	130
	N = 1	140
	NM = LOGF(M)	150
	JEND = 2*(M − 1)	160
	LOOPST	170
LP1)	N = N + .5	180
	J = J + 1	190
	ANG(J − 1) = −LOGF(N)/NM*THETA − 90	200
	IF (JEND − J) LP2, LP2, LP1	210
LP2)	LOOPND	220

5.4 INCLUSIVE SUBSCRIPTS

The convenience of subscripting variables in loops is often counteracted when strings of subscripted values must be specified individually. For example, *n* consecutive values of an array can be specified as

A(1),A(2),A(3),A(4), . . .,A(n)

It is not uncommon to require a string of values, such as in the tabulated cylinder definition of Section 6.2. In these cases, specifying an array name, a subscript, and a pair of parentheses for each occurrence of the array value is not only inconvenient but is susceptible to error. A convenience feature known as ''inclusive subscripts'' (not available in all APT systems) allows array values to be written in an abbreviated form. The string above would be written as

A(1,THRU,n)

The inclusive subscript feature is more general than indicated above. It allows specifying selected sequences of values, specifying values in an ascending or descending order by subscript value, and specifying values by spanning sequences of values in increments of the subscript value. The general format of the inclusive subscript specification is

$$name\left(\begin{Bmatrix} \text{ALL} \\ a, \text{THRU}, \begin{Bmatrix} b \\ \text{ALL} \end{Bmatrix} \end{Bmatrix} \left[, \begin{Bmatrix} \text{INCR} \\ \text{DECR} \end{Bmatrix}, c \right] \right)$$

where *name* is the name of the array being subscripted, *a* is the value of the lowest subscript, *b* is the value of the highest subscript, and *c* is the value of the increment for specifying intervals of subscripts. These values, which may be constants, scalar variables, or some combination thereof, are used to compute a sequence of subscripts for the array values desired. Since the subscripts are not explicitly given, collectively they are referred to as an implied list of subscripts.

Selection of the subscripts proceeds from the low value *a* to the high value *b* if the increment is positive, and from *b* to *a* if the increment is negative. The sign of the increment is determined from the combination of the sign of *c* and its modifier (INCR for increasing, DECR for decreasing). The first clause of the implied list specifies the range of subscript values from which the selection will be made. The second clause, which is optional, specifies the direction for selection and the increment for choosing the subscripts. The four principal combinations of these clauses are interpreted as follows.

1. ALL

All elements of the array are specified, from 1 through n, where n is determined from the limit given in the RESERV statement.

2.
$$\text{ALL}, \begin{Bmatrix} \text{INCR} \\ \text{DECR} \end{Bmatrix}, c$$

For *c* positive, the elements specified with INCR are 1, $1+c$, $1+2c$, . . . , r, where r is either the limit of the array or the last value computed that does not exceed the limit. For *c* positive, the elements specified with DECR are n, n-c, n-2c, . . . , r, where r is either 1 or the last value computed that is not less than 1.

3.
$$a, \text{THRU}, \begin{Bmatrix} b \\ \text{ALL} \end{Bmatrix}$$

All elements of the array from *a* through *b*, inclusive, are specified. If *b* equals the limit given in the RESERV statement, the effect is the same as if ALL was chosen.

4.
$$a, \text{THRU}, \begin{Bmatrix} b \\ \text{ALL} \end{Bmatrix}, \begin{Bmatrix} \text{INCR} \\ \text{DECR} \end{Bmatrix}, c$$

The elements are specified from the range described in interpretation 3 but chosen from this range as described in interpretation 2.

It is possible for *c* to be negative, thus leading to inverted logic for interpreting its effect. Since this action is contrary to the philosophy of making programs understandable, this option should not be exercised. The following are examples of inclusive subscripts with constants for *a*, *b*, and *c*. The array was declared with RESERV/A,7.

Form	Elements referenced
A(ALL)	A(1),A(2),A(3),A(4),A(5),A(6),A(7)
A(ALL,INCR,1)	A(1),A(2),A(3),A(4),A(5),A(6),A(7)
A(1,THRU,4)	A(1),A(2),A(3),A(4)
A(3,THRU,6)	A(3),A(4),A(5),A(6)
A(3,THRU,6,INCR,1)	A(3),A(4),A(5),A(6)
A(2,THRU,ALL,INCR,2)	A(2),A(4),A(6)
A(7,THRU,3)	A(7)
A(ALL,DECR,3)	A(7),A(4),A(1)
A(2,THRU,6,DECR,3)	A(6),A(3)
A(6,THRU,2,DECR,3)	A(2)

The material presented thus far provides limited opportunity for applying the inclusive subscript feature. Since it allows us to abbreviate the lists of subscripted values and because it conforms with our notion of program understandability because it adds to clarity, we will use it often in the material to come. Meanwhile, here are additional examples of its use:

L1 = LINE/A(1,THRU,4)

P5 = POINT/W(I,THRU,J)

PRINT/3,ANG(1,THRU,(J − 1))

L1 is defined with the first four values of array A. These values would have to be two pairs of *x-y* coordinates. P5 is defined by either two or three values from array W depending on the values of I and J. The PRINT statement is suitable for printing out the values for the angles computed in the part program segment of Section 5.3. This statement would follow line 220. When this statement is reached, J equals 9. Therefore, eight values are printed out. Note the need for enclosing the high value within parentheses since it is an expression.

5.5 EXAMPLES

Example 5.1: Part Program with Subscripted Line Definitions

Figure 5.2 shows a part whose boundary conforms with a grid structure but whose upper edge exhibits a repetitious character different from that of the lower edge. Furthermore, the character of the upper edge is dependent on whether the number of vertical lines is an even number [8 shown in Figure 5.2(a)] or an odd number [7 shown in Figure 5.2(b)]. The grid structure will simplify the line definitions since they may now be defined as subscripted variables within a loop. The repetitious nature of the upper and lower edges also implies a loop for commanding the tool. We will do this in a counterclockwise direction beginning from SETPT. However, special care must be taken to exit from the loop and properly contour the left edge of the part as required by either an even or an odd number of vertical lines.

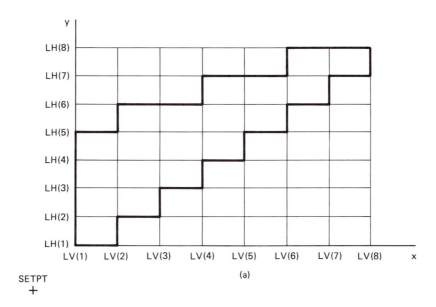

SETPT
+

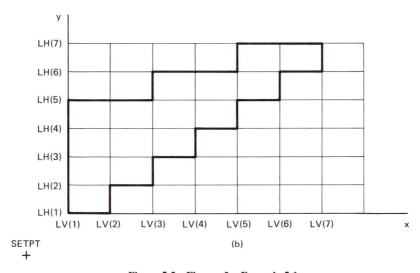

SETPT
+

Figure 5.2   Figures for Example 5.1.

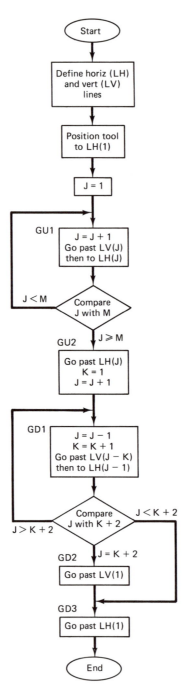

Figure 5.3   Flowchart for Example 5.1.

The flowchart of Figure 5.3 is for a part program to produce either of the parts shown in Figure 5.2. A part program written according to this flowchart is as follows:

```
          CUTTER/.2                                                    100
          RESERV/LH,30,LV,30                                           110
SETPT   = POINT/−1,−1,1                                                120
ORIG    = POINT/0,0                                                    130
LH(1)   = LINE/XAXIS                                                   140
LV(1)   = LINE/YAXIS                                                   150
          M = 8          $$ NUM OF LINES IN GRID                       160
          LOOPST                                                       170
          J = 0                                                        180
LD1)      J = J + 1                                                    190
          LH(J+1) = LINE/PARLEL,LH(1),YLARGE,J   $$ LOOP TO            200
          LV(J+1) = LINE/PARLEL,LV(1),XLARGE,J   $$ DEFINE             210
          IF (J−M) LD1, LD2, LD2     $$ VERT AND HORIZ LINES           220
LD2)      FROM/SETPT                                                   230
          GO/TO,LH(1)                                                  240
          J = 1                                                        250
GU1)      J = J + 1                                                    260
          TLRGT,GORGT/LH(J−1),PAST,LV(J)   $$ LOOP FOR LOWER EDGE      270
          GOLFT/LV(J),TO,LH(J)                                         280
          IF (J−M) GU1, GU2, GU2                                       290
GU2)      GOFWD/LV(J),PAST,LH(J)                                       300
          K = 1                                                        310
          J = J + 1                                                    320
GD1)      J = J − 1                                                    330
          K = K + 1                                                    340
          INDIRV/−1,0,0                                                350
          TLRGT,GOFWD/LH(J),PAST,LV(J−K)   $$ LOOP FOR UPPER EDGE      360
          GOLFT/LV(J−K),TO,LH(J−1)                                     370
          IF (J−K−2) GD3, GD2, GD1                                     380
GD2)      GORGT/LH(J−1),PAST,LV(1)  $$ EXECUTE IF M= ODD               390
          INDIRV/0,−1,0                                                400
GD3)      GOFWD/LV(1),PAST,LH(1)     $$ FINISH ALONG LV(1)             410
          LOOPND                                                       420
          GOTO/SETPT                                                   430
```

The grid structure is defined with subscripted vertical (LV) and horizontal (LH) lines. LH(1) and LV(1) are initially defined as the *x*- and *y*-axis, respectively. All others are defined parallel with these lines within a loop. The spacing between lines was arbitrarily chosen as one unit for both the LH and LV lines, thus producing square grid sections. This dimension is inserted with the scalar variable J (line 190). The part program establishes a part with an even number of vertical lines as determined by the value of the scalar variable M (line 160). Changing the value of M changes the part size with respect to the number of steps.

The motion commands take the tool counterclockwise around the part, alternately using the horizontal and vertical lines as check surfaces. The first loop takes the tool right along the lower edge, the second loop takes the tool left along the upper edge. The proper check surfaces are selected by the subscript values which become incremented, or decremented as the case may be, for each loop iteration. Special care must be used in computing the subscripts for the top edge so as to step left in the proper manner. This requires scalar variable K. Note that J is used as a subscript for the loop defining the lines and then redefined and used again as a subscript for the loops with the motion commands.

Example 5.2: Part Program with Subscripted Circle Definitions

Figure 5.4 shows a part that consists solely of tangent circles. From the geometry, there must always be an even number of circles. These are labeled in subscript manner. The circles whose centers lie on line LX are to be defined such that given the radius of the first circle, C(1), each subsequent circle must have a radius 1.25 times larger than its preceding tangent circle. Thus if C(1) has a radius of 0.5, then C(2) will have a radius of $(1.25)(0.5) = 0.625$. The radius of C(3) will be $(1.25)(0.625) = 0.7813$, and so on for the other circles. The radius of the largest circle (tangent to the end circles) is 1.25 times the sum of the radii of all circles whose centers are on LX. A part program to produce the part follows.

```
        RESERV/C,63                                              100
SETPT   = POINT/-1,-1,1                                          110
LX      = LINE/XAXIS                                             120
        LOOPST                                                   130
        R = .5                                                   140
        M = 8        $$ NUM OF CIRCLES (EVEN NUMBER)             150
        CR = 0       $$ FOR LOCATING CIRCLE CENTERS ON LX        160
        J = 0                                                    170
CR1)    J = J + 1      $$ LOOP TO DEFINE LINES                   180
        C(J) = CIRCLE/(CR+R),0,R                                 190
        CR = CR + 2*R                                            200
        R = 1.25*R                                               210
        IF (J + 1 - M) CR1, CR2, CR2                             220
CR2)    C(M) = CIRCLE/YSMALL,IN,C(1),IN,C(J),RADIUS,(1.25*CR/2)  230
        FROM/SETPT                                               240
        GO/ON,LX                      $$ POSITION THE TOOL FOR   250
        TLON,GORGT/LX,TO,C(1)     $$ CCW MOTION                  260
        INDIRV/0,-1,0                                            270
        J = 0                                                    280
CL1)    J = J + 1     $$ CUTTING LOOP                            290
        TLRGT,GOFWD/C(J),TANTO,C(J+1)                            300
        IF (J + 1 - M) CL1, CL2, CL2                             310
CL2)    GOFWD/C(M),TANTO,C(1)                                    320
        GOFWD/C(1),PAST,LX                                       330
        LOOPND                                                   340
        GOTO/SETPT                                               350
```

The part program is written such that given the radius of the first circle (line 140) and the total number of circles (line 150), the geometry of the part is completely specified. The circles whose centers are on line LX are defined in a loop (lines 180 through 230) with a radius whose value increases by the factor 1.25 (line 210). The last circle is defined in line 230. CR is a scalar variable with the *x*-coordinate of the center of the circles that are located on line LX and with the radius of the last circle C(M). The motion commands are executed

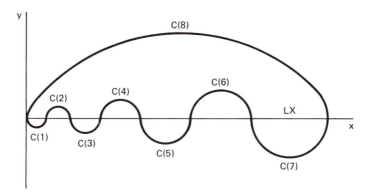

Figure 5.4   Geometry of Example 5.2.

in a loop (lines 290 through 320) that is followed by tool movement along circle C(M) back to the initial circle (line 330).

5.6 PRECAUTIONS IN USING SUBSCRIPTED VARIABLES

Part programs with arrays present few new challenges to checkout and debugging. To the list of program development and debugging aids already developed, we add these suggestions.

1. The RESERV statement specifies an upper limit to the size of the array. An error message is printed if the subscript exceeds this value. If possible, do not oversize the array so as to retain this built-in error checking feature.

2. The applications to which APT applies seldom require nesting of subscripts. Nesting, such as A(B(J)), complicates the program for readability and should be avoided by assigning the value of the innermost subscripted variable to a simple variable for use as a subscript. Although this introduces a new variable, it also tends to reduce errors during program preparation.

3. APT permits different data types to be assigned to different elements of an array. For example, one element may contain a scalar value, another may contain a line definition, another a circle definition, and so on. Such mixing of data types leads to confusion in identifying elements of an array during program checkout because the functional intent of the array will be disguised by neutral name selection. Therefore, restrict the use of each array to one data type.

4. Use the PRINT/3 statement to print out values of array elements during program checkout. The printout includes the subscript value. If variables are used as subscripts, this printout value can be correlated with that expected from the program, thus helping to isolate problems.

PROBLEMS

5.1 Rework Problem 4.8 using subscripted circle definitions for the circles of radius r_3.

5.2 Rework Problem 4.9 using subscripted line definitions for the polygon sides and elsewhere as needed.

5.3 Rework Problem 4.10 using subscripted circle definitions.

5.4 Rework Problem 4.11 using subscripted circle definitions.

5.5 Rework Problem 4.12 using subscripted circle and line definitions.

5.6 Figure P5.6 shows a part composed only of circular edges as determined by the radii r_1 and r_2, and n, the number of circular edges. Write a general-purpose APT part program that defines the surfaces as subscripted variables and that takes a tool counterclockwise around the part. For the computer run let $r_1 = 6$ in., $r_2 = 2$ in., $n = 8$, and the tool diameter $= \frac{1}{4}$ in.

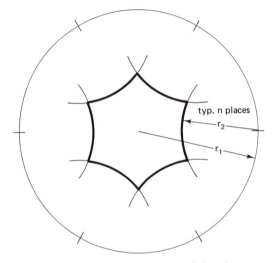

typ. n places

r_2

r_1

n = number of circular edges (n = 6 shown)   **Figure P5.6**

5.7 The part of Figure P5.7 has an edge with teeth determined by lines at angles α and β and by the dimension d_1, and an overall size determined by dimension d_2 and the number of teeth n. The length is determined by the number and size of teeth. Write a general-purpose APT part

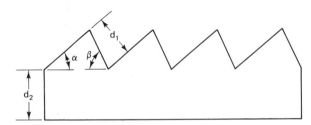

d_1

α β

d_2

n = number of teeth (n = 4 shown)   **Figure P5.7**

program to produce the part by defining the teeth with subscripted geometric entities and take a tool clockwise around the part. Run the part program with $\alpha = 40°$, $\beta = 65°$, $d_1 = 1$ in., $d_2 = 1.25$ in., $n = 6$, and a tool diameter of $\frac{1}{4}$ in.

5.8 The triangular-shaped part of Figure P5.8 has two symmetrical adjacent edges with an n number of semicircular arcs of radius r_1 that blend with the end circular transitions of radius r_2. Note that depending on whether n is an even number (solid semicircles only) or an odd number (with additional dashed semicircles) the end circular transition may lie on either side of the edge. Write a general-purpose APT part program with subscripted circle definitions and the logic to define the edges and to drive along the proper surface as determined by the value of n. Run the part program with $\alpha = 40°$, $r_1 = \frac{3}{4}$ in., $r_2 = \frac{3}{8}$ in., $d_1 = 10$ in., $d_2 = 1.75$ in., and tool diameter $= \frac{1}{4}$ in. Two runs are required to test the part program logic. Let $n = 4$ for one run and $n = 5$ for the other. Take the tool in a counterclockwise direction.

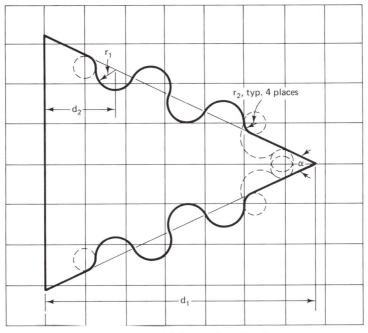

n = number of semicircles per edge with radius r_1
(n = 4 for solid; n = 5 for dashed)

Figure P5.8

5.9 Figure P5.9 shows a part with a variable number of slots in each long dimension edge as determined by the value of n. Write a general-purpose APT part program to define the slot edges with subscripted geometric entities and take a tool counterclockwise around the part. Run the part program with $n = 7$, $r_1 = \frac{3}{8}$ in., $d_1 = 1\frac{1}{2}$ in., $d_2 = 1\frac{1}{4}$ in., $d_3 = 2$ in., $d_4 = 2$ in., $d_5 = 1\frac{1}{2}$ in., $d_6 = 1.75$ in., and tool diameter $= \frac{3}{8}$ in.

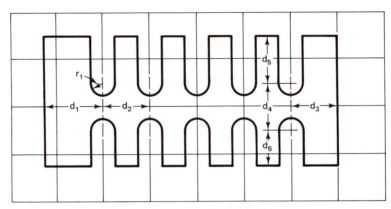

n = number of slots (n = 5 shown)

Figure P5.9

CHAPTER 6

General Conics, Loft Conics, and Tabulated Cylinders

The points, lines, and circles of Chapter 3 are the most used APT geometric figures. However, we may wish to describe figures that are of general conic form (ellipses, parabolas, hyberbolas), that are given only by the coordinates of a set of points, or that are described in analytic form for which no standard APT format exists. The latter two descriptions can be accommodated with the loft conic and tabulated cylinder features, respectively. In this chapter we discuss the general conic, loft conic, and tabulated cylinder features. As a result, we will greatly extend our ability to define geometric entities in APT.

6.1. GENERAL CONICS

The surface of revolution obtained from a line passing through and inclined with its axis of revolution is a pair of right circular cones whose vertices are on the axis as shown in Figure 6.1. When the cone is cut by a plane, the section formed is a circle, ellipse, parabola, or hyperbola, unless the plane passes through the vertex, in which case the section formed is either a point or a pair of intersecting lines. These sections, shown in Figure 6.2, are all referred to as **conic sections**. In APT they are called **general conics**. Those sections cut by planes not passing through the origin are designated as **regular conics**, while those cut by planes passing through the vertex are designated as **degenerate conics**. The regular conics characteristically contain no inflection point.

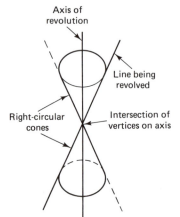

Figure 6.1   Generating right-circular cones.

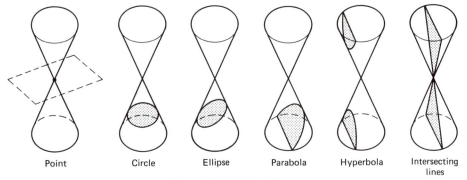

Figure 6.2   Conic sections.

6.1.1 Equation of a General Conic

The most general equation of second degree in two variables, x and y, is

$$Ax^2 + Bxy + Cy^2 + Dx + Ey + F = 0 \qquad (6.1)$$

where the coefficients are constants. If $B \neq 0$, the axis or axes of the conic section are rotated with respect to the coordinate axes x and y. Under this condition, it is shown in analytic geometry that the conic can be identified as follows:

1. A circle or ellipse if $B^2 - 4AC < 0$
2. A parabola if $B^2 - 4AC = 0$
3. A hyperbola if $B^2 - 4AC > 0$

However, the special case in which $B = 0$ occurs frequently (i.e., the cross-product term is missing). Under this condition, as shown in the following paragraphs, when this equation has a locus it is always a conic section and, if it is not a degenerate conic, it is

1. A circle if $A = C$ and A and C have the same sign
2. An ellipse if $A \neq C$ and A and C have the same sign
3. A parabola if A or C equal zero
4. A hyperbola if A and C have opposite signs

When using general conics in APT programs, we must have numeric values for the coefficients A through F in equation (6.1). From this form of the equation we may use the quadratic formula to solve for y as a function of x to get

$$y(x) = px + q \pm \sqrt{rx^2 + sx + t} \tag{6.2}$$

where p through t must each have numeric values. We may also solve equation (6.1) for x as a function of y to get

$$x(y) = p'y + q' \pm \sqrt{r'y^2 + s'y + t'} \tag{6.3}$$

where p' through t' must each have numeric values.

For APT applications, we would not algebraically manipulate one form of equation above to get another since there are APT defining formats for each of them. We always choose the most convenient format available to avoid algebraic manipulation that has the potential for introducing errors into our work. In APT, we have three formats for the general conic, each of the form

$$name = \text{GCONIC}/list$$

where *name* is the symbolic name given to the general conic being defined and *list* is a parameter list unique to each definition. We have these defining formats.

1. For equation (6.1):

   GCONIC/A,B,C,D,E,F

   where A through F are the coefficients in the equation.
2. For equation (6.2):

   GCONIC/P,Q,R,S,T

   where P through T are coefficients p through t in the equation.
3. For equation (6.3):

   GCONIC/P,Q,R,S,T,FUNOFY

   where P through T are coefficients p' through t' in the equation.

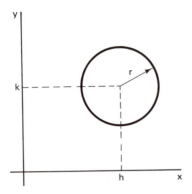

Figure 6.3 Geometry of a circle with center at (h, k).

APT assumes one of the formats based on the number of parameters, five or six, and the appearance, or not, of the keyword FUNOFY. The parameters must be numeric literals or scalar variables. All parameters must be present even though they may be zero. As for circles and lines, APT treats the general conics as cylindrical surfaces perpendicular to the xy-plane and infinite in the $\pm z$-direction. In this book we restrict our use of the general conic to the first format shown above.

6.1.2 Circle

Formats for defining circles in APT were presented in Chapter 3. Those definitions should be used whenever possible. However, an alternative for defining a circle is the GCONIC statement, especially when only the equation is available. The general equation for the circle is

$$(x - h)^2 + (y - k)^2 = r^2$$

where, as shown in Figure 6.3, h and k are the coordinates of the center and r is the radius. When expanded, this equation becomes

$$x^2 + y^2 - 2hx - 2ky + h^2 + k^2 - r^2 = 0$$

which is in the general conic form with the cross-product term zero. In this case,

$$
\begin{aligned}
A &= 1 & D &= -2h \\
B &= 0 & E &= -2k \\
C &= 1 & F &= h^2 + k^2 - r^2
\end{aligned}
$$

We identify the conic as a circle by noting that $A = C$ and A and C have the same sign.

6.1.3 Ellipse

The ellipse is defined as the locus of a point such that the sum of its undirected distances from two fixed points, the foci, is a constant (see Figure 6.4). If we place the center of

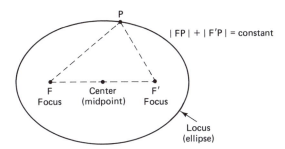

Figure 6.4   Definition of an ellipse.

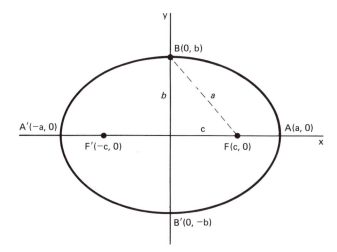

Figure 6.5   Geometry of an ellipse with center at (0, 0).

the ellipse at the origin of the coordinate system and if we align the ellipse such that its foci lie on one of the axes, from the geometry of Figure 6.5 we obtain the following ellipse equations:

$$\frac{x^2}{a^2} + \frac{y^2}{b^2} = 1 \qquad \text{(for the foci on the } x\text{-axis)}$$

$$\frac{x^2}{b^2} + \frac{y^2}{a^2} = 1 \qquad \text{(for the foci on the } y\text{-axis)}$$

where $a^2 = b^2 + c^2$.

The major axis of the ellipse is defined as the line through A-A′ with length $2a$, and the minor axis is defined as the line through B-B′ with length $2b$. Comparable equations for an ellipse with center at (h, k) are

$$\frac{(x - h)^2}{a^2} + \frac{(y - k)^2}{b^2} = 1 \qquad \text{(for major axis parallel with } x\text{-axis)}$$

$$\frac{(x - h)^2}{b^2} + \frac{(y - k)^2}{a^2} = 1 \qquad \text{(for major axis parallel}$$
with y-axis)

When the first of these equations is expanded the result is

$$b^2x^2 + a^2y^2 - 2b^2hx - 2a^2ky + b^2h^2 + a^2k^2 - a^2b^2 = 0$$

which is in the general conic form with the cross-product term zero. We observe that the coefficients of x^2 and y^2 will always have the same sign and we identify the conic as an ellipse by noting that $A \neq C$ and that A and C have the same sign. If c is made zero (see Figure 6.5), then $a^2 = b^2$ and the figure reduces to a circle.

The ellipse is a frequently occurring geometric figure. Generally, its dimensions are given rather than the coefficients of the general equation. So a special APT format is available for its definition.

name = ELLIPS/CENTER,*point,length1,length2,*θ

Name is the symbolic name given to the ellipse whose center is at *point*, whose semimajor axis length is *length1*, whose semiminor axis length is *length2*, and whose semimajor axis is inclined at an angle θ to the *x*-axis.

From the discussion above we note that the semimajor axis has length a and the semiminor axis has length b. The angle must be in degrees measured from the positive *x*-axis. An example appears in Figure 6.6. Other formats for an ellipse are given in Section 12.6.3. The ellipse is used in designing gears.

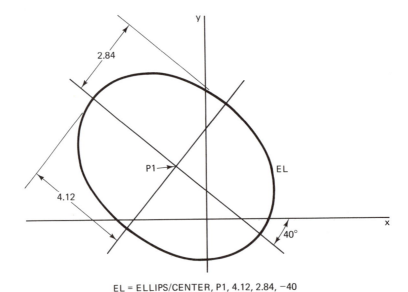

EL = ELLIPS/CENTER, P1, 4.12, 2.84, −40

Figure 6.6 Defining an ellipse with ELLIPS.

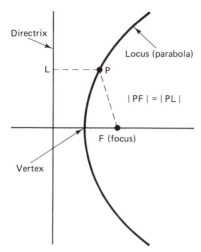

Figure 6.7   Definition of a parabola.

6.1.4 Parabola

The parabola is defined as the locus of a point such that the sum of its undirected distances from a fixed point, the focus, and a fixed line, the directrix, are equal (see Figure 6.7). We note that a vertex is defined as the point at which the parabola axis, the line about which the parabola is symmetrical, intersects the parabola. If we place the vertex of the parabola at the origin of the coordinate system and align the parabola axis coincident with either the x-axis or y-axis, we obtain the following parabola equations (see Figure 6.8):

$$y^2 = 2px \qquad \text{(for parabola axis on x-axis)}$$
$$x^2 = 2py \qquad \text{(for parabola axis on y-axis)}$$

where p is the distance from the focus to the directrix. From Figure 6.8 we see that the parabolas open to the right or upward. If the equations are $y^2 = -2px$ and $x^2 = -2py$, the parabolas would open to the left and downward, respectively.

Comparable equations for a parabola with vertex at (h, k) are

$$(y - k)^2 = 2p(x - h) \qquad \text{(for parabola axis parallel with x-axis)}$$

$$(x - h)^2 = 2p(y - k) \qquad \text{(for parabola axis parallel with y-axis)}$$

When the first of these equations is expanded, the result is

$$y^2 - 2px - 2ky + 2ph + k^2 = 0$$

which is in the general conic form with $A = B = 0$. A similar equation is obtained for a parabola whose axis is parallel with the y-axis. When in this form, we identify the conic as a parabola by noting that one of the coefficients A or C is always zero. The commonly encountered quadratic equation $y = ax^2 + bx + c$ is of parabolic form.

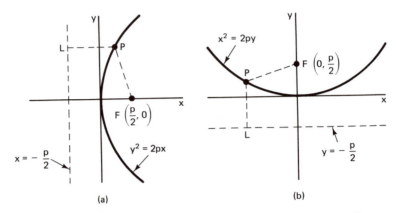

Figure 6.8   **Parabolas with axis parallel to (a) *x*-axis and (b) *y*-axis.**

Parabolas are used in designing mirrors, antennas, and reflectors in general because of their property of reflecting parallel rays to the focus or, in the reverse sense, of reflecting rays from the focus to be parallel with the axis.

6.1.5 Hyperbola

The hyperbola is defined as the locus of a point such that the difference of its undirected distances from two fixed points, the foci, is a constant (see Figure 6.9). We note that the point can be on either of two curves and still satisfy the definition. The formation of two curves is graphically illustrated in Figure 6.2. Because APT assumes these curves infinite and perpendicular to the *xy*-plane, these surfaces are also called **sheets**. If we place the center of the hyperbola at the origin of the coordinate system and if we align

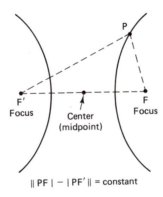

$$\| PF \| - | PF' \| = \text{constant}$$

Figure 6.9   **Definition of a hyperbola.**

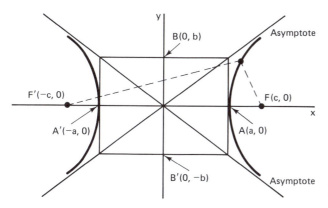

Figure 6.10 Geometry of the hyperbola with center at (0, 0).

the hyperbola such that its foci lie on one of the axes, from the geometry of Figure 6.10 we obtain the following hyperbola equations:

$$\frac{x^2}{a^2} - \frac{y^2}{b^2} = 1 \qquad \text{(for the foci on the x-axis)}$$

$$\frac{y^2}{a^2} - \frac{x^2}{b^2} = 1 \qquad \text{(for the foci on the y-axis)}$$

where $c^2 = a^2 + b^2$.

The transverse axis of the hyperbola is defined as the line through A-A′ with length $2a$, and the conjugate axis is defined as the line through B-B′ with length $2b$. The curves approach the asymptotes at distances far removed from the foci. The asymptotes in the figure have the equations $y = \pm(b/a)x$. Comparable equations for a hyperbola with center at (h, k) are

$$\frac{(x - h)^2}{a^2} - \frac{(y - k)^2}{b^2} = 1 \qquad \begin{array}{l}\text{(for transverse axis parallel} \\ \text{with x-axis)}\end{array}$$

$$\frac{(y - k)^2}{a^2} - \frac{(x - h)^2}{b^2} = 1 \qquad \begin{array}{l}\text{(for transverse axis parallel} \\ \text{with y-axis)}\end{array}$$

When the first of these equations is expanded, the result is

$$b^2x^2 - a^2y^2 - 2b^2hx + 2a^2ky + b^2h^2 - a^2k^2 - a^2b^2 = 0$$

which is in the general conic form with the cross-product term zero. A similar equation is obtained for a hyperbola whose transverse axis is parallel with the y-axis. In this form we identify the conic as a hyperbola by noting that A and C have opposite signs.

The hyperbola portrays a functional relationship between two variables whereby one of them is inversely related to the other. At times it is more convenient to describe

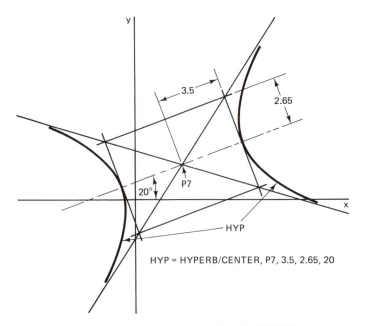

Figure 6.11 Defining a hyperbola with HYPERB.

the hyperbola in terms of its asymptotic characteristics as shown in Figure 6.11 rather than by its general equation. The following APT statement is provided for this purpose:

$$name \ = \ \text{HYPERB/CENTER}, point, length1, length2, \theta$$

Name is the symbolic name given to the hyperbola whose center is at *point*, whose half-transverse axis length is *length1*, whose half-conjugate axis length is *length2*, and whose transverse axis is inclined at an angle θ to the *x*-axis.

From the discussion above we note that the half-transverse axis has length *a* and the half-conjugate axis has length *b*. The angle must be in degrees measured from the positive *x*-axis. An example appears in Figure 6.11.

6.1.6 Examples

Two examples are presented. The first illustrates the use of the GCONIC and HYPERB for drive and check surfaces and shows that both sheets of the hyperbola are represented by the same equation. The second example shows the use of the ELLIPS definition and how, with appropriate computations, the GCONIC and ELLIPS may be used as check surfaces with the TANTO modifier.

Example 6.1: The Hyperbola and Parabola

Figure 6.12 shows a part whose boundary is formed by a parabola, both sheets of a hyperbola, and a straight line. The center of the hyperbola is located at the origin of the coordinate

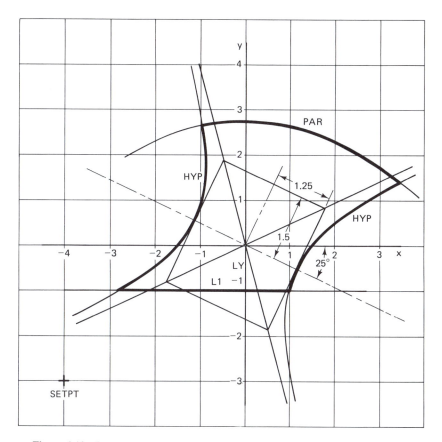

Figure 6.12   Part configuration composed of a parabola, both sheets of a hyperbola, and a line.

system and is described in terms of its transverse and conjugate axes lengths. The axis of the parabola is parallel with the y-axis and opens downward. Its vertex is located at (h, k) = $(0, 2.75)$ and $p = 4.5$. From these parameters the equation is

$$x^2 = -9(y - 2.75)$$

which, when expanded, becomes the general conic equation

$$x^2 + 9y - 24.75 = 0$$

The following is a suitable APT part program for taking a small-diameter tool counterclockwise around the part.

```
SETPT = POINT/−4,−3,0                                      100
L1    = LINE/XAXIS,−1                                      110
LY    = LINE/YAXIS                                         120
HYP   = HYPERB/CENTER,(POINT/0,0),1.25,1.5,−25            130
```

```
PAR    = GCONIC/1,0,0, 0,9, − 24.75              140
         FROM/SETPT                               150
         GO/TO,L1                                 160
         TLRGT,GORGT/L1,PAST,LY                   170
         GOFWD/L1,PAST,HYP                        180
         GOLFT/HYP,PAST,PAR                       190
         GOLFT/PAR,PAST,2,INTOF,HYP               200
         GOLFT/HYP,PAST,L1                        210
         GOTO/SETPT                               220
```

This example illustrates a type of problem that may develop when using general conics. First, it would be expected that lines 170 and 180 would be replaced with the statement

TLRGT,GORGT/L1,PAST,2,INTOF,HYP

However, we see that the left sheet of HYP intersects L1 at a small angle that may cause APT difficulty in determining the second intersection. When this part program was run with the IBM APT-AC system, the program determined the second intersection essentially at the location of the first intersection. Problems of this sort are not uncommon when using special curves such as introduced in this chapter. To avoid them it may be necessary to introduce an auxiliary check surface, LY in this case, then continue forward from that point. When the hyperbola angle was changed from − 25° to 0°, this problem did not occur because the left sheet was more nearly perpendicular to L1.

Example 6.2: Using ELLIPS and Defining a Parabola Tangent to a Circle

Figure 6.13 shows a part whose boundary is partly composed of an ellipse and a parabola that is tangent to a circle. The coordinate system is placed such that the center of C1 is at the origin. C2 is tangent to C1 and located with its center at (3, 0). The ellipse E1 is to be tangent to C2, its center x-coordinate is to have the value 1, its center y-coordinate lies on the line through the center of C2 as shown (this line makes an angle of 30° with the x-axis), and its semimajor axis length is given as 2.5. The parabola is located with its vertex at $(-5, 1.5)$ and is tangent to C1. A part program for this geometry is as follows:

```
SETPT = POINT/ − 6, − 1,2                         100
ORIG  = POINT/0,0                                 110
LX    = LINE/XAXIS                                120
C1    = CIRCLE/CENTER,ORIG,RADIUS,3.5             130
C2    = CIRCLE/CENTER,(POINT/3,0),RADIUS,.5       140
SMIN  = 2/COSF(30)                                150
SMINR = SMIN − .5                                 160
YE    = − SMIN*SINF(30)                           170
E1    = ELLIPS/CENTER,(POINT/1,YE),2.5,SMINR, − 60  180
XP    = − 2.2707                                  190
YP    = SQRTF(12.25 − XP*XP)                      200
P     = − YP*(XP + 5)/XP                          210
PAR   = GCONIC/1,0,0, 10,( − 2*P),(25 + 3*P)      220
L1    = LINE/(POINT/ − 5,1.5),ATANGL,90           230
        FROM/SETPT                                240
        GO/TO,LX                                  250
        TLRGT,GORGT/LX,PAST,E1                    260
        GOLFT/E1,TANTO,C2                         270
```

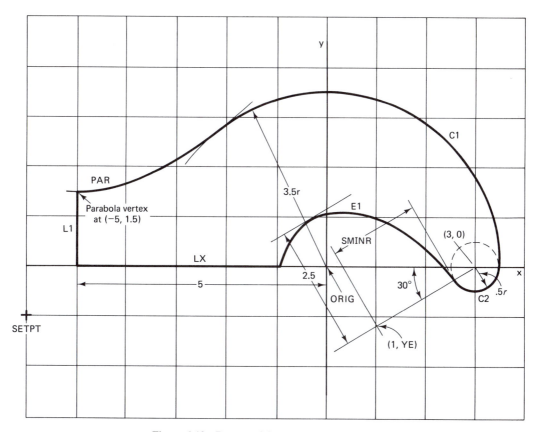

Figure 6.13 Part requiring computation of tangent points.

GOFWD/C2,TANTO,C1	280
GOFWD/C1,TANTO,PAR	290
GOFWD/PAR,PAST,L1	300
GOLFT/L1,PAST,LX	310
GOTO/SETPT	320

We must compute the ellipse semiminor axis length and the y-coordinate of its center. From the geometry it is relatively easy to compute the semiminor axis length SMINR, as shown in lines 150 and 160. The y-coordinate of the center is computed in line 170 and the ellipse is defined in line 180.

The computations for the parabola are more involved. The equation of a parabola with vertex at $(-5, 1.5)$ is

$$(x + 5)^2 = 2p(y - 1.5) \tag{6.4}$$

or

$$x^2 + 10x - 2py + 25 + 3p = 0$$

This equation is in the form needed to use the GCONIC format, but we do not yet have a value for p. Since the parabola is to be tangent to circle C1, the procedure will be to equate the equations for the slopes of the parabola and circle (since the slopes must be equal at the point of tangency), compute the coordinates of the tangent point, and substitute these coordinate values into an expression for p.

The equation of a circle with center at (0, 0) and radius = 3.5 is

$$x^2 + y^2 = 12.25 \tag{6.5}$$

Differentiating this equation, we get

$$2x\,dx + 2y\,dy = 0$$

from which the slope of the circle is

$$\frac{dy}{dx} = -\frac{x}{y} \tag{6.6}$$

Differentiating the equation of the parabola [equation (6.4)] gives

$$2(x + 5)\,dx = 2p\,dy$$

from which its slope is

$$\frac{dy}{dx} = \frac{x + 5}{p} \tag{6.7}$$

Upon equating the slopes [equations (6.6) and (6.7)], we have

$$\frac{x + 5}{p} = -\frac{x}{y}$$

from which

$$p = \frac{-y(x + 5)}{x} \tag{6.8}$$

To obtain a numeric value for p it is necessary to have values for the coordinates of the tangent point. To get these coordinate values we first substitute equation (6.8) into (6.4), giving

$$(x + 5)^2 = \frac{-2y(x + 5)(y - 1.5)}{x}$$

or

$$x^2 + 5x = -2y^2 + 3y \tag{6.9}$$

From (6.5), the equation of the circle, we solve for y in terms of x ($y = \sqrt{12.25 - x^2}$) and substitute this into (6.9), giving

$$x^2 + 5x = -2(12.25 - x^2) + 3\sqrt{12.25 - x^2}$$

which reduces to

$$x^4 - 10x^3 - 15x^2 + 245x + 490 = 0$$

This is a quartic equation which, when solved independently of the APT part program, gives us the x-coordinate of the tangent point as $x = -2.2707$. We then use this value to compute the tangent point y-coordinate value from the equation of the circle, line 200, and then use both coordinate values to compute p, line 210, from equation (6.8). The value of p is now available for use in the GCONIC statement to compute the values of the parameters E and F as shown in line 220.

The motion statements take the tool around the part in a counterclockwise direction. Note that a TANTO modifier is used when going from the ellipse E1 to C2 and again when going from C1 to the parabola PAR.

6.2 LOFT CONICS

The general conics of Section 6.1 were described to APT in terms of a mathematical expression—the coefficients of the most general equation of second degree in two variables. We now address the problem of conveying to the APT system information from which it can construct the equation of a general conic, other than by its coefficients.

Often, a smooth curve is needed as a transition from one surface to another, perhaps for technical reasons but maybe also for its esthetic value. Whether an ellipse, parabola, or hyperbola is used is generally immaterial. In fact, its shape may be derived from a smoothly curved plastic device known as a spline, French curve, or ship's curve. By design, some of these devices closely approximate the shape of varying general conics. This same device may be used to fit a smooth curve through plotted data points. Conversely, deriving data points from such a curve is also a widely accepted practice.

The loft conic to be introduced is a mathematical counterpart of the physical spline. The physical spline is a long, narrow, flexible strip of wood or plastic that is shaped by hand to form a smooth curve and held in place by weighted supports. This device was used in a special attic or loft to lay out the contours of ships. The term "lofting with ship's curves" was applied to this process. If we let the APT system mathematically form smooth curves of general conic shape through data points, we can define the curve as a **loft conic**. Internal to the APT system, the result will be the same as if we had defined a general conic via the GCONIC statement.

6.2.1 LCONIC Formats

A curve must exist so that we may define the loft conic. Given this curve, it is then necessary to define points on the curve, or to define some combination of points and slopes for the curve as described below. The assumption is that we do not know the equation of the curve, for if we did, we certainly would define it with the GCONIC statement. The APT formats for the loft conic are

$$name = \text{LCONIC/} \begin{Bmatrix} \text{5PT,}point,point,point,point,point \\ \text{4PT1SL,}point,slope,point,point,point \\ \text{3PT2SL,}point,slope,point,slope,point \end{Bmatrix}$$

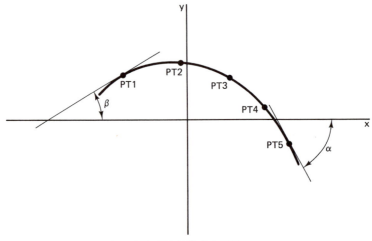

LCON1 = LCONIC/5PT, PT1, PT2, PT3, PT4, PT5
LCON1 = LCONIC/4PT1SL, PT5, TANF(−ALFA), PT3, PT2, PT1
LCON1 = LCONIC/3PT2SL, PT1, TANF(BETA), PT5, TANF(−ALFA), PT3

Figure 6.14   Curve defined by each of the three loft conic formats.

where *name* is the symbolic name given to the loft conic being defined, *point* is either the symbolic name for a point or its *x-y* coordinates, and *slope* is the slope of the curve at the *point* immediately preceding the slope value. Either all symbolic points or all *x-y* coordinates must be used; they cannot be mixed in any of the formats. Note that five independent conditions are needed to define the loft conic.

The first format defines the loft conic with five points. The points do not have to be given in the order of their appearance on the curve. The second format defines the loft conic with four points and the slope of the curve at the first point listed, which must be an endpoint on the curve. The third format defines the loft conic with three points and the slopes of the curve at the first two points listed, respectively, which must be endpoints on the curve. The third point is an in-between point. These formats are illustrated in Figure 6.14.

An equivalent infinite value for the slope should be chosen if the tangent to the curve is parallel with the *y*-axis; 1,000,000 is such a value. Numerical values for the point coordinates must be accurately determined and entered with adequate precision so that the APT system can derive a general conic representative of the curve being defined. Special problems arise with the LCONIC definition as discussed in Section 6.2.2.

6.2.2 Examples

The first example shows the geometry from various loft conics specified with points only. The second example shows contouring along loft conics and reveals problems that may result from their use.

Example 6.3: Notes on Loft Conic Geometry

Curves drawn by a French curve closely follow a general conic shape when defined as a loft conic. Several such curves are shown in Figure 6.15. Complete curves are shown (except for the hyperbola) by plotting the cut vectors produced by the APT system while driving on each curve. The check surfaces are not shown. The points specified in the LCONIC definition are also plotted. These LCONICs are

LCON1 = LCONIC/5PT,1.5, − 1, 1.4, − .5, 1.55,.5, 1.8,1.25, 2,1.75
LCON2 = LCONIC/5PT, − .25, − .85, .25, − .35, .5,.5, 0,1.5, − .85,2.1
LCON3 = LCONIC/5PT, − .5,2.25, .75,2.1, 1.35,1.75, 2,1.25, 2.5,.75
LCON4 = LCONIC/5PT, − 1.5, − 2, .75,1.2, 0,1, 1.25,1.45, 2,2.25

The points for LCON1 produce an hyperbola (both sheets are shown). The other curves are ellipses. Care must be taken when driving onto the curves to avoid inadvertent stopping at invisible check surfaces—portions of the general conic outside the endpoints of the defining segments.

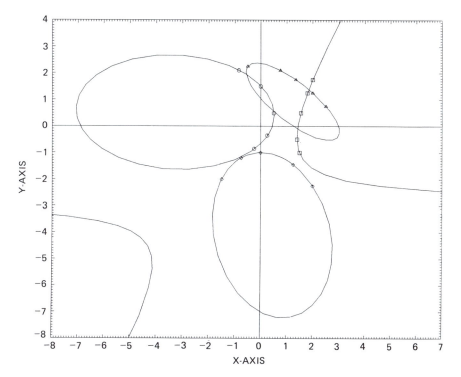

Figure 6.15   Selection of curves produced from loft conic definitions.

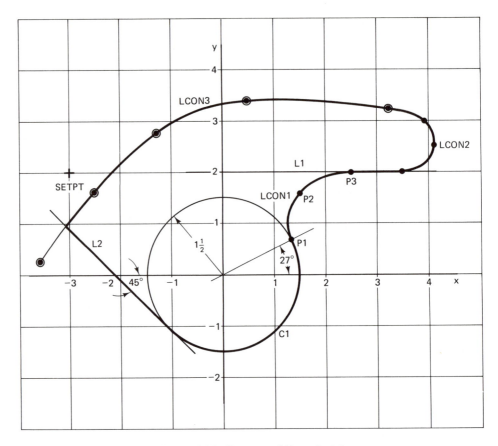

Figure 6.16   Geometry of Example 6.4.

Example 6.4: Contouring Along LCONICs

Figure 6.16 shows a part comprised partly of loft conics. Contouring is counterclockwise beginning at SETPT. The following is a suitable part program segment.

C1	= CIRCLE/0,0,1.5	100
L1	= LINE/XAXIS,2	110
L2	= LINE/XSMALL,TANTO,C1,ATANGL,−45	120
LT	= LINE/3.25,3.25,ATANGL,80,XAXIS	130
P1	= POINT/RTHETA,XYPLAN,1.5,27	140
P2	= POINT/1.5,1.58	150
P3	= POINT/2.5,2	160
LCON1	= LCONIC/3PT2SL,P1,TANF(27-90),P3,0,P2	170
LCON2	= LCONIC/4PT1SL,3.5,2, 0, 4.13,2.5, 3.92,3, 3.25,3.25	180
LCON3	= LCONIC/5PT,−3.6,.25, −2.5,1.6, −1.25,2.75, .5,3.4, 3.25,3.25	190
	FROM/SETPT	200
	GO/PAST,L2	210

TLRGT,GOLFT/L2,TANTO,C1	220
GOFWD/C1,TANTO,LCON1	230
GOFWD/LCON1,TANTO,L1	240
GOFWD/L1,TANTO,LCON2	250
GOFWD/LCON2,ON,LT	260
GOFWD/LCON3,PAST,L2	270
GOTO/SETPT	280

Each loft conic definition of Section 6.2.1 is used in this part program. When necessary, the slope is computed (line 170); otherwise, it is obvious from Figure 6.16.

The TANTO check surface modifier must be used with care. Tangent conditions at endpoints of loft conics on simple curves cause little problem in contouring when the tangents are made equal. However, it is difficult to make two loft conics tangent to each other at a point, as is the case of LCON2 and LCON3. To avoid problems in driving from LCON2 to LCON3, line LT was defined through the common point of these two loft conics and made nearly normal to them. The tool is driven onto this auxiliary check surface to avoid the tangent condition. If the two curves have nearly equal slopes at this point, there should be no problem in continuing the contouring with a new drive surface, such as in lines 260 and 270. But an out-of-tolerance condition should be anticipated. It is common to use an auxiliary check surface in this manner.

6.3 TABULATED CYLINDERS

The APT geometric definitions presented thus far allow the part programmer to describe what may be referred to as "regular" geometric figures, that is, those easily described from their geometry or with the general conic equation. Certainly, for many applications this much capability is adequate. However, this restriction must be removed so that part programs may be written to accommodate geometric figures that may be known only as follows.

1. Analytically as a function of two variables but whose equation is not in the form of a general conic, such as the curves in Figure 6.17
2. Analytically in the form of parametric equations, such as those shown in Figure 6.18
3. As a physical model (probably a drawing) that cannot be identified as one of the standard APT surfaces (circle, plane, general conic, etc.)
4. In tabular form as a set of points represented by their x- and y-coordinate values

These geometric figures may all be represented in APT as tabulated cylinders.

6.3.1 The Tabulated Cylinder Definition

In this chapter the tabulated cylinder is discussed for the xy-plane only. Other definitions for three-dimensional applications are given in Section 12.6.4.

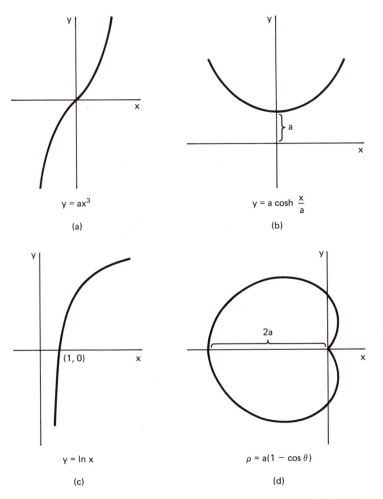

$y = ax^3$

(a)

$y = a \cosh \dfrac{x}{a}$

(b)

$y = \ln x$

(c)

$\rho = a(1 - \cos \theta)$

(d)

Figure 6.17   Curves with equations not in the form of a general conic: (a) cubical parabola; (b) catenary; (c) logarithmic; (d) cardioid.

For two-dimensional applications, the APT tabulated cylinder is a curve in the xy-plane defined by a set of points whose coordinate values are given. The fact that the APT system processes this curve as if it were a cylindrical surface, that is, the surface is generated by moving a line parallel with the z-axis (the generatrix) along the curve represented by the set of points (the directrix), and that the curve is defined as a set of points, leads to the term **tabulated cylinder**. As for previous cylindrical surfaces, the tabulated cylinder is considered to be infinite in the $\pm z$-direction. This representation is depicted in Figure 6.19.

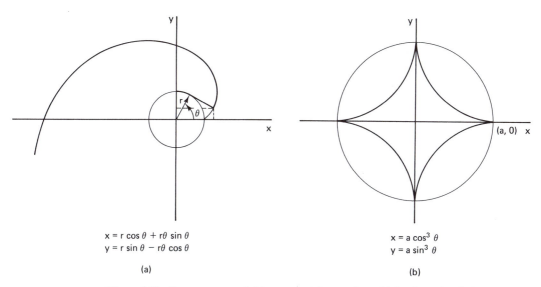

$x = r \cos \theta + r\theta \sin \theta$
$y = r \sin \theta - r\theta \cos \theta$

(a)

$x = a \cos^3 \theta$
$y = a \sin^3 \theta$

(b)

Figure 6.18   Curves represented by parametric equations: (a) involute of a circle; (b) hypocycloid of four cusps.

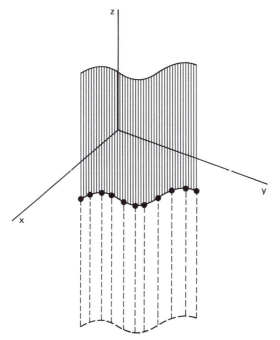

Figure 6.19   Set of points in the xy-plane represented as a tabulated cylinder.

The APT system processes the tabulated cylinder by fitting smooth curves through the given points. The resulting curves are used for interpolating between points as the tabulated cylinder is referenced in motion commands.

The APT format for defining the tabulated cylinder is as follows:

$$name = \text{TABCYL/} \left\{ \begin{array}{l} \text{NOX} \\ \text{NOY} \\ \text{NOZ} \\ \text{RTHETA} \\ \text{THETAR} \end{array} \right\} ,\text{SPLINE,[TRFORM,}matrix\text{,]}data$$

where *name* is the symbolic name given to the tabulated cylinder being defined, *data* is the list of points used to define the curve (see below), the first modifier identifies the coordinate-pair values given in *data*, while the second modifier, SPLINE, selects the curve-fitting option that produces the smoothest curve through the given points. Other curve-fitting options are available but none of them are discussed here. The optional TRFORM,*matrix* couplet is discussed in Example 7.2 and in Section 12.6.4.

Coordinate-pair values in the *xy*-plane can be supplied as follows (NOX and NOY are discussed in Section 12.6.4).

NOZ   = x,y values in the rectilinear coordinate system
RTHETA = r,θ values in the polar coordinate system
THETAR = θ,r values in the polar coordinate system

Angle θ must be in degrees measured from the positive *x*-axis. Distance r is measured from the origin of the coordinate system. The coordinate-pair values must be given in the order shown for the modifier.

The precise form for the list of points given in *data* is as follows:

$$a_1,b_1 \left[, \left\{ \begin{array}{l} \text{SLOPE,}s_1 \\ \text{NORMAL,}\alpha_1 \end{array} \right\} \right] ,a_2,b_2, \ \ldots \ ,a_k,b_k \left[, \left\{ \begin{array}{l} \text{SLOPE,}s_k \\ \text{NORMAL,}\alpha_k \end{array} \right\} \right]$$

where the a_i,b_i coordinate-pair values are determined by the modifier as discussed above. A slope constraint may be imposed on the tabulated cylinder at one, or both, of the endpoints. When supplied, the slope value applies to the immediately preceding point. It may be given as the tangent value s_i of the curve at that point, in which case the qualifier SLOPE must precede the value, or it may be given as the angle α_i of the normal to the curve at that point, in which case the qualifier NORMAL must precede the value. The angle must be in degrees measured from the positive *x*-axis. The slope constraint is needed to achieve a tangent condition between the tabulated cylinder and another curve.

When the NOZ modifier is used the list of points may include *x*-*y* pairs, point symbols, nested point definitions, or any combination of these forms. The points must be listed in succession beginning with either end of the tabulation and must be separated by commas. The modifier NOZ signifies that the *z*-coordinate value of symbolically named points is to be ignored; that is, the desired curve is the projection of the points on the *xy*-plane.

The following examples each define a tabulated cylinder of six points.

TAB1 = TABCYL/NOZ,SPLINE, − .1,3.2, .3,2.2, 1.1,1.5, 2.2,1.1, $
        3.1,.1, 3.8, − 1.4

TAB2 = TABCYL/NOZ,SPLINE,X1,Y1, X2,Y2, X3,Y3, X4,Y4, $
        X5,Y5, X6,Y6

TAB3 = TABCYL/NOZ,SPLINE,X1,Y1, .3,2.2, X3,Y3, 2.2,1.1, $
        X5,Y5, 3.8, − 1.4

TAB4 = TABCYL/NOZ,SPLINE,X1,3.2, X2,2.2, X3,1.5, X4,1.1, $
        X5,.1, X6, − 1.4

TAB5 = TABCYL/NOZ,SPLINE,PT1,PT2,PT3,PT4,PT5,PT6

TAB6 = TABCYL/NOZ,SPLINE,PT1,X2,Y2, 1.1,1.5, X(4),Y(4), $
        PT(5), 3.8, − 1.4

It is assumed that each of the examples above defines a tabulated cylinder through the same set of points. These points are listed in the definition consecutively in the order in which they occur along the curve. Since it does not matter from which end of the curve the points are listed, TAB5, for example, could also have been defined properly as follows:

TAB5 = TABCYL/NOZ,SPLINE,PT6,PT5,PT4,PT3,PT2,PT1

When the points are subscripted symbols, it is convenient to define the tabulated cylinder via the inclusive subscript feature described in Section 5.4. For example, we may have

TAB7 = TABCYL/NOZ,SPLINE,PT(1,THRU,6)

The tabulated cylinder must be defined with at least four points. The maximum number of points allowed is governed by the APT implementation. If a curve contains more points than can be accommodated by one TABCYL statement, as many adjoining TABCYLs as required may be used to define the curve. Adjoining TABCYLs should be overlapped by at least three points so as to blend the curves smoothly. Because TABCYLs require a large amount of computer storage space, they should not be defined in nests, looping regions, or macros in order to conserve memory space for the remainder of the part program. Instead, they should be assigned a symbolic name in a separate defining statement and then referenced elsewhere as required.

6.3.2 Using TABCYLs

Some understanding of the construction of tabulated cylinders is helpful when writing part program motion statements to avoid conflicting and ambiguous geometric conditions when TABCYLs are referenced. It is especially important to understand the construction in the endpoint region.

The tabulated cylinder defined with the SPLINE curve-fitting routine consists of a collection of cubic polynomials joining the tabulated points and a straight-line extension at each endpoint. Thus the tabulated cylinder is not infinite in the xy-plane nor is it limited

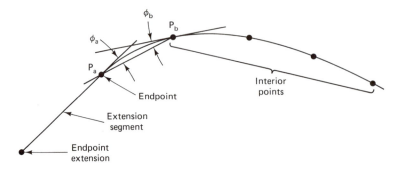

Figure 6.20   Tabulated cylinder extension segment construction.

to the confines of the endpoints. Instead, it extends beyond each endpoint by the amount of extension provided by the APT system.

When specified, the spline routine is used to compute the coefficients of the polynomials. The basic algorithm requires that curvature and slope computations be made and adjusted such that the resulting curve smoothly joins the points, a process called **fairing**. Further details are provided in Section 6.3.4. However, unless an endpoint slope is specified, the computations for each extension are based only on the geometry of an endpoint and its adjacent point and the slope of the polynomial at that adjacent point.

Figure 6.20 shows the construction in the vicinity of an endpoint. A straight-line segment joins endpoint P_a with its adjacent interior point P_b. Angle ϕ_b is computed from the slope of the polynomial at P_b and the slope of the line segment joining P_a with P_b. Then a straight line is passed through P_a to form an equilateral triangle with the equal angles ϕ_a and ϕ_b occurring at the vertices P_a and P_b, respectively. A point is computed on this straight line 10 units of length beyond endpoint P_a and in the direction opposite from P_b. This becomes the endpoint extension. The extension segment is the line joining the endpoint with the endpoint extension.

APT considers the possibility that the extension segments intersect each other as shown in Figure 6.21. If the computation reveals that intersection has occurred, each extension is shortened so that its length is short of the intersecting point by 1 percent of the distance from the endpoint to the intersecting point.

TABCYL references in motion commands must be considered with respect to the tabulated cylinder geometry described above. Tabulated cylinders may be used as drive

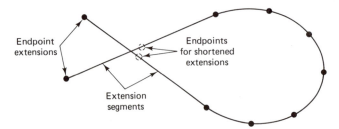

Figure 6.21   Intersecting extension segments.

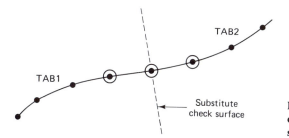

Figure 6.22   **Tangent TABCYLs with overlapping points (encircled) and a substitute check surface.**

and check surfaces. When used in GO and OFFSET startup commands, the tool may be checked anywhere on the tabulated cylinder, not just at a point of the definition. Also, one TABCYL may be used as a check surface when driving along another, and motion may be commanded in either direction along the TABCYL, that is, it is not restricted by the order in which the points are listed in the TABCYL definition. However, certain precautions must be observed when specifying check surfaces so as to avoid confusing the APT system during computation of a check point.

The geometry of Figure 6.22 shows where possible computational difficulties may be encountered. Two TABCYLs define one curve with overlap at the three encircled points. When driving along one TABCYL to a point tangent to the other, the APT system may become confused as to which tangent point to select from among alternative points. The problem can be resolved by introducing a new check surface and driving onto it. This substitute check surface should be constructed to pass through the middle point of the three overlapping points and be nearly orthogonal to the tabulated cylinders at that point. Then, instead of checking on the TABCYL with a TANTO modifier, the new check surface should be referenced with an ON modifier and motion continued from that point with another command. Note the substitute check surface in Figure 6.22.

Problems similar to the above may also arise when attempting to go tangent from a line to a TABCYL or from a TABCYL to a line, or for that matter with any type of curve in which a tangency condition is desired. These problems may also be resolved by introducing substitute check surfaces.

Other problems may arise when a tabulated cylinder defines a closed curve, such as shown in Figure 6.23(a), where the replacement geometry shows two overlapping TABCYLs and a substitute check surface. Similarly, a tabulated cylinder defining a self-intersecting curve should be replaced by the equivalent of two overlapping TABCYLs and a substitute check surface, as shown in Figure 6.23(b).

The location of the extension segments must be carefully considered during preparation of motion commands since they may inadvertently become the check surface. The geometry of Figure 6.24 shows how an incorrect stopping location may result. The 2,INTOF modifier is the remedy. Also, the extension segment must not be overlooked when a line is tangent to a tabulated cylinder at the end point. The use of a substitute check surface is necessary to avoid computational difficulty when either the line or TABCYL is the drive surface and the other is the check surface.

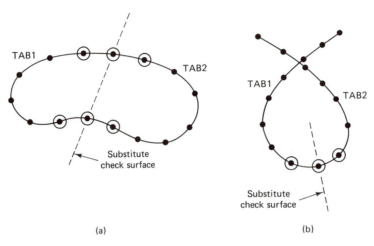

Figure 6.23   Two tabulated cylinders and a substitute check surface defining (a) a closed curve and (b) a self-intersecting curve. Overlapping points of the TABCYLs are shown encircled.

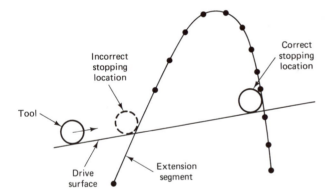

Figure 6.24   Inadvertently stopping on an extension segment.

Finally, caution must be exercised to avoid redundant points (multiply occurring values) in TABCYL definitions. These redundant points may not necessarily cause the APT system to fail during computation of the polynomials and movement of the tool, but some unwanted inflection may be introduced in the cut vector sequence, especially in the region of the redundant point.

Apart from the obvious case of inadvertently including redundant points in a TABCYL definition, such redundancy may also occur when computing points for a curve which is represented by different equations for adjoining segments of the curve. Such a curve might be one that is represented by the following two equations:

$$y = \frac{x^2}{2} \qquad\qquad 0 \leqslant x \leqslant 1$$

$$y = -\frac{x^2}{4} + \frac{3x}{2} - \frac{3}{4} \qquad 1 \leqslant x$$

Each equation properly represents the curve in the region indicated, but care must be taken so as not to recompute the point at the boundary of the two regions. Computations of this nature would typically involve subscripted scalar variables in loops where such redundancy is easily introduced through improper loop initialization.

6.3.3 Example

This example illustrates the tabulated cylinder feature. The part shown in Figure 6.25 consists of line L1, curve CURV1, involute of a circle (INVOL) given by the parametric equations of Figure 6.18(a) but rotated clockwise 20°, and line L4 at an angle of 75°

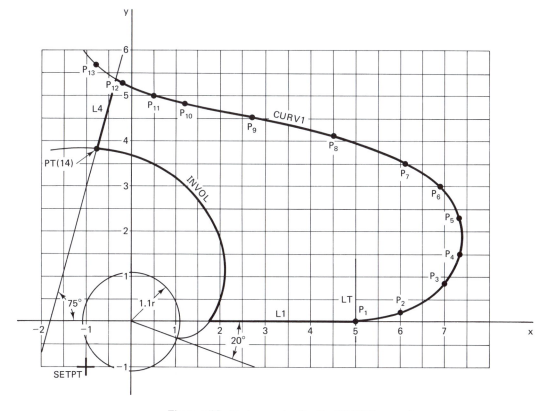

Figure 6.25   Part geometry for the TABCYL example.

with the *x*-axis and passing through a point of the involute. A part program to produce
the part is as follows:

```
SETPT = POINT/ - 1, - 1                                          100
L1      = LINE/XAXIS                                             110
LT      = LINE/YAXIS,5                                           120
CURV1 = TABCYL/NOZ,SPLINE,5,0,SLOPE,0, 6,.2, 7,.85,    $        130
                  7.32,1.5, 7.3,2.3, 6.9,3, 6.1,3.5,    $        140
                  4.5,4.1, 2.7,4.5, 1.2,4.85, .5,5,      $        150
                  - .2,5.3, - .8,5.7                             160
          RESERV/PT,15                                           170
          LOOPST                                                 180
          I = 1                                                  190
          THETA = 0                                              200
          RADUS = 1.1                                            210
          DTHETA = 15                                            220
          PHI = - 20                                             230
          SINPH = SINF(PHI)                                      240
          COSPH = COSF(PHI)                                      250
LA1)      SINTH = SINF(THETA)                                    260
          COSTH = COSF(THETA)                                    270
          THETR = THETA/57.2957795                               280
          XP = RADUS*(COSTH + THETR*SINTH)                       290
          YP = RADUS*(SINTH - THETR*COSTH)                       300
          XI = XP*COSPH - YP*SINPH                               310
          YI = XP*SINPH + YP*COSPH                               320
          PT(I) = POINT/XI,YI                                    330
          I = I + 1                                              340
          THETA = THETA + DTHETA                                 350
          IF (I - 16) LA1, LA2, LA2                              360
LA2)  LOOPND                                                     370
INVOL = TABCYL/NOZ,SPLINE,PT(ALL)                                380
L4      = LINE/PT(14),ATANGL,75                                  390
          FROM/SETPT                                             400
          GO/TO,L1                                               410
          TLRGT,GORGT/L1,ON,LT                                   420
          GOFWD/CURV1,PAST,L4                                    430
          GOLFT/L4,PAST,INVOL                                    440
          GOLFT/INVOL,PAST,L1                                    450
          GOTO/SETPT                                             460
```

The tabulated cylinder CURV1, lines 130 through 160, is defined by reading from
Figure 6.25 the coordinate values for various points selected on the curve and entering
them as numeric literals in the definition. The selected points are marked on the curve.

CURV1 is made tangent to the x-axis. A description for fairing CURV1 is given in Section 6.3.4.

The involute will be defined as a tabulated cylinder after first computing a set of points along the curve according to its parametric equations. These equations are

$$x' = r \cos \theta + r\theta \sin \theta$$

$$y' = r \sin \theta - r\theta \cos \theta$$

Angle θ is defined as shown in Figure 6.18(a) and must be expressed in radians because of the multiplication. The radius of the circle forming the involute is 1.1 (line 210). Computation of the x and y coordinate values for a set of points on the involute as θ varies from its initial value of 0° (line 200) in 15° increments (line 220) is performed in lines 260 through 300. The computation in line 280 converts θ from degrees to radians. Since the involute is actually rotated clockwise by 20° as given by ϕ in Figure 6.25, a transformation must be performed on the x' and y' coordinate values. This transformation is performed in lines 310 and 320 according to the equations

$$x = x' \cos \phi - y' \sin \phi$$

$$y = x' \sin \phi + y' \cos \phi$$

where ϕ is the angle of rotation ($-20°$ as given in line 230). These x and y coordinate values are stored in array PT (line 330) for later use. Fifteen points are computed for the involute as determined by the test performed in line 360. The points of the involute are computed as θ varies by 15° increments (line 350).

The involute INVOL is defined as a tabulated cylinder in line 380 using the inclusive subscript feature. This is a good application for that feature. Line L4 (line 390) is defined as passing through point 14 of the involute at an angle of 75°. The motion commands to take the tool counterclockwise around the part are given in lines 400 through 460 with a substitute check surface LT (line 420) used at the tangent point of L1 and CURV1.

6.3.4 Interpreting the TABCYL Printout

The APT system prints out information on each TABCYL in the part program. This information is obained from the canonical form entries for each TABCYL. The information is not of academic interest only; it is to be used by the part programmer to identify and correct irregular features in the curve. Figure 6.26 is a printout for the spline fit to the involute of Figure 6.25. It was produced by the IBM APT-AC system. The printout from other APT processors may differ from that shown here. This discussion is limited to interpretation of those printout values useful for the purpose stated above.

There are three main parts to Figure 6.26:

A. A listing of curvature values and a curvature plot
B. A printout of the rotation matrix elements for transformation of the TABCYL points to the uvw coordinate system

ISN 00029 INVOL   = TABCYL/NOZ,SPLINE,PT(ALL)

	.200000	.800000	1.80000	00003900	2.80000	3.80000

CURVATURE

1	3.54349
2	3.54407
3	1.47338
4	1.14080
5	.832851
6	.677951
7	.564552
8	.485606
9	.424977
10	.378492
11	.340528
12	.309654
13	.285140
14	.258259
15	0.0

CURVATURE

Ⓐ

DATA STORAGE = 125
ROTATION MATRIX

1.00000000	0.0	-0.0
-0.0	1.00000000	-0.0
0.0	0.0	1.00000000

Ⓑ

NUMBER OF POINTS = 17

Ⓒ

U ①	V ②	A ③	B ④	LENGTH ⑤	MAX ⑥	MIN ⑦
-8.67608748	2.01559039	0.0	0.0	10.0000000	0.0	0.0
1.03366188	-.376222157	.980334626 D-14	1.78372957	.037624691	-0.0	-.016778068
1.07071511	-.382754795	-3.22850863	1.84468342	.112778648	-0.0	-.036851303
1.18330269	-.376195551	-.306636177	.755804250	.187951397	-0.0	-.031466120
1.36205453	-.318113630	-.204327270	.587352702	.263157037	-0.0	-.033355476
1.55308705	-.175353460	-.079853521	.427971765	.338303589	-0.0	-.032778853
1.81042835	.075176469	-.047694681	.348610908	.413430554	-0.0	-.032965344
1.99989444	-.442693499	-.028421521	.290020383	.488657743	-0.0	-.032890643
2.10405566	.920120803	-.018688135	.249441508	.563835065	-0.0	-.039937936
2.07782119	1.48334521	-.012916158	.218254124	.639012473	-0.0	-.032892275
1.88410106	2.09228646	-.009229931	.194374958	.714189941	-0.0	-.032942330
1.49898325	2.69374431	-.006966876	.174648381	.789367451	-0.0	-.038579243
.915892466	3.22582147	-.004923462	.159002431	.864544993	-0.0	-.032987798
.148274484	3.62356673	-.005099721	.145372612	.939722558	-0.0	-.032701038
-.769526690	3.82535653	0.0	.132256354	1.01490014	-0.0	-.033658245
-1.79338772	3.77944683	0.0	0.0	10.0000000	0.0	0.0
-11.6234686	1.99820928					

Figure 6.26 Sample printout for a SPLINE definition of a TABCYL.

184

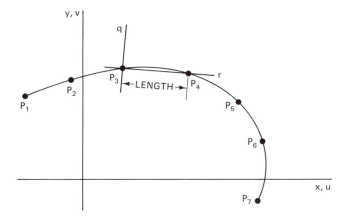

Figure 6.27   Coordinate systems placement for TABCYL cubic spline computations.

C. A table of TABCYL point coordinates together with information on the cubic equations that define the curve between points

Details on the entries in each part are discussed below, after which we will show their use in fine tuning the TABCYL entries.

1. The modifiers NOZ, RTHETA, and THETAR (see Section 6.3.1) signify that the points of the TABCYL definition are to be projected onto the xy-plane, that is, their z values are to be stripped off for purposes of defining the TABCYL. Modifiers NOX and NOY, discussed in Section 12.6.4, result in projection of the points onto the yz-plane and zx-plane, respectively. If the optional TRFORM couplet is used, this transformation occurs before projection. The original points, after the optional transformation, are in the xyz-coordinate system. To aid subsequent processing, the TABCYL routine transforms the points from the xyz-coordinate system to the uvw-coordinate system. This transformation, separate from any TRFORM transformation, is a rotation only that orients the w-axis in the same direction as the generatrix. The directrix is projected onto the uv-plane. Without the TRFORM couplet, correspondence between points in the xyz and uvw systems for the modifiers NOZ, RTHETA, and THETAR are $u = x$, $v = y$. These axes are shown in Figure 6.27. For the involute, the x-y values of array PT together with the endpoint extensions become the u-v values of part C, columns 1 and 2, in Figure 6.26. The r-θ values for TABCYLs defined in polar coordinates are their x-y equivalents when printed out as u-v values.

2. Part B shows the number of entries in this TABCYL canonical form storage (125 in Figure 6.26), the nine elements of the xyz-to-uvw rotation matrix (unity, or identity, matrix shown in Figure 6.26), and the number of defined points plus two extension points in the TABCYL ($15 + 2 = 17$ for Figure 6.26).

3. The cubic equation that represents the TABCYL between two points is defined in a local coordinate system, the rq-coordinates of Figure 6.27. This coordinate system

is relocated to each point pair of the TABCYL, including the endpoint extensions. Its origin is at the first point of the point pair in the order that the points are given in the TABCYL definition. The cubic equation for a point-pair TABCYL interval is given by the equation

$$q = Ar^3 + Br^2 + Cr$$

where coefficients A and B are listed in columns 3 and 4 of part C and coefficient C is computed from

$$C = -(A*LENGTH^2 + B*LENGTH)$$

where LENGTH, the distance between the two points, is listed in column 5 of part C. LENGTH is the distance between points P_3 and P_4 in Figure 6.27, for that point pair. In part C of Figure 6.26, LENGTH = 10 for the first point pair (the first endpoint extension and the first INVOL defined point). LENGTH = 10 also for the last point pair (the last INVOL defined point and its endpoint extension). The distance between INVOL points 2 and 3 (lines 3 and 4, respectively) is given as LENGTH = .112778488. The resulting cubic equation for this point pair is

$$q = -3.22850863r^3 + 1.84688342r^2 - 0.1672253592r$$

4. A measure of the deviation of the cubic equation from the r-axis between points P_i and P_{i+1} is given by the values for MAX and MIN, columns 6 and 7 in Figure 6.26. These values are multiplier values. The actual deviations are MAX*LENGTH and MIN*LENGTH as shown in Figure 6.28. If the cubic equation for a segment has an inflection point between P_i and P_{i+1}, both MAX and MIN are nonzero. For INVOL points 2 and 3 (lines 3 and 4, respectively, in Figure 6.26), MAX = 0 and MIN = -0.036851303. Therefore, the cubic equation passes through a minimum value for that interval. To show this to be true, we will differentiate the expression for q in paragraph 3 above and solve for the r-coordinate of the minimum, then substitute this value into the equation for q to calculate the deviation. This value will be compared with that obtained from Figure 6.26.

The first derivative of q with respect to r is

$$\frac{dq}{dr} = -9.68552589r^2 + 3.69376684r - 0.1672253592$$

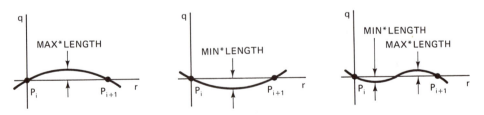

Figure 6.28   **TABCYL cubic spline fit between two points.**

which, when normalized and set to zero, becomes

$$r^2 - 0.3813697761r + 0.0172654909 = 0$$

Upon solving this equation with the quadratic formula, we get

$$r = 0.3288703997, 0.0524993764$$

The first value exceeds LENGTH, so is disregarded. The second value is the r-coordinate of the minimum. When substituted into the equation for q, we get for the deviation,

$$q = -0.0041560342$$

From Figure 6.26 we get

$$\text{Deviation} = \text{MIN*LENGTH} = (-.036851303)*(.112778488)$$
$$= -.0041560342$$

The two values are equal.

5. The curvature values at each defined point of the TABCYL and a curvature plot are given in part A. The plot is to a coarse scale. The shape of a curve at a point (its flatness or sharpness) depends on its rate of change of direction. This rate is called the **curvature** at the point and, when the equation of the curve is given in rectangular coordinates, is computed from

$$K = \frac{d^2y/dx^2}{\left[1 + (dy/dx)^2\right]^{3/2}}$$

The significance of the curvature for the APT part programmer is discussed below. For now, we will illustrate its computation for the second point of INVOL (line 2, part A, Figure 6.26). First, we need the second derivative of q with respect to r. This we get by differentiating the first derivative of paragraph 4.

$$\frac{d^2q}{dr^2} = -19.37105178r + 3.69376884$$

At the point in question $(r, q) = (0, 0)$. Letting $r = 0$ to get the first and second derivatives values and substituting these values into the expression for K, we get

$$K = \frac{3.69376684}{\left[1 + (0.1672253592)^2\right]^{3/2}} = 3.544071022$$

This value checks with that printed in Figure 6.26.

6. The TABCYL canonical forms are not the same for all APT processors. The exact form should be obtained from the reference manual. In any case, the printout(s) include most of the information from the canonical form. Such is the case for the IBM APT-AC system. This includes, from Figure 6.26, the number of data storage locations

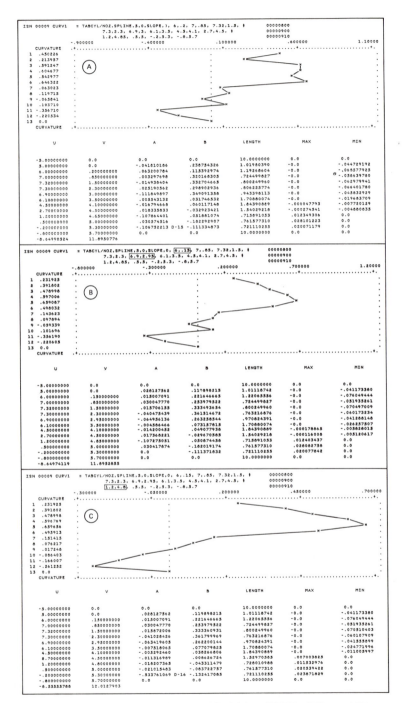

Figure 6.29   TABCYL printouts while fairing CURV1 of Figure 6.25.

and the rotation matrix elements of part B and all entries in part C. The remaining two entries, not shown in Figure 6.26, are of little interest to the part programmer.

We now return to the purpose of this section: to identify and correct irregular features in the curve. Fortunately, the APT system has made nearly all the computations we need. It is up to us to interpret them geometrically in the context of fairing the curve. Three features of the fitted curve are reflected in the values of the TABCYL printout.

The MAX and MIN Values

The number of inflection points in the original curve is generally evident on inspection of the curve, as is their approximate location. This number should agree with that determined from the MAX and MIN values. Each alternation of zero values between MAX and MIN columns identifies an inflection point. There are none in Figure 6.26, but there is one in Figure 6.29, the TABCYL printout for CURV1 of Figure 6.25. There should be only one inflection point in CURV1, somewhere near $(x, y) = (2, 4.6)$.

Nonzero MAX and MIN values in the same segment, such as shown in Figure 6.29(a), identify the condition shown in Figure 6.28(c). An inflection point within a segment is normally not desired unless it happens to be the inflection point of the curve. Such reversals of curvature lead to waviness in the curve. This is caused by improper location of the points and may indirectly result from too few or too many points in the region of the curve where such reversals occur.

Large-magnitude individual MAX or MIN values should be compared to the average value. If the difference is substantial, the large value indicates a bulge in the fitted curve. The actual value can be determined by multiplying the MAX or MIN value by LENGTH. Further interpretations of this bulge should be in conjunction with curvature discussed in the next paragraph.

The Curvature Plot

When a connect-the-dot procedure is performed on the curvature plot, fluctuations in the curvature are clearly revealed. To understand the importance of this, we first look at the physical meaning of curvature.

An object following the path of a curve is subject to an acceleration normal to the curve with a value dependent on the curvature at the point. The greater the curvature, the greater the acceleration for a given speed. To avoid sudden changes in acceleration, the curvature should be made gradual. For example, a straight section of railroad track is connected to a circular section by a transition section that helps pass smoothly from the infinite curvature of the straight section to the finite curvature of the circular track. The acceleration of the cars when passing abruptly from the straight section to the circular section would be hard on the equipment and unpleasant for the passengers. Wing profiles are designed to avoid sudden changes in acceleration of the air particles flowing past the wing. Similarly, cams are designed to avoid abrupt changes in acceleration of the activated components. Unless directed otherwise, numerically controlled machine tools may be subjected to similar acceleration components. These affect the actual tool path, which is a reflection of the dynamics of the machine-tool/controller combination.

The curvature plot reveals inflection points (polarity changes in curvature values). These have already been discussed. Frequent fluctuations in the curvature may be caused by poorly fitting points in the region of such curvature changes. Depending on the shape of the TABCYL curve, not all curvature fluctuations may be undesired. For example, straight lines tangent to circles are allowed APT geometric constructions that will result in curvature effects analogous to the railroad example discussed above. Since such effects are permitted, TABCYL curvature changes should be somewhat liberally interpreted.

Endpoint Extensions and Extension Segment Slopes

The coordinates of the TABCYL endpoint extensions appear on the printout. These points can be located on the part drawing and used to discover potential problems when driving onto the TABCYL. Simple computations will yield the slopes of the extension segments, which can be used to examine transition conditions between two drive surfaces.

In fairing a curve, we strive to eliminate the most irregular features of the curvature plot by fine adjustments in the location of certain points. An acceptable fit is obtained

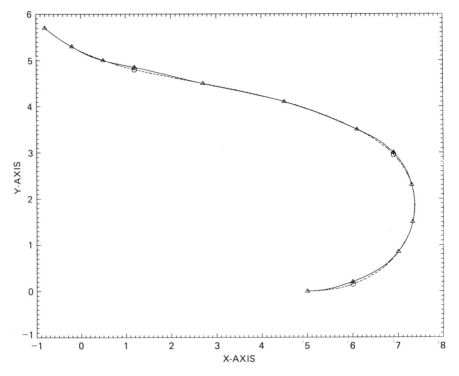

Figure 6.30   Before and after fairing CURV1 of Figure 6.25

when the curve formed by connecting the points of the curvature plot is void of fluctuation. Adjustments made in point location affect the curvatures of adjacent points also, so only gross irregularities should be worked on first. A systematic procedure should be followed that includes beginning at one end of the curve and working toward the other end. Iterations in following the procedure surely will be required. A purely cut-and-try approach should be avoided.

A fairing procedure will be applied to the first TABCYL fit to CURV1 as shown in Figure 6.29(a). The curvature of point 2 is not large enough, while that of point 6 is too large. We will move point 2 from (6, 0.2) to (6, 0.15), thus making points 1, 2, and 3 depart more from a straight line. We will move point 6 from (6.9, 3) to (6.9, 2.95), thus bringing points 5, 6, and 7 more toward a straight line. These changes produced the curvature plot of Figure 6.29(b), where the first seven points no longer cause severe fluctuation in the plot.

The curvature for point 10 should be made negative, or at least more nearly like that of its immediate neighbors. We will do this by moving the point from (1.2, 4.85) to (1.2, 4.8) so as to remove an unwanted inflection in its vicinity. This resulted in the curvature plot of Figure 6.29(c). The MAX and MIN values show one inflection point in the curve and the curvature plot is now quite regular, so we will accept this set of points for CURV1. The TABCYL curves for Figure 6.29(a) and (c) are shown in Figure 6.30.

6.4 DEBUGGING PROCEDURES FOR GCONICS, LCONICS, AND TABCYLS

The geometric entities of this chapter are not often encountered. This is fortunate since they can cause time-consuming and frustrating debugging sessions. It is natural for us to expect an APT processor running on a large computer to produce perfect results. But APT does have its problems when we command tool motion along the curves of this chapter. When we consider the alternative, it's best that we do what we can to allow the APT processor to compute the cut vectors we need. The following suggestions are offered to aid the debugging process.

6.4.1 Preliminaries

We cannot overemphasize the need to follow good programming practices so as to get into execution (computation of cut vectors) as soon as possible. This means that we must eliminate syntax and semantic errors quickly. Here are suggestions that apply to all curves of this chapter.

1. The statements describing the geometry of this chapter can be quite complicated. Syntax errors can be easily made, especially when many points are used to define a curve. Omitting commas and neglecting to include the statement continuation symbol are common errors. Statements must be proofread thoroughly.

2. Assuming that good programming practices are observed (tool-to-surface modifiers specified, explicit check surfaces specified, etc.), multiple computer runs for checkout should be expected. The APT processor may find an incorrect check surface, or no check surface at all. This may be caused by impossible geometry (perhaps bad values in the geometric definitions), contradictory geometry (expected tangencies that aren't), or by arithmetic difficulties within the APT processor. Often, the result is the computation of a large number of cut vectors, limited by the default maximum value for a cut sequence (those produced by a single motion command). Unless a large number are expected, the remedy is to redefine downward the default maximum value via the NUMPTS statement. Its format is

   NUMPTS/n

   where n is an integer. This is a modal statement, remaining in effect until countermanded by another NUMPTS statement. It can be used to increase or decrease the default maximum value. For debugging purposes, it should be placed at the beginning of the part program and set for a small value, say 50 cut vectors.

3. Fatal diagnostic messages are often issued but warning messages may be expected. Serious problems exist when the number of cut vectors calculated exceeds the NUMPTS value in effect, if the check surface cannot be located while moving along the drive surface (assuming they intersect), or if the tool is moving in the direction opposite that from which it was commanded. Some of these problems may defy resolution but suggestions for confronting them are usually related to the geometry. The paragraphs that follow provide insight to possible dilemmas.

6.4.2 General Conics

These factors should be considered to ease the debugging task with GCONICs.

1. The shape of the general conic is known when defined, but some curve segments may not be visible on the drawing. These should be visualized, especially the image sheet of the hyperbola, to avoid stopping at unexpected check surfaces.

2. Difficulties in reaching the proper GCONIC check surface may arise from too loose a tolerance. This may be overcome with an auxiliary check surface, such as used with the hyperbola of Example 6.1.

3. The general conic equation can be retrieved from a printout of its canonic form. However, its usefulness for debugging purposes is suspect. Because the canonic form differs for various APT processors, it is not described in this book.

6.4.3 Loft Conics

The factors discussed above for general conics, and the following, apply to loft conics.

1. The shape of the general conic derived from the loft conic points may differ drastically from that expected. This was illustrated in Example 6.3. It is recom-

mended that curves such as in Figure 6.15 be prepared for LCONICs for which stopping and contouring is giving difficulty.

2. Driving the tool tangent to a loft conic should be done only after defining it with a slope parameter and only in conjunction with a substitute check surface at the tangent point.

6.4.4 Tabulated Cylinders

Identifying and correcting problems with tabulated cylinders has already been discussed in Section 6.3. However, here are reminders and other factors to consider.

1. Extension segments can become unwanted check surfaces. Their location may be deceptive when many TABCYLs are used in a local area. It is recommended that endpoint extensions be located on the drawing at least once to confirm or deny their potential for causing problems.

2. The tool should never be driven onto a TABCYL with a TANTO check surface modifier. Instead, substitute check surfaces should be used.

3. The coordinates of each TABCYL should be verified via sight check of the printout for proper or approximate expected value. Constructions such as discussed in Section 6.3.2 should be incorporated in the part program design.

4. Fairing of the curve should be done according to the procedure of Section 6.3.4, to avoid computational difficulty while driving along the TABCYL.

PROBLEMS

6.1. Figure P6.1 shows an x-y coordinate system on which appears various lines, a circle tangent to two lines, a parabola, and a hyperbola. Define the appropriate surfaces and write the motion commands to take a $\frac{1}{2}$-in.-diameter tool counterclockwise around the part outline. Locate the set point at (4, 0).

6.2. Repeat Problem 6.1 but for the part outline of Figure P6.2.

6.3. Repeat Problem 6.1 but for the part outline of Figure P6.3.

6.4. Figure P6.4 shows a collection of curves defined by equations written with respect to the axes shown. Write an APT part program to take a $\frac{1}{4}$-in.-diameter tool counterclockwise around the part.

6.5. Figure P6.5 shows a collection of lines, parabolas, and an hyperbola. One line passes through the focus of the upper parabola and makes an angle with the x-axis as shown. The other line also passes through the focus but is tangent to the lower parabola. The equations are written with respect to the axes shown. Write an APT part program to take a $\frac{1}{4}$-in.-diameter tool counterclockwise around the part.

6.6. A line and two curves form the boundary of the part shown in Figure P6.6. Define each of the curves as a loft conic and write an APT program to take a $\frac{1}{4}$-in.-diameter tool clockwise around the part. Locate the set point at (3, -2).

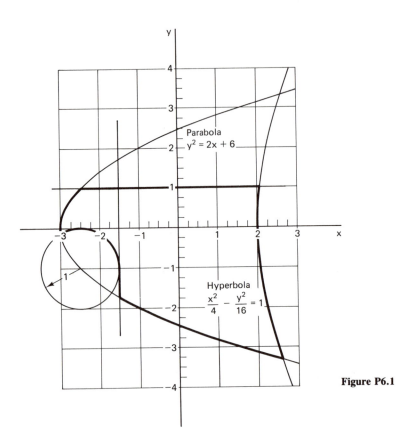

Figure P6.1

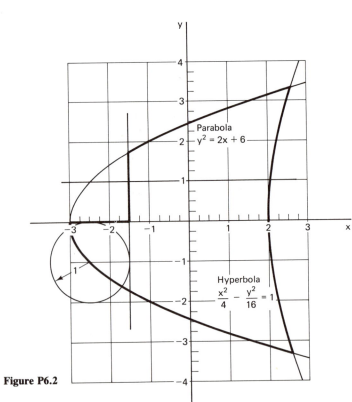

Figure P6.2

194

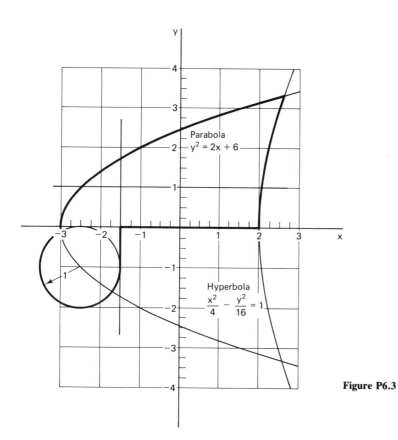

Figure P6.3

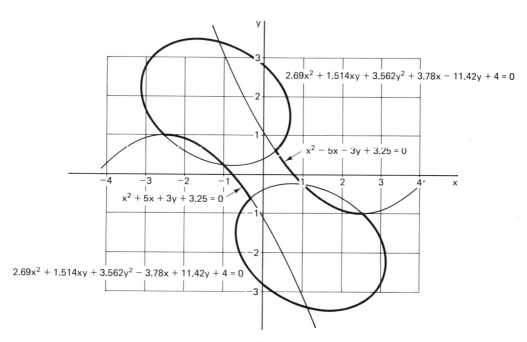

Figure P6.4

195

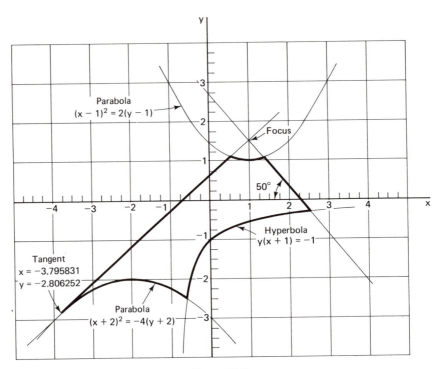

Figure P6.5

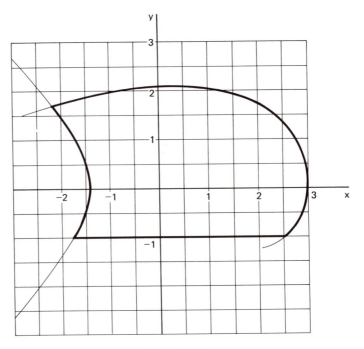

Figure P6.6

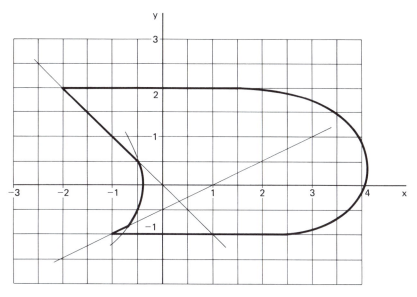

Figure P6.7

6.7. The part outline of Figure P6.7 is formed from four lines and two curves. Define each of the curves as a loft conic and write an APT part program to take a ½-in.-diameter tool clockwise around the part.

6.8. A line, a circular segment, and two curves form the boundary of the part shown in Figure P6.8. Define each of the curves as a loft conic and write an APT part program to take a ½-in.-diameter tool counterclockwise around the part.

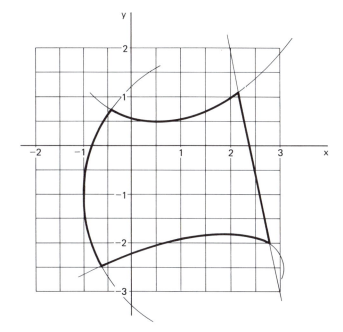

Figure P6.8

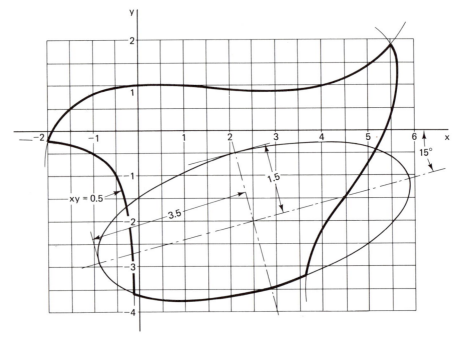

Figure P6.9

6.9. Figure P6.9 shows a part whose boundary is an hyperbola, an ellipse, and two curves of arbitrary character defined only by their geometry as drawn on the grid of the figure. Using the axes of the figure and the part dimensions shown, define the arbitrary curves as tabulated cylinders. Write an APT part program to take a ½-in.-diameter tool counterclockwise around the part.

6.10. Figure P6.10 shows a part whose boundary is a line parallel with the *x*-axis, a circle tangent to both the *x* and *y* axes, a line passing through the point (−3, −1) and tangent to the circle, and two curves of arbitrary character defined only by their geometry as drawn on the grid of the figure and where one of them is tangent to the circle. Define the arbitrary curves as tabulated cylinders and write an APT part program to take a ½-in.-diameter tool counterclockwise around the part.

6.11. Figure P6.11 shows a part whose boundary is arbitrarily shaped except for a semicircular edge. Define the arbitrarily shaped portion as a tabulated cylinder and write an APT part program to take a ¼-in.-diameter tool counterclockwise around the part.

6.12. Figure P6.12 shows a part whose boundary is a portion of a circle, a straight line, a cubic equation, and an arbitrarily shaped curve. Define as APT tabulated cylinders the arbitrarily shaped curve and the cubic equation, where the points for the cubic equation are to be computed algebraically in a loop and placed in a subscripted variable to be used in the tabulated cylinder definition. Then write an APT part program to take a ½-in.-diameter tool counterclockwise around the part.

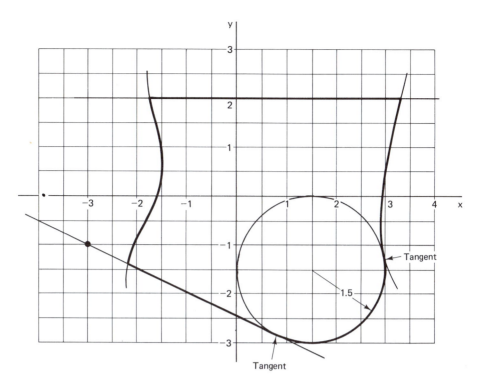

Figure P6.10

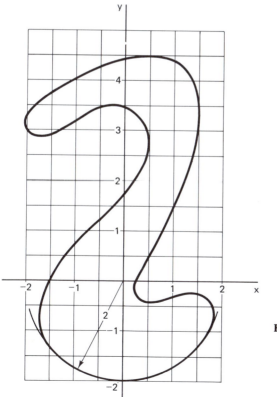

Figure P6.11

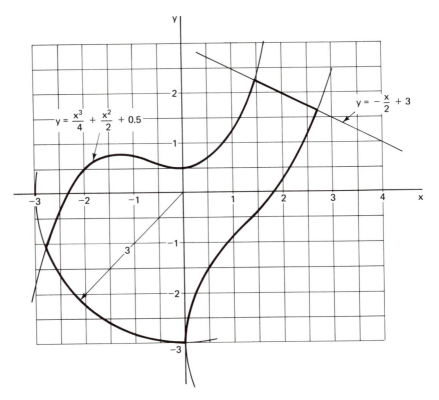

$y = \dfrac{x^3}{4} + \dfrac{x^2}{2} + 0.5$

$y = -\dfrac{x}{2} + 3$

Figure P6.12

6.13. Figure P6.13 shows a part whose boundary is a circle, an ellipse, a straight line, and a cosine curve. Define the cosine curve as a tabulated cylinder, where the points for the curve are to be computed algebraically in a loop and placed in a subscripted variable to be used in the tabulated cylinder definition. Then write an APT part program to take a ½-in.-diameter tool counterclockwise around the part.

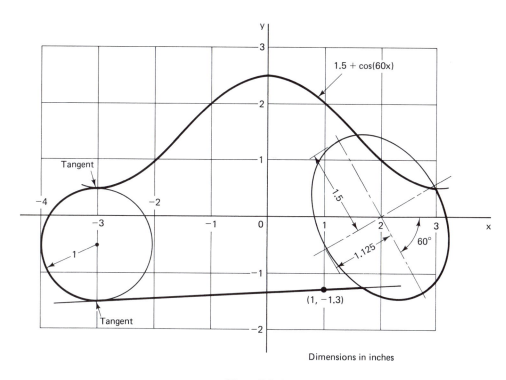

Figure P6.13

CHAPTER **7**

Coordinate Transformations, COPY and TRACUT

At times we must describe a part or portion of a part in a coordinate system different from one that is more convenient for us. We may resolve this issue by performing coordinate transformations. We do this in APT by defining transformation matrices and specifying them in the REFSYS, COPY, and TRACUT statements. Their use is described in this chapter.

The requirement for maintaining the tool within tolerance of a drive and part surface while contouring often leads to awkward sequences of tool movement for positioning the tool before executing the desired tool commands. Inhibiting the output of these undesired cut vectors to the CL-file is the purpose of the DNTCUT-CUT statements. Their use is also described in this chapter.

7.1 COORDINATE TRANSFORMATIONS

For the most part, the applications of APT involve some physical object that can be represented in the Cartesian coordinate system by the coordinate values of its geometrical features. For many simple objects a representation with respect to one coordinate system is sufficient. However, sometimes it is desirable or necessary to use different coordinate systems to describe different parts of the object or to otherwise perform some operations on the coordinate values so as to relocate them within the same coordinate system. These operations are referred to in general as **coordinate transformations** and are most conveniently represented in matrix form. Various APT statements allow transformations to be specified in matrix form. This feature relieves the part programmer from performing the mathematical operations to transform coordinate values. The requirement, of course, is that the part programmer be able to communicate the required transformation in matrix

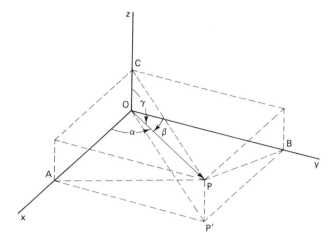

Figure 7.1 Geometry for defining the direction cosines.

form. The following paragraphs present the essentials for defining such matrices in APT. A mathematical presentation is given to lay groundwork for Chapters 8 and 12.

7.1.1 Direction Cosines and Vector Components

Figure 7.1 shows a point P in the Cartesian coordinate system. This point can be connected to the origin O with a straight line as shown. The directed line segment from the origin to the point, line OP, is referred to as a **vector** and will be denoted as **p**. It has a direction and magnitude (length). From the projection of this vector onto the xy-plane, we can obtain its components OA and OB (signed quantities) in the x and y directions, respectively, by noting the intercepts on the x and y axes. Similarly, we can obtain its component OC in the z-direction by projection onto the zx- or yz-plane. The three components OA, OB, and OC are sufficient to completely describe the vector, which can be represented as the triplet of numbers $\mathbf{p} = \{OA, OB, OC\}$ in which it is understood that the components are ordered along the x, y, and z axes, respectively.

We obtain the magnitude of **p** as follows:

$$OP' = \sqrt{OA^2 + OB^2}$$
$$OP = \sqrt{OP'^2 + OC^2} = \sqrt{OA^2 + OB^2 + OC^2}$$

Therefore,

$$|\mathbf{p}| = \sqrt{OA^2 + OB^2 + OC^2}$$

By defining α, β, and γ as the angles from the x, y, and z axes, respectively, to the vector **p**, we obtain the trigonometric relationships

$$OA = OP \cos \alpha$$
$$OB = OP \cos \beta$$
$$OC = OP \cos \gamma$$

which lead us to the definition of the **direction cosines**:

$$\cos \alpha = \frac{OA}{OP}$$

$$\cos \beta = \frac{OB}{OP}$$

$$\cos \gamma = \frac{OC}{OP}$$

We note the identity:

$$
\begin{aligned}
1 &= \frac{OP^2}{OP^2} \\
&= \frac{OA^2 + OB^2 + OC^2}{OP^2} \\
&= \frac{OA^2}{OP^2} + \frac{OB^2}{OP^2} + \frac{OC^2}{OP^2} \\
&= \cos^2 \alpha + \cos^2 \beta + \cos^2 \gamma
\end{aligned}
$$

Let the vector **p** be translated to some other position in the coordinate system such that it now begins at $P_1(x_1, y_1, z_1)$ and ends at $P_2(x_2, y_2, z_2)$, as shown in Figure 7.2. This new vector **p′** is equivalent to **p** if the following relationships are maintained:

$$a = OA = x_2 - x_1$$
$$b = OB = y_2 - y_1$$
$$c = OC = z_2 - z_1$$

The vector components have been redefined as a, b, and c to emphasize the fact that they are independent of the origin of the coordinate system. Thus two vectors whose components are identical are considered as identical vectors. The components are also

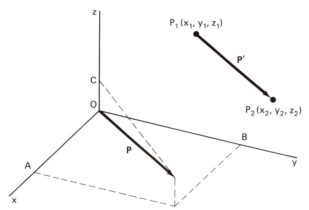

Figure 7.2 Geometry for defining the direction numbers.

referred to as **direction numbers**. From them we obtain the direction cosines:

$$\cos \alpha = \frac{a}{\sqrt{a^2 + b^2 + c^2}}$$

$$\cos \beta = \frac{b}{\sqrt{a^2 + b^2 + c^2}}$$

$$\cos \gamma = \frac{c}{\sqrt{a^2 + b^2 + c^2}}$$

We can now represent any vector as a triplet of direction numbers (e.g., $\mathbf{v} = \{a, b, c\}$) in which it is understood that the numbers are ordered along the x, y, and z axes, respectively, and the vector can begin anywhere, not necessarily at the origin of the coordinate system. This is the same interpretation that must be given to the vector specified by the minor section of the INDIRV statement. There, of course, the origin of the vector is fixed by the tool-end location and the vector components are with respect to that point.

7.1.2 Transformation Matrices

The transformation of coordinate values is accomplished most conveniently through matrix operations. The principal operation requires that the coordinate values of the point to be transformed be operated on (multiplied) by the **transformation matrix**. We will proceed with the derivation of this transformation matrix by developing the relationships necessary to transform a vector from one three-dimensional coordinate system to another three-dimensional coordinate system. We initially assume that the origin of each coordinate system is at the same point and that the geometric vector to be transformed begins at this point and terminates at some point P. With this constraint there can be no translation of the geometric vector since the origins always remain fixed. Translation will be considered later.

The vector to be transformed is shown in the geometry of Figure 7.3 where, for convenience in mathematical notation during the derivation, the axes nomenclature is different from that previously used. The geometric vector is shown in two coordinate systems, primed and unprimed, for which the axes are in the directions labeled by the $\hat{\mathbf{e}}'_i$ or $\hat{\mathbf{e}}_i$. The geometric vector is assumed to be known in terms of the primed coordinate system. Its components in terms of the unprimed coordinate system are to be determined.

Let the unprimed system be the standard Cartesian coordinate system with axes in the directions of the three basic mutually orthogonal unit vectors.

$$\hat{\mathbf{e}}_1 = \{1, 0, 0\}$$
$$\hat{\mathbf{e}}_2 = \{0, 1, 0\}$$
$$\hat{\mathbf{e}}_3 = \{0, 0, 1\}$$

The components of the geometric vector that lie along these unit vectors may be represented as the elements of the three-dimensional numeric vector

$$\mathbf{x} = \{x_1, x_2, x_3\}$$

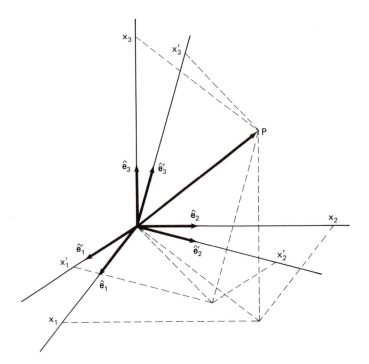

Figure 7.3   Geometric vector represented in two different coordinate systems.

This representation corresponds to the relationship

$$\mathbf{x} = \{x_1, x_2, x_3\} = x_1\hat{\mathbf{e}}_1 + x_2\hat{\mathbf{e}}_2 + x_3\hat{\mathbf{e}}_3 = \sum_{k=1}^{3} x_k\hat{\mathbf{e}}_k \qquad (7.1)$$

This vector is also represented in the primed system whose coordinate axes lie in the directions of the **linearly independent** unit vectors $\hat{\mathbf{e}}_1'$, $\hat{\mathbf{e}}_2'$, $\hat{\mathbf{e}}_3'$. Linear independence means that none of these unit vectors can be derived from any linear combination of the other two. The primed unit vectors are related to the unprimed unit vectors by the equations

$$\hat{\mathbf{e}}_1' = q_{11}\hat{\mathbf{e}}_1 + q_{21}\hat{\mathbf{e}}_2 + q_{31}\hat{\mathbf{e}}_3$$
$$\hat{\mathbf{e}}_2' = q_{12}\hat{\mathbf{e}}_1 + q_{22}\hat{\mathbf{e}}_2 + q_{32}\hat{\mathbf{e}}_3 \qquad (7.2)$$
$$\hat{\mathbf{e}}_3' = q_{13}\hat{\mathbf{e}}_1 + q_{23}\hat{\mathbf{e}}_2 + q_{33}\hat{\mathbf{e}}_3$$

or

$$\hat{\mathbf{e}}_k' = \sum_{j=1}^{3} q_{jk}\hat{\mathbf{e}}_j \qquad k = 1, 2, 3 \qquad (7.3)$$

where the q_{jk} are the direction cosines (since the $\hat{\mathbf{e}}_i'$ are unit vectors) as shown in Figure 7.4.

If we denote as $\mathbf{x}'$ the numeric vector comprising the components x_1', x_2', x_3' of the geometric vector along the primed axes, we have

$$\mathbf{x}' = \{x_1', x_2', x_3'\} = x_1'\hat{\mathbf{e}}_1' + x_2'\hat{\mathbf{e}}_2' + x_3'\hat{\mathbf{e}}_3' = \sum_{k=1}^{3} x_k'\hat{\mathbf{e}}_k' \qquad (7.4)$$

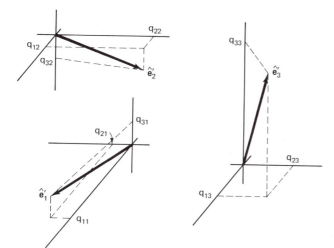

Figure 7.4 Resolving the primed unit vectors into components in the direction of the unprimed unit vectors.

Now, substituting equation (7.3) into (7.4) we obtain

$$\mathbf{x}' = \sum_{k=1}^{3} x_k' \left(\sum_{j=1}^{3} q_{jk}\hat{\mathbf{e}}_j \right) = \sum_{j=1}^{3} \left(\sum_{k=1}^{3} q_{jk}x_k' \right) \hat{\mathbf{e}}_j = \mathbf{x} \qquad (7.5)$$

Of course, we already know that $\mathbf{x}' = \mathbf{x}$, since they represent the same geometric vector, in different coordinate systems to be sure. However, since the vectors $\hat{\mathbf{e}}_j$ are mutually orthogonal, their respective coefficients in equations (7.1) and (7.5) must be equal. Therefore,

$$x_j = \sum_{k=1}^{3} q_{jk}x_k' \qquad j = 1, 2, 3 \qquad (7.6)$$

or, expanding,

$$x_1 = q_{11}x_1' + q_{12}x_2' + q_{13}x_3'$$
$$x_2 = q_{21}x_1' + q_{22}x_2' + q_{23}x_3'$$
$$x_3 = q_{31}x_1' + q_{32}x_2' + q_{33}x_3'$$

Placing these equations in matrix form, we have

$$\begin{bmatrix} x_1 \\ x_2 \\ x_3 \end{bmatrix} = \begin{bmatrix} q_{11} & q_{12} & q_{13} \\ q_{21} & q_{22} & q_{23} \\ q_{31} & q_{32} & q_{33} \end{bmatrix} \begin{bmatrix} x_1' \\ x_2' \\ x_3' \end{bmatrix} \qquad (7.8)$$

which, in matrix notation, is

$$\mathbf{x} = \mathbf{Q}\mathbf{x}' \qquad (7.9)$$

where $\mathbf{Q}$ is referred to as the transformation matrix. This matrix is the **transpose** of the coefficient matrix in equations (7.2), that is, row i of equations (7.2) becomes column i

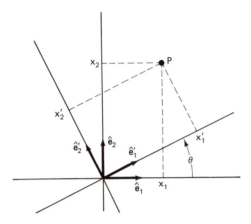

Figure 7.5   Point *P* represented in two different coordinate systems.

of equation (7.8). Each column of $\mathbf{Q}$ contains the components (direction cosines) of a primed unit vector along the unprimed coordinate axes.

In order to interpret equation (7.9), consider the two-dimensional example of Figure 7.5, where point P is given in the primed system with orthogonal axes. The primed system is rotated through an angle θ with respect to the unprimed system, also with orthogonal axes. The coordinates of P, after rotation, with respect to the unprimed system are to be obtained.

First, we obtain the direction cosines,

$$\hat{e}_1' = q_{11}\hat{e}_1 + q_{21}\hat{e}_2 = \cos\theta\hat{e}_1 + \sin\theta\hat{e}_2$$
$$\hat{e}_2' = q_{12}\hat{e}_1 + q_{22}\hat{e}_2 = -\sin\theta\hat{e}_1 + \cos\theta\hat{e}_2$$

In the form of equation (7.8), we have then,

$$\begin{bmatrix} x_1 \\ x_2 \end{bmatrix} = \begin{bmatrix} \cos\theta & -\sin\theta \\ \sin\theta & \cos\theta \end{bmatrix} \begin{bmatrix} x_1' \\ x_2' \end{bmatrix}$$

If the point P is located at $(x_1', x_2') = (2, 1.5)$ and if $\theta = 30°$, then

$$\begin{bmatrix} x_1 \\ x_2 \end{bmatrix} = \begin{bmatrix} 0.866 & -0.5 \\ 0.5 & 0.866 \end{bmatrix} \begin{bmatrix} 2 \\ 1.5 \end{bmatrix} = \begin{bmatrix} (0.866)(2) + (-0.5)(1.5) \\ (0.5)(2) + (0.866)(1.5) \end{bmatrix} = \begin{bmatrix} 0.982 \\ 2.299 \end{bmatrix}$$

The arithmetic is that implied by equations (7.7). The result is that the coordinates of P with respect to the unprimed system are at $(x_1, x_2) = (0.982, 2.299)$. The coordinates of any other point in the primed system may be obtained in terms of the unprimed system by making the appropriate substitutions for x_1' and x_2'. Note that the origin, point $(0, 0)$, is transformed into itself (i.e., it is **invariant** under this transformation).

We originally represented P as the endpoint of a vector from the origin. That same interpretation can be used in this example where we see that it is necessary only to consider the vector endpoint in determining its new coordinates. If the vector had endpoints at P_1 and P_2, its new endpoint coordinates would be obtained by transforming P_1 and P_2 individually.

Returning to equation (7.9), we might choose to interpret it according to either of the following ways, depending on the physical circumstances in which the geometric vector was derived:

1. The transformation matrix **Q** is used to relate the components of the geometric vector along the axes of one coordinate system to the components of the same geometric vector along the axes of another coordinate system.
2. The transformation matrix **Q** is used to transform a geometric vector into another one within the same coordinate system.

Inasmuch as each x_i of equation (7.7) is a linear combination of the x_i', the set of equations (7.7) is viewed as a **linear transformation** in which the set of numbers $\{x_1', x_2', x_3'\}$ is transformed into the set of numbers $\{x_1, x_2, x_3\}$. Furthermore, since these equations contain no constant term, they are known as **homogeneous linear equations**. The linear transformation matrices **Q** have the property of leaving the origin fixed.

Occasionally, it is necessary not only to perform a linear transformation as described above, but also to translate the origin and other points the same distance in a specified direction. This transformation, referred to as an **affine transformation**, is represented by the set of **nonhomogeneous linear equations**,

$$x_j = \sum_{k=1}^{3} q_{jk}x_k' + c_j \qquad j = 1, 2, 3 \tag{7.10}$$

or

$$\begin{aligned}
x_1 &= q_{11}x_1' + q_{12}x_2' + q_{13}x_3' + c_1 \\
x_2 &= q_{21}x_1' + q_{22}x_2' + q_{23}x_3' + c_2 \\
x_3 &= q_{31}x_1' + q_{32}x_2' + q_{33}x_3' + c_3
\end{aligned} \tag{7.11}$$

which, in matrix notation becomes

$$\mathbf{x} = \mathbf{Q}\mathbf{x}' + \mathbf{c} \tag{7.12}$$

where **Q** is the transformation matrix as previously derived and **c** is a column vector whose elements c_j are the directed distances, representing the translation of a point, parallel with their respective axes, $\hat{e}_j$. The affine transformation equation (7.12) includes the linear transformations (with $\mathbf{c} = \mathbf{0}$, the zero vector) and the translation only (with $\mathbf{Q} = \mathbf{I}$, where **I** is the **identity**, or **unity**, matrix; a square matrix with principal diagonal element values each equal to one and off-diagonal elements each equal to zero).

Equations (7.11) may also be represented in matrix form as

$$\begin{bmatrix} x_1 \\ x_2 \\ x_3 \\ 1 \end{bmatrix} = \begin{bmatrix} q_{11} & q_{12} & q_{13} & c_1 \\ q_{21} & q_{22} & q_{23} & c_2 \\ q_{31} & q_{32} & q_{33} & c_3 \\ 0 & 0 & 0 & 1 \end{bmatrix} \begin{bmatrix} x_1' \\ x_2' \\ x_3' \\ 1 \end{bmatrix} \tag{7.13}$$

where a trivial fourth equation, $1 = 1$, has been added, which places no conditions on the relationship between the coordinates of a point and those of its transformed and translated point. In APT this matrix is presented in its simplified form as

$$\begin{bmatrix} q_{11} & q_{12} & q_{13} & c_1 \\ q_{21} & q_{22} & q_{23} & c_2 \\ q_{31} & q_{32} & q_{33} & c_3 \end{bmatrix} \tag{7.14}$$

7.1.3 Matrix Definitions

The APT general form of matrix definition is

[Statement-label)]   Symbolic-name = MATRIX/minor-section

If the statement is labeled and reexecuted, either the canonic form of matrix definition must be used or the symbolic name must be subscripted and a different index used for each execution; otherwise, a multiply defined geometric statement error will result. The rules for symbol naming apply here also. The minor section is completed according to any one of several different formats provided for the convenience of the part programmer. A selection of matrix definitions is given in Table 7.1. Also shown in the table are the matrices that the formats generate.

When required in the minor section, scalar values can be numeric literals, scalar variables (subscripted or nonsubscripted), arithmetic expressions, or some mixture of these. Angles are scalar values expressed in degrees. They are positive when measured about the reference axis in a counterclockwise direction.

7.1.4 Matrix Multiplication

We showed in Section 7.1.2 that a vector $\mathbf{x}'$ can be transformed into another vector $\mathbf{x}$ with the matrix equation

$$\mathbf{x} = \mathbf{Q}\mathbf{x}'$$

where $\mathbf{Q}$ is the linear transformation matrix (translation is not included). Let us suppose that the vector $\mathbf{x}'$ is to be transformed into the vector $\mathbf{x}$ with matrix $\mathbf{V}$, and that the vector $\mathbf{x}$ is to be transformed into the vector $\mathbf{r}$ with the matrix U. The required operations to accomplish this are as follows:

$$\mathbf{x} = \mathbf{V}\mathbf{x}'$$
$$\mathbf{r} = \mathbf{U}\mathbf{x}$$

We can substitute the first equation into the second and obtain

$$\mathbf{r} = \mathbf{U}\mathbf{V}\mathbf{x}'$$

which is a matrix equation of the form

$$\mathbf{r} = \mathbf{W}\mathbf{x}$$

TABLE 7.1 MATRIX DEFINITIONS

1	By the coefficients of the transformation equations:

MATRIX/$a_1, b_1, c_1, d_1, a_2, b_2, c_2, d_2, a_3, b_3, c_3, d_3$
   or
MATRIX/CANON, $a_1, b_1, c_1, d_1, a_2, b_2, c_2, d_2, a_3, b_3, c_3, d_3$

$$\begin{bmatrix} a_1 & b_1 & c_1 & d_1 \\ a_2 & b_2 & c_2 & d_2 \\ a_3 & b_3 & c_3 & d_3 \end{bmatrix}$$

The parameters are the elements of the linear transformation matrix and the translation elements as shown in equation (7.14) of section 7.1.2. All elements must be included even though they may be zero. When used with nonzero translation elements, the linear transformation is always performed first, then followed by the translation.

      MAT2 = MATRIX/A1, 0, −JA, 0, 0, B2, C2, 0, 0, B3, C3, 0
      MATR = MATRIX/(D*A1), 0, 0, T1, 0, (D*B2), 0, T2, 0, 0, (D*C3), T3
      MMA = MATRIX/CANON, R(1, THRU, 12)

2	As a translation only:

MATRIX/TRANSL, d_1, d_2, d_3

The matrix defined performs translation only. The parameters d_1, d_2, and d_3 are the same d_i as defined in Matrix-1. They are the increments of translation in the x, y, and z directions, respectively.

      M33 = MATRIX/TRANSL, 10, −4.5, −1
      M16 = MATRIX/TRANSL, X1, Y1, Z1
      M79 = MATRIX/TRANSL, 0, Y(3), 0

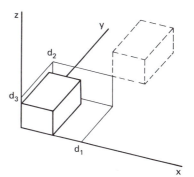

3	As a rotation about a coordinate axis:

$$\text{MATRIX/} \begin{Bmatrix} \text{XYROT} \\ \text{YZROT} \\ \text{ZXROT} \end{Bmatrix}, \theta$$

The matrix defined performs rotation about a coordinate axis only. Rotation through an angle θ occurs about the axis whose character designation is not included in the modifier. For example, YZROT causes rotation about the x-axis. A positive θ causes CCW rotation when viewed toward the origin from the positive direction of this axis.

$$\begin{bmatrix} \cos\theta & -\sin\theta & 0 & 0 \\ \sin\theta & \cos\theta & 0 & 0 \\ 0 & 0 & 1 & 0 \end{bmatrix}$$

XYROT
θ pos

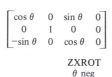

      MTR1 = MATRIX/XYROT, 55
      MTR2 = MATRIX/YZROT, −ANG
      MTR3 = MATRIX/ZXROT, WW(4)

$$\begin{bmatrix} 1 & 0 & 0 & 0 \\ 0 & \cos\theta & -\sin\theta & 0 \\ 0 & \sin\theta & \cos\theta & 0 \end{bmatrix}$$

YZROT
θ pos

$$\begin{bmatrix} \cos\theta & 0 & \sin\theta & 0 \\ 0 & 1 & 0 & 0 \\ -\sin\theta & 0 & \cos\theta & 0 \end{bmatrix}$$

ZXROT
θ neg

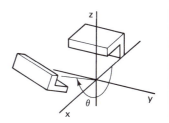

TABLE 7.1 (CONTINUED)

4	As a scale factor:

MATRIX/SCALE, *k*

The matrix defined performs scaling (multiplying by a constant) of each coordinate value by the same amount, that given by the value of *k*. For $k > 1$ there is uniform stretching, for $0 < k < 1$ there is uniform compression, and for $k = 1$ (unit matrix) there is no change in the coordinate values. For $k < 0$, scaling and reflection through the origin occurs.

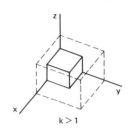

$$k > 1$$

MA = MATRIX/SCALE, 10
MB = MATRIX/SCALE, D
MC = MATRIX/SCALE, (PR + S1)

$$\begin{bmatrix} k & 0 & 0 & 0 \\ 0 & k & 0 & 0 \\ 0 & 0 & k & 0 \end{bmatrix}$$

5	As the product of two matrices:

MATRIX/*matrix1*, *matrix2*

The matrix defined is the result of multiplying *matrix1* by *matrix2*. For the multiplication, the matrices are ordered from left to right as given in the definition. See the discussion on matrix multiplication in section 7.1.4.

$$\text{MAT} = \text{MATRIX/M1, M2} \longrightarrow \text{MAT} = \text{M1} \cdot \text{M2}$$

The result is the same as if *matrix2* were performed first, then followed by *matrix1*.

6	As a mirror image about one or more coordinate planes:

$$\text{MATRIX/MIRROR,} \left\{\begin{matrix} \text{XYPLAN} \\ \text{YZPLAN} \\ \text{ZXPLAN} \end{matrix}\right\} \left[\left\{\begin{matrix} \text{XYPLAN} \\ \text{YZPLAN} \\ \text{ZXPLAN} \end{matrix}\right\}, \left\{\begin{matrix} \text{XYPLAN} \\ \text{YZPLAN} \\ \text{ZXPLAN} \end{matrix}\right\}\right]$$

MATRIX/MIRROR, YZPLAN

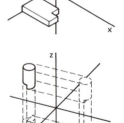

The matrix defined functions as a mirror through one, two, or three coordinate planes (a pair of axes define a coordinate plane), depending upon the number of modifiers chosen. The end result is independent of the order of the modifiers. Duplicate modifiers function as a single modifier.

A given modifier changes the sign of the coordinate value whose character designation is not included in the modifier. For example YZPLAN changes the sign of the *x*-coordinate values [numerically, the point (x, y, z) = (−3, −1, 4) is transformed to (3, −1, 4)]. This definition generates the following matrix, where a, b, and c are each +1 unless changed by the modifiers as follows:

YZPLAN, $a = -1$
ZXPLAN, $b = -1$
XYPLAN, $c = -1$

$$\begin{bmatrix} a & 0 & 0 & 0 \\ 0 & b & 0 & 0 \\ 0 & 0 & c & 0 \end{bmatrix}$$

MATRIX/MIRROR, XYPLAN, ZXPLAN
or
MATRIX/MIRROR, ZXPLAN, XYPLAN

7	As a mirror image about a line or a plane:

$$\text{MATRIX/MIRROR,} \left\{\begin{matrix} line \\ plane \end{matrix}\right\}$$

The matrix defined functions as a mirror through the plane defined by *line* (which is a plane perpendicular to the xy-plane) or through *plane*, an arbitrary plane in space as defined in Chapter 8. In general, the generated matrix performs both a linear transformation and a translation.

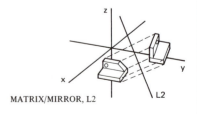

MATRIX/MIRROR, L2

212

where $\mathbf{W} = \mathbf{UV}$ is the resultant transformation matrix. The application of successive transformations through matrix multiplication is shown in linear algebra theory. This result is adopted here without proof. Let us first review matrix multiplication before constructing transformation matrices in this manner.

The product of two matrices $\mathbf{A}$ and $\mathbf{B}$, each having three rows and three columns, yields another 3×3 matrix $\mathbf{C} = \mathbf{AB}$ whose elements are computed from the following:

$$c_{ij} = \sum_{k=1}^{3} a_{ik}b_{kj} \qquad i = 1, 2, 3, \quad j = 1, 2, 3$$

where the a_{ij}, b_{ij}, and c_{ij} are the elements of row i and column j of the matrices $\mathbf{A}$, $\mathbf{B}$, and $\mathbf{C}$, respectively. The double subscript notation was introduced in Section 7.1.2. The equation sums products of elements from row i of matrix $\mathbf{A}$ with the elements of column j of matrix $\mathbf{B}$ as illustrated below for the case where $i = 3$ and $j = 2$.

$$\begin{bmatrix} a_{11} & a_{12} & a_{13} \\ a_{21} & a_{22} & a_{23} \\ a_{31} & a_{32} & a_{33} \end{bmatrix} \begin{bmatrix} b_{11} & b_{12} & b_{13} \\ b_{21} & b_{22} & b_{23} \\ b_{31} & b_{32} & b_{33} \end{bmatrix} = \begin{bmatrix} c_{11} & c_{12} & c_{13} \\ c_{21} & c_{22} & c_{23} \\ c_{31} & c_{32} & c_{33} \end{bmatrix}$$

$$a_{31}b_{12} + a_{32}b_{22} + a_{33}b_{32} = c_{32}$$

Implicit in this operation is ordering of the matrices. Matrix multiplication is in general not commutative; that is, the product $\mathbf{AB}$ is not necessarily the same as the product $\mathbf{BA}$. The matrices do not have to be square as illustrated above, but the number of columns in the first matrix must equal the number of rows in the second matrix. Common applications require the multiplication of a transformation matrix (square) by a column vector, as developed in Section 7.1.2.

Now, let it be required to transform a point by first rotating it by $\theta = +30°$ in the xy-plane and then mirroring it about the x-axis. The point is initially located at $P = (5, 4, 3)$, as shown in Figure 7.6. The rotation matrix is

$$\begin{bmatrix} \cos\theta & -\sin\theta & 0 \\ \sin\theta & \cos\theta & 0 \\ 0 & 0 & 1 \end{bmatrix}$$

which transforms the point as follows:

$$\begin{bmatrix} 0.866 & -0.5 & 0 \\ 0.5 & 0.866 & 0 \\ 0 & 0 & 1 \end{bmatrix} \begin{bmatrix} 5 \\ 4 \\ 3 \end{bmatrix} = \begin{bmatrix} 2.33 \\ 5.964 \\ 3 \end{bmatrix}$$

The operation above rotates the point to P_1 as shown in Figure 7.6, from which it is then mirrored to P_2 by the following operation:

$$\begin{bmatrix} 1 & 0 & 0 \\ 0 & -1 & 0 \\ 0 & 0 & 1 \end{bmatrix} \begin{bmatrix} 2.33 \\ 5.964 \\ 3 \end{bmatrix} = \begin{bmatrix} 2.33 \\ -5.964 \\ 3 \end{bmatrix}$$

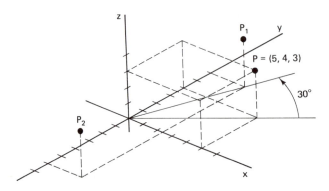

Figure 7.6   Rotating, then mirroring a point.

This is the desired result. However, the two transformation matrices can be combined into one as follows:

$$\begin{bmatrix} 1 & 0 & 0 \\ 0 & -1 & 0 \\ 0 & 0 & 1 \end{bmatrix} \begin{bmatrix} 0.866 & -0.5 & 0 \\ 0.5 & 0.866 & 0 \\ 0 & 0 & 1 \end{bmatrix} = \begin{bmatrix} 0.866 & -0.5 & 0 \\ -0.5 & -0.866 & 0 \\ 0 & 0 & 1 \end{bmatrix}$$

which when multiplied by the column vector representing the point will accomplish the same results. Alternatively, we could select matrix formats from Table 7.1 and let the APT processor do all the work with these statements.

> MAT1 = MATRIX/XYROT,30
> MAT2 = MATRIX/MIRROR,YZPLAN
> MAT3 = MATRIX/MAT2,MAT1

Above we rotated first, then mirrored. This implied an ordering of the transformation matrices. If the matrices were applied in the reverse order we would have the points shown in Figure 7.7, where P_1 results from mirroring P and P_2 from rotating P_1. These operations would be combined as follows:

$$\begin{bmatrix} 0.866 & -0.5 & 0 \\ 0.5 & 0.866 & 0 \\ 0 & 0 & 1 \end{bmatrix} \begin{bmatrix} 1 & 0 & 0 \\ 0 & -1 & 0 \\ 0 & 0 & 1 \end{bmatrix} = \begin{bmatrix} 0.866 & 0.5 & 0 \\ 0.5 & -0.866 & 0 \\ 0 & 0 & 1 \end{bmatrix}$$

which would transform the point to

$$\begin{bmatrix} 0.866 & 0.5 & 0 \\ 0.5 & -0.866 & 0 \\ 0 & 0 & 1 \end{bmatrix} \begin{bmatrix} 5 \\ 4 \\ 3 \end{bmatrix} = \begin{bmatrix} 6.33 \\ -0.964 \\ 3 \end{bmatrix}$$

We could define the matrix by the statement

> MAT4 = MATRIX/MAT1,MAT2

The example above demonstrates that matrix multiplication is not necessarily commutative.

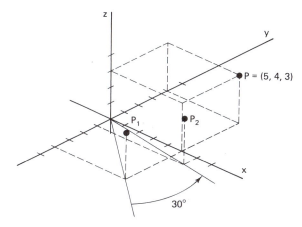

Figure 7.7   Mirroring, then rotating a point.

7.2 TRANSFORMING GEOMETRIC DEFINITIONS

We must address problems related to parts of complex geometry. Until now we used the APT computational capability to determine coordinate values, lengths, slopes, and so on, so that we could define the appropriate part geometry. It is desirable to avoid this computation by rearranging the part geometry through transformations. For this purpose APT has the REFSYS convenience feature, which allows the part programmer to describe the geometry in a convenient local coordinate system, then transform it to the part base coordinate system.

Because tabulated cylinders cannot be transformed by the REFSYS feature, we take this opportunity to use the TRFORM,*matrix* couplet as a substitute for the REFSYS transformation. Since it is unlikely that we can avoid all auxiliary computation, we introduce the OBTAIN statement to allow us to extract canonic parameters from geometric definitions. The philosophy is to avoid hand computation and to let the computer do the computing needed.

7.2.1 The REFSYS Statements

The reference system feature allows the part programmer to specify a transformation matrix for transformation of a set of APT geometric definitions bounded by a mating pair of REFSYS statements. The structure of this feature is as follows:

REFSYS/*matrix*   invoking statement
        :   statements subject to geometric
        :   transformation by *matrix*
REFSYS/NOMORE   countermanding statement

REFSYS/*matrix* is a modal command, remaining in effect until countermanded by the REFSYS/NOMORE or another REFSYS/*matrix* statement. This permits only one level of referencing; nesting is not allowed. *Matrix* must be a named matrix even though it may be a nested definition in the invoking statement. Except for the tabulated cylinder, the matrix transformation applies to all geometric definitions introduced thus far. The TABCYL can be transformed in the manner described in Example 7.2 and in Section 12.6.4.

The APT system "tags" each geometric definition within the REFSYS mating pair with the name *matrix*. No transformations are performed until these definitions are referenced in a different coordinate system. Transformation occurs prior to cut vector computation, but only as needed.

7.2.2 Geometric Interpretation of the REFSYS Transformation

The REFSYS feature applies to geometry in a local area of the part. We will refer to the part coordinate system as the base coordinate system and the coordinate system of a local area as a local coordinate system. Part geometry of a local area is a candidate for the REFSYS transformation if it requires awkward and complicated geometric descriptions when defined in the base coordinate system but which can easily be described given a choice of coordinate system. Usually, only a small number of geometric definitions are subject to a given REFSYS transformation.

Figure 7.8 shows a part with four local areas (marked A through D), each of which could be described with the REFSYS feature. Suitable local coordinate axes are marked on the part. The transformations for two-dimensional parts perform individually, or in combination, rotation or translation. Computation may be needed to locate the origin of the local axis coordinate system. A scaling matrix should be avoided for REFSYS purposes and a matrix should not be redefined, perhaps inadvertently with the CANON/ON statement (because as a tag, its redefined value would lead to errors for symbols referenced outside the REFSYS area).

The matrix of the REFSYS feature may be interpreted in one of the two ways described in Section 7.1.2. It describes what must be done to the geometry of the local system to place it properly within the base system. Should there be any doubt that the direction of the transformation is correct, the canonic form of a simple geometric feature, say a point, should be printed out and the values checked for proper location within each system. This can be done with the PRINT/3,--- statement when used in both coordinate systems.

7.2.3 Extracting Values from Canonic Forms

The convenience afforded the part programmer by the variety of definition formats for a given entity is slightly offset by the uncertainty in knowing the parameters that define the entity. Although the canonic-form definition of Section 3.7 allows us to modify parameters, thus redefining the entity, often we must extract values from the canonic form for use elsewhere. The parameters are scalar values and can be used in calculations

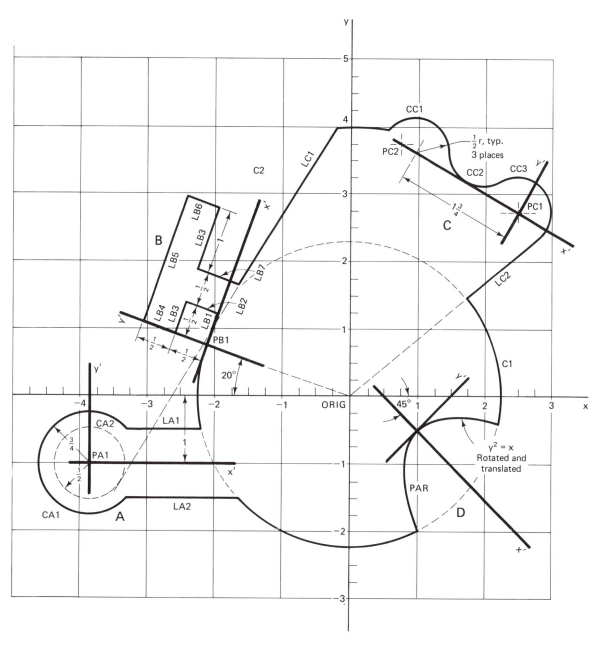

Figure 7.8   Candidate REFSYS local areas.

supporting geometric definitions, transformation matrices, and so on. The OBTAIN statement is used for extracting these parameters. Its format is

OBTAIN,*surface*/*name*,*list*

where *surface* is the entity type from which parameters are to be extracted, *name* is the symbolic name of the instance of entity type, and *list* is a selection of scalar variables into which the parameters are placed. The variables are delimited in *list* by commas.

In principle, the OBTAIN statement functions like the arithmetic assignment statement. The variables in *list* receive values from the canonic form of *name*. Since canonic form entries are ordered values (see the discussion of Section 3.7.1), the variables in *list* are position sensitive. The first parameter of the canonic form is assigned to the leftmost variable, the remainder to the rest of the variables in a left-to-right manner. Fewer than the number of parameters in the canonic form for *surface* may be extracted, but the number of variables in *list* cannot exceed the number of parameters in the canonic form. Consecutive commas are used to skip over parameters not being extracted but trailing consecutive commas are not needed. The inclusive subscript feature may be used in *list*.

Examples:

OBTAIN,POINT/P1,X1,Y1

Only the *x* and *y* coordinates of P1 are extracted.

OBTAIN,CIRCLE/C1,C1X,C1Y,,,,,C1R

The *x* and *y* coordinates of the center of C1 and its radius are extracted.

OBTAIN,CIRCLE/C1,C1PAR(1,THRU,7)

All parameters of C1 are extracted.

7.2.4 Examples

Example 7.1 applies the REFSYS feature to the geometry of Figure 7.8. Supporting calculations require canonic-form parameters that are extracted with the OBTAIN feature. Example 7.2 addresses some of the problems related to transforming tabulated cylinders.

Example 7.1: Setting Up the REFSYS Matrix

The principal geometry of Figure 7.8 that can be defined in the base coordinate system includes the origin, labeled ORIG, and circles C1 and C2 centered there. C2 is needed to establish the local area A coordinate system, while C1 is needed for local area B. We can write the following code:

```
ORIG   = POINT/0,0,0                              100
C1     = CIRCLE/CENTER,ORIG,RADIUS,2.25           110
C2     = CIRCLE/CENTER,ORIG,RADIUS,4              120
$$     **** LOCAL AREA A ****
```

```
PA1     = POINT/XSMALL,INTOF,(LINE/XAXIS, – 1),C2        130
        OBTAIN,POINT/PA1,PA1X,PA1Y                       140
MAT1    = MATRIX/TRANSL,PA1X,PA1Y,0                       150
        REFSYS/MAT1                                       200
CA1     = CIRCLE/0,0,.75                                  210
CA2     = CIRCLE/0,0,.5                                   220
LA1     = LINE/YLARGE,TANTO,CA2,ATANGL,0                  230
LA2     = LINE/YSMALL,TANTO,CA2,ATANGL,0                  240
        REFSYS/NOMORE                                     250
```

ORIG, C1, C2 and PA1 are defined in the base coordinate system. PA1 locates the origin of the local area A coordinate system. From it, with the OBTAIN statement, we extract the x-y coordinates for defining the translation matrix MAT1. Then we define the geometry within the REFSYS region.

The origin of the local area B coordinate system is at the intersection of line LB4 with C1. These coordinates are computed in a manner analogous to those for local area A. We have the following code:

```
$$      **** LOCAL AREA B ****                           300
LB4     = LINE/ORIG,ATANGL, – 20,XAXIS                    305
PB1     = POINT/XSMALL,INTOF,LB4,C1                       310
        OBTAIN,POINT/PB1,PB1X,PB1Y                        315
MAT2    = MATRIX/XYROT,70                                 320
MAT3    = MATRIX/TRANSL,PB1X,PB1Y,0                       325
        REFSYS/(MAT4 = MATRIX/MAT3,MAT2)                  330
LB1     = LINE/XAXIS                                      335
LB2     = LINE/YAXIS,.5                                   340
LB3     = LINE/XAXIS,.5                                   345
LB5     = LINE/XAXIS,1                                    350
LB6     = LINE/YAXIS,2                                    355
LB7     = LINE/YAXIS,1                                    360
        REFSYS/NOMORE                                     365
```

From the geometry we determine the angle of rotation (70°) and with the OBTAIN statement we extract the coordinates of the origin of local area B. The matrix of the REFSYS statement must be named (MAT4 in line 330). MAT4 is the product of two matrices, where MAT2, the linear transformation (rotation), is performed first, then followed by the translation (MAT3).

The origin of the local area C coordinate system is PC1, read off the grid paper as $(x, y) = (2.5, 2.75)$. We will calculate the rotation angle because we also know the coordinates of PC2. The following code applies:

```
$$      **** LOCAL AREA C ****                           400
MAT5    = MATRIX/XYROT,ATAN2F( – 1,1.75)                  410
MAT6    = MATRIX/TRANSL,2.5,2.75,0                        420
        REFSYS/(MAT7 = MATRIX/MAT6,MAT5)                  430
CC3     = CIRCLE/0,0,.5                                   440
CC1     = CIRCLE/ – 1.75,0,.5                             450
CC2     = CIRCLE/YLARGE,OUT,CC1,OUT,CC3,RADIUS,.5         460
        REFSYS/NOMORE                                     470
```

The equation of local area D defines a parabola with its vertex at the origin and with it symmetrical about and opening in the direction of the positive x-axis. In local area D its vertex is at $(x, y) = (1, -0.5)$ and its axis is at a 45° angle to the x-axis. The following code defines it as a general conic:

$$	**** LOCAL AREA D ****	500
COS45	= COSF(-45)	510
SIN45	= SINF(-45)	520
MAT8	= MATRIX/CANON,COS45,$-$SIN45,0,1,SIN45,COS45, $	530
	0,$-$.5,0,0,1,0	540
	REFSYS/MAT8	550
PAR	= GCONIC/0,0,1,$-$1,0,0	560
	REFSYS/NOMORE	570

We finish the geometric descriptions with the following two definitions in the base coordinate system:

LC1 = LINE/RIGHT, TANTO,CA2, LEFT, TANTO,C1
LC2 = LINE/ORIG, RIGHT,TANTO,CC3

With respect to the origin of the local areas, ORIG appears to have coordinates as given in the following table:

Local area	ORIG	
	x	y
A	3.87298	1
B	0	-2.25
C	-0.80623	-3.62802
D	-1.06066	-0.35355

It is usually difficult to determine the equation for a general conic whose axis does not lie on, or is not parallel with, a coordinate axis. The equation for the parabola, obtained from a PRINT/3,PAR statement placed after the local area D REFSYS/NOMORE statement is

$$0.5x^2 + xy + 0.5y^2 - 1.2071x + 0.2071y + 1.18566 = 0$$

Example 7.2: Transforming Tabulated Cylinders

With the subject of matrix definitions and transformations covered and with the REFSYS feature discussed, we return to the subject of transforming tabulated cylinders. A good vehicle for this purpose is the involute of Section 6.3.3, where, originally computed from its parametric equations, it was rotated clockwise 20°. We will define the involute first with the TRFORM,*matrix* couplet in the TABCYL definition, then with the REFSYS feature.

The statement numbering of the part program in Section 6.3.3 is retained. It will be modified as needed but only statements beginning with line 200 will be repeated here. The plan is to rotate tabulated cylinder INVOL 20° clockwise with a rotation matrix. Upon deleting redundant statements and modifying the TABCYL statement, we get the following part program segment:

```
          LOOPST                                               180
          I = 1                                                190
          THETA = 0                                            200
          RADUS = 1.1                                          210
          DTHETA = 15                                          220
LA1)      SINTH = SINF(THETA)                                  260
          COSTH = COSF(THETA)                                  270
          THETR = THETA/57.2957795                             280
          XP = RADUS*(COSTH + THETR*SINTH)                     290
          YP = RADUS*(SINTH − THETR*COSTH)                     300
          PT(I) = POINT/XP,YP                                  330
          I = I + 1                                            340
          THETA = THETA + DTHETA                               350
          IF (I − 16) LA1, LA2, LA2                            360
LA2)      LOOPND                                               370
          MATR = MATRIX/XYROT, − 20                            375
          INVOL = TABCYL/NOZ,SPLINE,TRFORM,MATR,PT(ALL)        380
          OBTAIN,POINT/PT(14),X14,Y14                          382
          REFSYS/MATR                                          384
          P14 = POINT/X14,Y14                                  386
          REFSYS/NOMORE                                        388
          L4 = LINE/P14,ATANGL,75                              390
```

We simplified the code for computing the points of the involute. It was only necessary to define matrix MATR and specify its use in the TABCYL statement. The points of array PT are not rotated until INVOL is referenced in another statement. Consequently, we cannot define L4 with PT(14). Instead, we extract the coordinates of PT(14) with the OBTAIN statement, tag point P14 for rotation by defining it within a REFSYS region, then use it outside the REFSYS region so that it can be rotated prior to defining L4.

Had it not been for L4, the part program segment above would have been a clean way to write the program. This next approach will rotate the points first, then define the involute and L4.

```
          LOOPST                                               180
          I = 1                                                190
          THETA = 0                                            200
          RADUS = 1.1                                          210
          DTHETA = 15                                          220
MATR      = MATRIX/XYROT, − 20                                 224
          REFSYS/MATR                                          226
LA1)      SINTH = SINF(THETA)                                  260
          COSTH = COSF(THETA)                                  270
          THETR = THETA/57.2957795                             280
          XP = RADUS*(COSTH + THETR*SINTH)                     290
          YP = RADUS*(SINTH − THETR*COSTH)                     300
          PT(I) = POINT/XP,YP                                  330
          I = I + 1                                            340
          THETA = THETA + DTHETA                               350
          IF (I − 16) LA1, LA2, LA2                            360
```

LA2)	LOOPND	370
	REFSYS/NOMORE	375
INVOL	= TABCYL/NOZ,SPLINE,PT(ALL)	380
L4	= LINE/PT(14),ATANGL,75	390

The only geometric definitions within the REFSYS bounding region are points PT. Since the points are tagged for rotation when used outside the REFSYS region, INVOL and L4 must be defined outside the REFSYS region.

7.3 COPY

The part programmer may have to describe a part with symmetrical features or one that has repetition of geometry. In such applications the straightforward application of the APT geometric definitions thus far presented may require slight modification in statements of similar form but perhaps applied with much repetition. Of course, loops could be written within which geometric definitions can be modified and appropriate motion commands incorporated to produce the part. Both of these approaches would accomplish the same result, calculation of the proper set of cut vectors, but through varying demands on the part programmer with respect to ability, time, and effort. APT provides the COPY convenience feature (executed in Section 3 of the APT system) to assist the part programmer in accomplishing the results described above but with considerably less effort (i.e., it avoids the necessity for rewriting the instructions). The use of COPY is sometimes clearly evident and its application becomes almost trivial. At other times, considerable imagination is required to identify suitable transformation matrices to enable the COPY feature to be incorporated. COPY differs from REFSYS in that it transforms cut vectors, not geometric definitions.

7.3.1 Fundamental Usage and Statement Formats

When the APT system is executing a part program that does not include the COPY feature it writes auxiliary commands (cutter diameter, tolerances, postprocessor commands, etc.) and cut vectors onto the CL-file for use by the postprocessor section. This output is generated on a one-to-one basis as the commands are executed. The COPY feature allows the part programmer to cause a section of CL-file data to be repeated (copied) under transformation on the CL-file without reexecuting the part program segments that caused the data to be originally produced. An expanded CL-file is the result.

To use the COPY feature, the part programmer must tag with INDEX statements the section of CL-file data to later be copied onto the CL-file at the location designated by the COPY statement. The structure of the COPY feature is as follows:

INDEX/k

   .
   .
   .

INDEX/k,NOMORE

   .
   .
   .

COPY/k, . . .

The INDEX statements are bounding statements containing the same index number, k (numeric literal or scalar variable), that is referenced in a subsequent COPY statement. Index k bears no relation to the number of copies, nor does it connote anything other than identification for a section of CL-file data. The number k cannot appear in more than one INDEX bounding pair within the same part program.

The COPY statement designates the relative location within the CL-file, where the copied data will be inserted to generate the expanded CL-file. The structure defined above allows for multiple reference of the bounded statements for copying purposes. Figure 7.9 illustrates the effect that the COPY feature has on the data within the CL-file. The INDEX/k,NOMORE statement may be omitted when followed immediately by the COPY statement referencing the same k. The section of program thus bounded cannot again be referenced by another COPY statement in the manner illustrated by the lower diagram of Figure 7.9.

The following two sequences produce equivalent results:

INDEX/7	INDEX/7
.	.
.	.
.	.
INDEX/7,NOMORE	COPY/7, . . .
COPY/7, . . .	

The COPY statement specifies the number of copies to be made and the transformation they are to undergo during the copy operation. The syntax for a representative sample of COPY statements is as follows:

$$\text{COPY}/k, \begin{Bmatrix} \text{XYROT}, \theta \\ \text{TRANSL}, dx, dy, dz \\ \text{SAME} \\ \text{MODIFY}, matrix \end{Bmatrix}, n$$

Parameter k references a previously defined INDEX bounding pair while parameter n (a nonnegative integer numeric literal or scalar variable) defines the number of copies of CL-file data to be generated. Parameter n specifies a quantity *in addition* to that produced from the original pass through the INDEX bounding-pair statements.

The choice of transformation (modifier) for the COPY statement is governed by the geometry of the part and the motion commands included within the INDEX bounding pair. Each modifier is now discussed in conjunction with an example whose geometry suggests its use.

COPY Using the XYROT Modifier

Figure 7.10 shows a part of angular symmetry if the origin of the coordinate system is placed at the center of the part. It is redundant to describe the geometry and motion commands for more than one octant of the part. We could write the part geometry and motion commands for one octant within a loop, then reexecute these statements seven more times with appropriate geometry redefinition so as to generate a complete set of cut vectors. This requires computation of cut vectors from new geometric definitions and motion commands for each octant, an APT Section 2 operation. The saving is the time

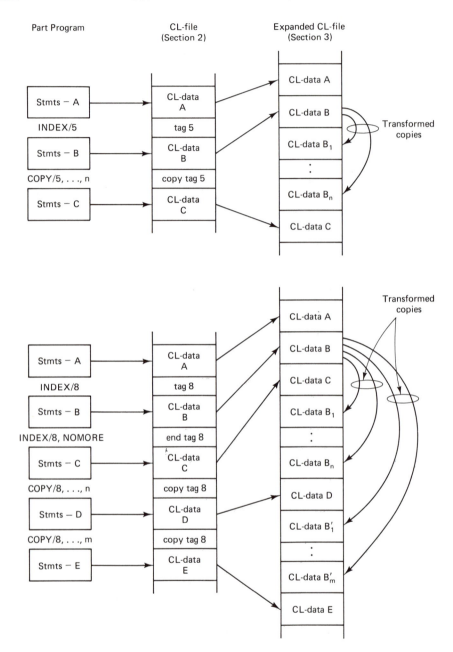

Figure 7.9 Two examples showing the effect the COPY statement has on the data within the CL-file to generate an expanded CL-file.

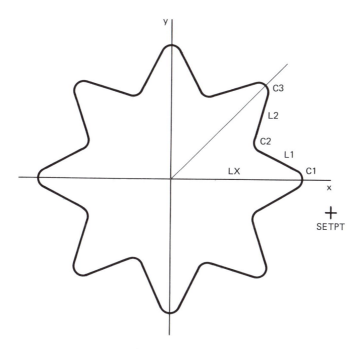

Figure 7.10   Part geometry suggesting the XYROT modifier in a COPY statement.

and effort in not preparing the geometric definitions and motion commands for the other seven octants.

The symmetry of the part suggests that the cut vectors for one octant be computed as mentioned above and the cut vectors for the other octants be derived from these by a linear transformation, which for this part would be a rotation in the xy-plane. It is more economical of part programmer time and of computer time to transform the cut vectors in this manner (APT Section 3 operation) than to recompute them from geometric definitions and motion commands (the preparation of which leads to errors). The XYROT modifier in a COPY statement will be used for the linear transformation to be made through specification of the angle of rotation (θ, in degrees, is positive when measured counterclockwise). The motion commands to produce the part shown in Figure 7.10 are as follows:

FROM/SETPT	100
GO/TO,LX	110
TLLFT,GOLFT/LX,TO,C1	120
TLRGT,GORGT/C1,ON,LX	130
INDEX/1	140
GOFWD/C1,TANTO,L1	150
GOFWD/L1,TANTO,C2	160

GOFWD/C2,TANTO,L2	170
GOFWD/L2,TANTO,C3	180
GOFWD/C3,ON,LA	190
COPY/1,XYROT,45,7	200
GOTO/SETPT	210

Geometric definitions were prepared for the octant labeled in Figure 7.10. The tool is initially brought from SETPT to a position on LX tangent to C1. These motion commands are not included within the INDEX/1 bounding-pair (COPY is the termination statement for the sequence) so the resulting cut vectors will not be copied, especially the coordinates of the tool end when the tool is on LX and tangent to C1. The cut vectors computed from this point along the contour to and including that cut vector computed when the tool is on LA and tangent to C3 are within the INDEX/1 bounding pair and will be used to compute, in a counterclockwise direction because of the positive angle specified, the cut vectors for the other seven octants. The cut vectors are transformed and copied in the same order in which they were computed within the INDEX/1 bounding pair. When the COPY is satisfied, the tool is still located where placed by the last motion statement within the INDEX/1 bounding pair. In this case it is on LA and tangent to C3. For this example the tool will not really be located there when the part is produced (it will be located on LX and tangent to C1) and it can be returned immediately to SETPT as shown. If further contouring were to occur after the COPY operation, the tool would have to be brought to within tolerance of the new drive surface from its location on LA and tangent to C3. In this example the tool will not cut into the part on its way back to SETPT.

One must understand that the cut vectors produced will direct the tool in a continuous manner in a counterclockwise direction along the contour as required. The same cut vector sequence cannot be produced by copying clockwise (specifying a negative angle) since this will result in a counterclockwise sequence of cut vectors within octants that are transformed in a clockwise direction and written on the CL-file in this order. The tool motion from one octant to its adjacent octant will be undesired, as will the tool location upon completion of the COPY.

COPY Using the TRANSL Modifier

Figure 7.11 shows a part with repetition of geometry. The repetitious geometry, together with the appropriate motion commands, could be defined within a loop. As for the previous example, we will write the geometric definitions and motion commands for an element of the part repetition and then compute the remaining cut vectors through transformation with translation only. We will use the COPY statement with the TRANSL modifier.

The TRANSL modifier parameters specify incremental movement for each copy operation with the signed quantities dx, dy, dz, each of which signifies movement in a positive axis direction when positive and in a negative axis direction when negative. Translation is cumulative with each copy operation. For the first copy the original cut vectors are translated by an amount dx, dy, dz, for the second copy by an amount $2dx$,

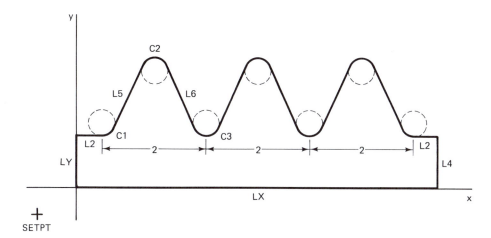

Figure 7.11   Part geometry suggesting the TRANSL modifier in a COPY statement.

$2dy$, $2dz$, and so on through the *n*th copy, which results in the original cut vectors being translated by an amount of ndx, ndy, ndz. The motion commands to produce the part of Figure 7.11 are as follows:

FROM/SETPT	100
GO/TO,LY	110
TLLFT,GOLFT/LY,PAST,L2	120
GORGT/L2,TANTO,C1	130
INDEX/7	140
GOFWD/C1,TANTO,L5	150
GOFWD/L5,TANTO,C2	160
GOFWD/C2,TANTO,L6	170
GOFWD/L6,TANTO,C3	180
GOFWD/C3,TANTO,L2	190
COPY/7,TRANSL,2,0,0,2	200
GOFWD/L2,PAST,L4	210
GORGT/L4,PAST,LX	220
GORGT/LX,PAST,LY	230
GOTO/SETPT	240

The element of repetition is selected such that tool movement is along C1-L5-C2-L6-C3 (clockwise around the part), beginning at the point where L2 and C1 are tangent and ending at the point where L2 and C3 are tangent. When the COPY operation is complete the tool is commanded along L2 into position to finish the part. The cut vectors under the TRANSL modifier are transformed and written onto the CL-file in the same sequence as computed during the original pass through the INDEX/7 bounding pair. The motion commands are written to avoid undesirable tool movement between adjacent sequences because of the COPY operation.

COPY Using the SAME Modifier

Sometimes we wish to write onto the machine control tape identical sets of cut vectors. Perhaps in the production of a large number of identical parts it would be expedient to continue sequentially through the tape as each part is produced rather than to rewind the tape for each part. Of course, this requires postprocessor commands to stop the director while the workpiece is being changed. Sequential processing would continue to the end of the machine control tape, at which time the tape must be rewound. In another application, the first of two sets of identical cut vectors may be used for a rough cut with an undersized tool, the second for the finish cut with the proper tool. Of course, postprocessor commands would be required to effect the tool change.

Part programs for the applications above require no transformation of cut vectors and can be written to use the SAME modifier in a COPY statement (the effect is the same as if a unit matrix is used for the transformation). Fifty copies of the simple part in Figure 7.12 can be produced with the following motion commands:

FROM/SETPT	100
INDEX/4	110
GO/TO,C1	120
TLRGT,GORGT/C1,TANTO,C2	130
GOFWD/C2,TANTO,C3	140
GOFWD/C3,TANTO,LX	150
GOFWD/LX,PAST,C1	160
GOTO/SETPT	170
COPY/4,SAME,50	180

The motion commands take the tool in a counterclockwise direction around the part and return it to SETPT. The cut vectors produced by the program are written onto the CL-file, then copied onto the CL-file without change 50 more times. Postprocessor commands must be added to stop the machine so that the completed part may be replaced with a new workpiece.

The case where rough/finish cuts are made with two different diameter tools but with the same set of cut vectors can be addressed with the THICK statement. This is a modal statement that allows a thickness to be added to one or more of the surfaces in startup or motion commands. Its format is

THICK/*thk-ps,thk-ds,thk-cs*

where *thk-ps*, *thk-ds*, and *thk-cs* are scalar values for the thickness of the part, drive, and check surface, respectively. Multiple check surface thicknesses may be allowed for a given APT processor. A positive thickness causes excess material to be left on the cutter side of the surface; a negative thickness causes undercutting of the part. Other factors

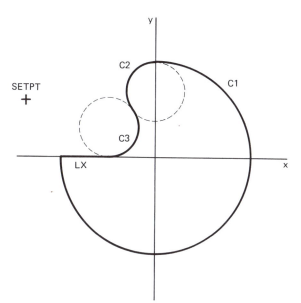

Figure 7.12   Part geometry for illustrating the use of the SAME modifier in a COPY statement.

must be considered when using this statement. Because we have no need for it in this book, the APT reference manual should be consulted regarding its application. A form of the statement is reflected in the POLYCONIC defining statement (see Section 12.3).

COPY Using the MODIFY Modifier

The part may contain repetitious geometry not addressed by the transformations of the preceding COPY formats. These cases may be handled by constructing a matrix and referencing it in the MODIFY form of the COPY statement. The matrix is constructed as described in Section 7.1 and its symbol included in the COPY statement for *matrix*. The part in Figure 7.13 contains a repeated geometric element that can be transformed by a matrix performing rotation and translation.

The left and right edges of the part have identical geometries. It is possible that a center of rotation could be found such that the XYROT modifier form of COPY statement could be used. However, the coordinate system location facilitates description of the geometry and preparation of the matrix will not be difficult. Because the geometry of the top edge differs from that of the bottom edge, thus requiring intervening cut vectors between the original and copied cut vectors, the terminator of the INDEX/k bounding pair will be the INDEX/k,NOMORE statement.

Since the y-axis is also drive surface LY, the cut vectors of the right edge will be rotated 180° about the origin, then translated to the left edge. The matrix for the COPY

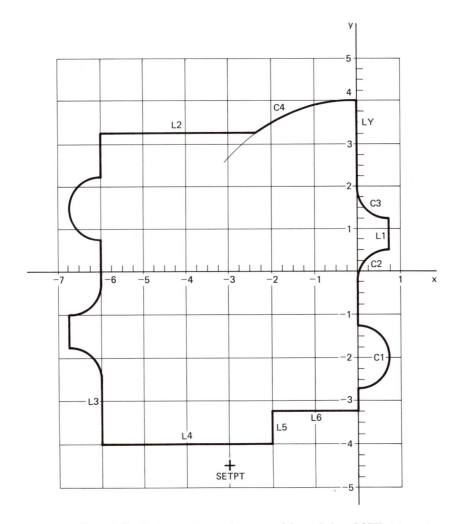

Figure 7.13 Part geometry requiring a special matrix in a COPY statement.

statement and the motion commands to take a tool counterclockwise around the part of Figure 7.13 are as follows:

$$
\begin{array}{lr}
\text{MAT1 = MATRIX/XYROT,180} & 100 \\
\text{MAT2 = MATRIX/TRANSL}, -6, -.5, 0 & 110 \\
\text{MAT3 = MATRIX/MAT2,MAT1} & 120 \\
\quad \text{FROM/SETPT} & 130 \\
\quad \text{GO/PAST,L5} & 140 \\
\end{array}
$$

TLRGT,GOLFT/L5,TO,L6	150
GORGT/L6,PAST,LY	160
INDEX/9	170
GOLFT/LY,TO,C1	180
GORGT/C1,TO,LY	190
GORGT/LY,TANTO,C2	200
GOFWD/C2,PAST,L1	210
GOLFT/L1,PAST,C3	220
GOLFT/C3,TANTO,LY	230
INDEX/9,NOMORE	240
GOFWD/LY,PAST,C4	250
GOLFT/C4,TO,L2	260
GORGT/L2,PAST,L3	270
COPY/9,MODIFY,MAT3,1	280
GOLFT/L3,PAST,L4	290
GOLFT/L4,PAST,L5	300
GOTO/SETPT	310

For each of the four examples above, the tool was positioned before the COPY statement so that it could be commanded after the cut vectors were copied without generating unwanted cut vectors. The tool position does not change when a COPY statement is executed. The straight-line drive surfaces prevented computation of unwanted cut vectors. Inhibiting unwanted cut vectors can be handled by the technique described in Section 7.4.

7.3.2 Nested COPY

Each COPY statement can reference only one INDEX bounding pair, and several COPY statements may reference the same INDEX bounding pair. Also, a COPY statement may be contained within another INDEX bounding pair. This latter condition is known as nested COPY, which is analogous to nested looping. Whereas nested loops cause statements within the loop to be executed during each pass through the loop, nested COPY statements cause multiple copies to be made under transformation for each nest. Possible nesting cofigurations are shown in Figure 7.14 for the case of nonoverlapping COPY and for a maximum of three levels of nesting, the usual limitation on nesting.

A COPY nesting contouring example is shown in Figure 7.15(a). This figure has a repetitious element (the arm) for which cut vectors can be copied twice with appropriate translation. However, the arm itself [Figure 7.15(b)] has a repetitious element whose cut vectors can also be copied with translation. All motion commands for the arm can be included within the INDEX bounding pair of one COPY with some of these commands also within the INDEX bounding pair of another COPY, thus forming a nested COPY part program. Additional commands are required to position the tool and to finish the

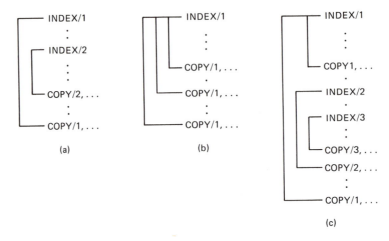

Figure 7.14 Possible nonoverlapping COPY nesting configurations.

part after the copying operations are completed. The motion commands to take a tool counterclockwise around the part are as follows:

FROM/SETPT	100
GO/TO,LX	110
INDEX/1	120
TLRGT,GORGT/LX,PAST,L5	130
GOFWD/LX,TANTO,C2	140
GOFWD/C2,TANTO,C3	150
INDEX/2	160
GOFWD/C3,TANTO,L3	170
GOFWD/L3,TANTO,C4	180
GOFWD/C4,TANTO,L4	190
GOFWD/L4,TANTO,C5	200
GOFWD/C5,TANTO,L2	210
COPY/2,TRANSL,$-1.5,0,0,1$	220
GOFWD/L2,TANTO,C6	230
GOFWD/C6,TANTO,L5	240
COPY/1,TRANSL,(5*COSF(60)/SINF(60)),5,0,2	250
GOFWD/L5,PAST,L6	260
GOLFT/L6,PAST,L7	270
GOLFT/L7,TANTO,C7	280
GOFWD/C7,TANTO,LX	290
GOFWD/LX,PAST,L5	300
GOTO/SETPT	310

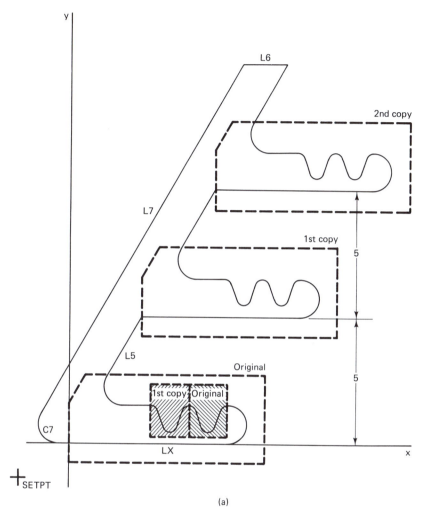

Figure 7.15 (a) Tool endpoints which are copied by nesting; (b) (see next page) details of the arm shown in Figure 7.15(a).

The commands for one complete arm are contained within the INDEX/1 bounding pair. However, some of these commands are also within the INDEX/2 bounding pair, thus forming the nested COPY structure of Figure 7.14(a). Since contouring is to continue after each COPY operation, one must ensure that the tool will be within tolerance of a suitable drive surface at all times. For the innermost COPY, the tool must be within tolerance of L2, a condition guaranteed by positioning the tool on L2 with the beginning and ending commands for the repetitious element (C3 is tangent to L2). For the outermost

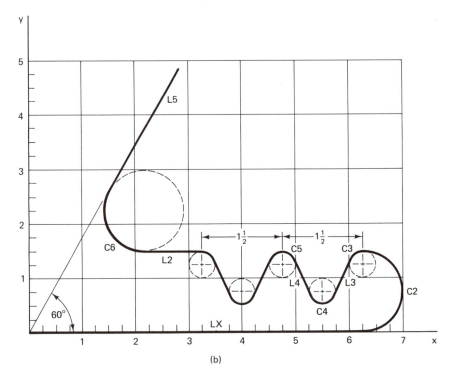

(b)

Figure 7.15 (Continued)

COPY, the tool must be within tolerance of L5, a condition guaranteed by initially positioning the tool on L5 with a PAST modifier and ending the COPY by driving onto L5 from C6 with a TANTO modifier. Figure 7.15(a) shows the areas for which the cut vectors are obtained with the COPY feature.

Some APT systems permit overlapping COPY, such as shown in Figure 7.16. Use of this capability is contrary to program clarity and understanding. Thus its application is discouraged.

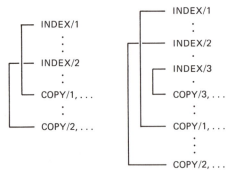

Figure 7.16   Some overlapping COPY configurations.

7.4 INHIBITING OUTPUT TO THE CL-FILE

We stated in Chapter 1 that the part geometry is resolved in APT Section 1, all cutter locations are determined in Section 2, and all coordinate transformations and copies are made in Section 3. This means that all cut vectors are calculated before any transformations are applied and thus are determined with reference to only one coordinate system, subject to REFSYS transformations of course. Therefore, when preparing motion statements whose cut vectors will require Section 3 processing, one must carefully develop all motion commands in only one coordinate system. Special attention must be given to statements whose cut vectors will interface with those obtained via transformation and copying.

The examples of Section 7.3 were constructed to avoid the interfacing problems. The part program for Figure 7.10 had SETPT located such that startup was no problem. Also, the ending motion command took the tool in a straight-line motion back to SETPT without cutting across the part. This ending motion did not require a drive surface, so the tolerance relationships were unimportant. The part program for Figure 7.11 required tool movement after the COPY operation along drive surface L2, for which it was already within tolerance as it completed the last command within the INDEX bounding pair (C3 tangent to L2). Again, no tool repositioning was required and the part was not unintentionally cut into. The part program for Figure 7.12 required no special tool movement during startup and ending operations. The part program for Figure 7.13 had the tool positioned to continue contouring along L3. The part program for Figure 7.15 required tool movement after the COPY operations in a manner similar to that required for Figure 7.11.

The geometry of a part may not be so simple that the foregoing problems can always be avoided. Additional commands may be required to position the tool for the next sequence of commands. Furthermore, cut vectors produced by these additional commands may not even be desired on the CL-file. Thus the writing of cut vectors on the CL-file must be inhibited while manipulating the tool in this manner. This is done with the reserved words DNTCUT (don't cut) and CUT, which would appear in a part program as follows:

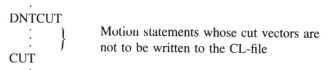

The DNTCUT statement signals the APT system to process the statements that follow (compute cut vectors) but not to write them onto the CL-file or cause them to appear on the CL-printout. This is a modal command and remains in effect until countermanded by the CUT statement. The CUT statement signals the APT system to resume outputting the cut vectors to the CL-file and on the CL-printout for all commands that follow it. Upon encountering a CUT statement, the first cut vector which the APT system will write is the last one computed within the range of the DNTCUT-CUT statements. This cut vector is derived from the motion command immediately preceding the CUT

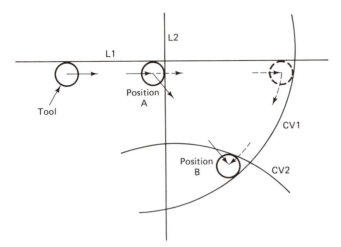

Figure 7.17   Geometry for illustrating DNTCUT-CUT statements.

statement and provides the terminal coordinates for tool motion originating at the coordinates of the previous cut vector output, which immediately preceded DNTCUT. Thus, when the CUT statement is encountered, the effect is to position the tool as desired but not have it follow a path given by cut vectors for intermediate positions. Following the CUT statement, all cut vectors are output in the usual manner.

The following part program segment will take the tool directly from position A to position B of Figure 7.17 without following the dashed-line path. The tool cannot be commanded directly from position A to position B since there is no drive surface for this purpose. The DNTCUT-CUT statements permit tool-to-drive-surface relationships to be maintained yet a direct path to result.

```
····/L1,TO,L2
DNTCUT
GOFWD/L1,TO,CV1
GORGT/CV1,PAST,CV2
CUT
    ·
    ·
    ·
```

7.5 TRANSFORMING CUT VECTORS VIA TRACUT

The features introduced in this chapter substantially increase the part programmer's capability for applying APT to parts of complex geometry. The power of the APT language should be apparent by now. Much of this power arises from transformations in the part program. We can transform geometry (REFSYS) and cut vectors for parts with symmetrical or repetitious geometry (COPY), but we cannot yet transform cut vectors for parts of arbitrary geometry. It may be that the workpiece cannot be mounted on the machine according to the desires of the part programmer, that is, the coordinate system chosen by the part programmer is inappropriate for the machine. In this case, all cut vectors for the part should be transformed to the machine coordinate system. It may be desired to scale part dimensions upward or downward (to allow for material characteristics, to graphically display the part to a more suitable scale, etc.). Or it may be more convenient to transform a set of cut vectors which are not copies. The TRACUT feature is provided for general cut vector transformation purposes.

7.5.1 Principles of the TRACUT Feature

The TRACUT statement causes cut vectors to be transformed from one coordinate system to another. Two formats for statements of this feature are usually available:

$$\text{TRACUT}/\begin{Bmatrix} matrix \\ \text{NOMORE} \end{Bmatrix} \qquad \text{TRACUT/LAST,} \begin{Bmatrix} matrix \\ \text{NOMORE} \end{Bmatrix}$$

The statement requires *matrix* to be defined, or its definition nested within the statement. Geometric definitions are not transformed by *matrix*. The case where cut vectors are subject to both TRACUT and COPY transformations is discussed in Section 7.5.2. When applied without the LAST modifier, the TRACUT feature is structured as follows:

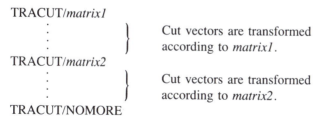

$$\left.\begin{array}{l} \text{TRACUT}/matrix \\ \qquad \vdots \\ \text{TRACUT/NOMORE} \end{array}\right\} \quad \begin{array}{l} \text{Statements whose cut vectors are} \\ \text{transformed according to } matrix \end{array}$$

Cut vectors transformed are generated within the TRACUT bounding-pair statements. The TRACUT/*matrix* statement is modal and remains in effect until countermanded by the TRACUT/NOMORE statement or by another TRACUT/*matrix* statement (see below). Multiple, but not nested nor multiplicative, transformations may be structured as follows:

$$\begin{array}{l} \text{TRACUT}/matrix1 \\ \qquad \vdots \quad \left.\right\} \quad \begin{array}{l} \text{Cut vectors are transformed} \\ \text{according to } matrix1. \end{array} \\ \text{TRACUT}/matrix2 \\ \qquad \vdots \quad \left.\right\} \quad \begin{array}{l} \text{Cut vectors are transformed} \\ \text{according to } matrix2. \end{array} \\ \text{TRACUT/NOMORE} \end{array}$$

The second TRACUT statement countermands the first. It, in turn, is countermanded by the TRACUT/NOMORE statement.

If the ranges of two sets of TRACUT bounding-pair statements overlap, where one set must include the LAST modifier, transformation within the overlapping ranges is cumulative. Modifier LAST effects a second transformation that is applied after the transformation required by the TRACUT bounding pair without LAST. The TRACUT/ LAST,*matrix* statement is modal and can be countermanded only by the TRACUT/

LAST,NOMORE statement or by another TRACUT/LAST,*matrix* statement. A selection of examples is given in Table 7.2 to reveal the functioning of the two forms of TRACUT bounding-pair statements.

The coordinate transformations of the TRACUT feature are not made until Section 3 of the APT system is executed. No transformations will be made if there are Section 2 errors. The transformation matrix appears in the CL-printout at the location where the TRACUT statement is encountered.

Since all tool motion commands reference part geometry described in only one coordinate system, care must be used with the TRACUT feature to maintain proper tool-to-surface relationships to avoid Section 2 errors. The DNTCUT-CUT feature is useful

TABLE 7.2

TRACUT/LAST structure	Equivalent TRACUT structure
⎯ TRACUT/LAST,MAT1 ⋮	TRACUT/MAT1 ⋮
⎯TRACUT/MAT2 ⋮	TRACUT/(MATRIX/MAT1,MAT2) ⋮
⎣TRACUT/NOMORE ⋮	TRACUT/MAT1 ⋮
⎯ TRACUT/LAST,NOMORE	TRACUT/NOMORE
⎯TRACUT/LAST,MAT1 ⋮	TRACUT/MAT1 ⋮
⎯TRACUT/MAT2 ⋮	TRACUT/(MATRIX/MAT1,MAT2) ⋮
⎯TRACUT/LAST,NOMORE ⋮	TRACUT/MAT2 ⋮
⎣TRACUT/NOMORE	TRACUT/NOMORE
⎯TRACUT/MAT1 ⋮	TRACUT/MAT1 ⋮
⎯TRACUT/LAST,MAT2 ⋮	TRACUT/(MATRIX/MAT2,MAT1) ⋮
⎣TRACUT/NOMORE ⋮	TRACUT/MAT2 ⋮
⎯TRACUT/LAST,NOMORE	TRACUT/NOMORE
⎯ TRACUT/LAST,MAT1 ⋮	TRACUT/MAT1 ⋮
TRACUT/MAT2 ⋮	TRACUT/(MATRIX/MAT1,MAT2) ⋮
TRACUT/MAT3 ⋮	TRACUT/(MATRIX/MAT1,MAT3) ⋮
⎣TRACUT/LAST,MAT4 ⋮	TRACUT/(MATRIX/MAT4,MAT3) ⋮

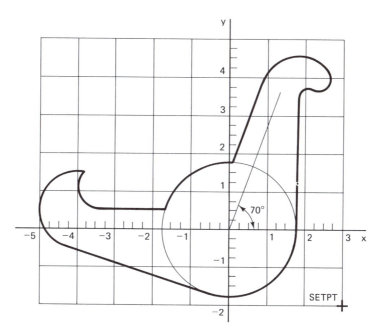

Figure 7.18 **Part geometry suggesting the use of common geometric definitions for TRACUT application.**

when positioning the tool with continuous motion commands to avoid unwanted cut vectors.

Part program simplification will result for the part of Figure 7.18 if the part is partitioned into two sections, each of which can be partially described with a common set of geometric definitions. Then, with a suitable transformation matrix in a TRACUT statement, cut vectors for one section can be transformed into place to complete the part. An APT part program to do this is as follows:

SETPT	= POINT/3, −2,2	100
	REFSYS/(MAT1 = MATRIX/TRANSL, −4,.5,0)	110
ORIGA	= POINT/0,0,0	120
LXA	= LINE/XAXIS	130
LYA	= LINE/YAXIS	140
C1	= CIRCLE/CENTER,ORIGA,RADIUS,1	150
C2	= CIRCLE/XLARGE,LYA,YLARGE,LXA,RADIUS,.5625	160
C3	= CIRCLE/TANTO,LXA,YLARGE,ORIGA,RADIUS,.25	170
C4	= CIRCLE/XSMALL,OUT,C3,IN,C1,RADIUS,.3	180
	REFSYS/NOMORE	190
ORIGB	= POINT/0,0,0	200
C5	= CIRCLE/CENTER,ORIGB,RADIUS,1.75	210
L1	= LINE/RIGHT,TANTO,C3,LEFT,TANTO,C5	220

L2	= LINE/RIGHT,TANTO,C1,RIGHT,TANTO,C5		230
L3	= LINE/YSMALL,TANTO,C1,ATANGL,0		240
L5	= LINE/ORIGB,ATANGL, − 55,XAXIS		250
L6	= LINE/ORIGB,ATANGL,55,XAXIS		260
MAT2	= MATRIX/XYROT, − 110		270
	FROM/SETPT		280
	GO/TO,L5		290
	TLRGT,GORGT/L5,TO,C5	\$\$ MOTION	300
	TLLFT,GOLFT/C5,TANTO,L2	\$\$ COMMANDS FOR	310
	GOFWD/L2,TANTO,C1	\$\$ THE LEFT ARM	320
	GOFWD/C1,PAST,C2		330
	GORGT/C2,TANTO,LXA		340
	GOFWD/LXA,TO,C5		350
	GOLFT/C5,ON,L5		360
	DNTCUT		370
	TLRGT,GOBACK/C5,ON,L6		380
	TRACUT/MAT2		390
	CUT		400
	TLLFT,GOBACK/C5,TO,L3	\$\$ MOTION	410
	GOLFT/L3,TANTO,C1	\$\$ COMMANDS FOR	420
	GOFWD/C1,TANTO,C4	\$\$ THE UPPER ARM	430
	GOFWD/C4,TANTO,C3		440
	GOFWD/C3,TANTO,L1		450
	GOFWD/L1,TANTO,C5		460
	GOFWD/C5,ON,L6		470
	TRACUT/NOMORE		480
	GOTO/SETPT		490

The geometric symbols appear in Figure 7.19, where both the left and upper arms of the part are described on the same sketch. The REFSYS feature is conveniently applied here. The part program shows the tool commanded clockwise around the part, transforming the cut vectors for the upper arm (lines 410 through 470) with a rotation matrix in the TRACUT statement. Inhibited tool positioning is necessary to generate a continuous sequence of cut vectors before transformation is performed.

7.5.2 Transforming Copied Cut Vectors

We have not yet demonstrated how the TRACUT matrix transforms cut vectors generated by the COPY feature. We find no restriction on applying COPY, TRACUT, and TRACUT/ LAST simultaneously to a set of motion statements. Indeed, any combination of these three features may be used within the same part program. No rules of application are needed for APT Section 2 processing since untransformed cut vectors only are computed there. But conflicts may arise in Section 3 processing when multiple transformations apply to a set of cut vectors (transformations are matrix operations which, in general,

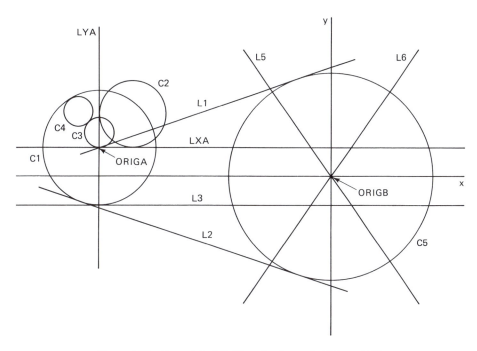

Figure 7.19   Geometric definitions for the part of Figure 7.18.

are not commutative). To avoid such conflicts, matrix operations are applied hierarchically as follows:

TRACUT   (highest priority)
COPY
TRACUT/LAST   (lowest priority)

We first illustrate the hierarchical order with the TRACUT-COPY structures of Figure 7.20. In part (a), cut vectors from statement groups A, B, and C are first transformed

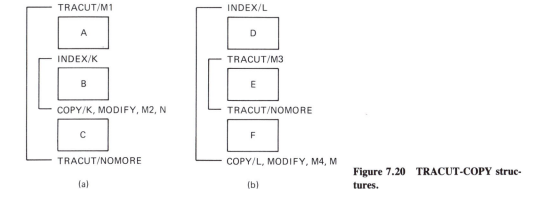

(a)   (b)

Figure 7.20   TRACUT-COPY structures.

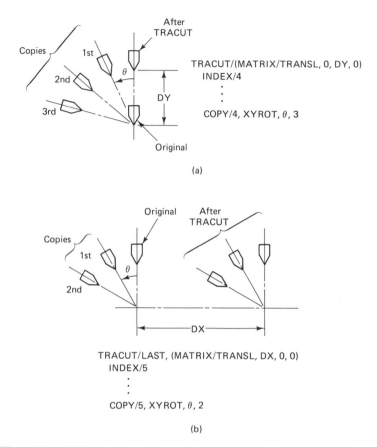

Figure 7.21   **COPY transformations under (a) TRACUT, and (b) TRACUT/ LAST.**

according to matrix M1. Then the transformed cut vectors from statement group B are again transformed N times according to matrix M2. In part (b) cut vectors from statement group E are first transformed according to matrix M3. Then these transformed cut vectors and those from statement groups D and F are transformed M times according to matrix M4.

The TRACUT/LAST feature applies a transformation delayed in a manner analogous to that described in Section 7.5.1. This unique characteristic is illustrated in Figure 7.21, where it is contrasted with the TRACUT feature. It is not difficult to generalize the hierarchical ordering to multiple levels, such as shown in Figure 7.22. A part program with a similar structure appears in Example 8.3.

The choice of transformation structure is guided by the geometry of the part and the ability of the part programmer to visualize the outcome when applying alternative transformation structures. Selection criteria must include clarity in understanding the

```
         TRACUT/LAST, . . .
           TRACUT/. . .
            INDEX/. . .
                 ⋮
                 ⋮
            COPY/. . .
         TRACUT/NOMORE
       TRACUT/LAST, NOMORE
```

Figure 7.22   Multiple levels of cut vector transformation.

sequences of matrix transformations and the need to position the tool for continued contouring to avoid Section 2 errors.

7.6 COPING WITH TRANSFORMATIONS WHILE DEBUGGING PART PROGRAMS

Part complexity determines whether or not matrix transformations are used in a part program. Transformation features add power to the APT language and permit a mature approach to be applied in part program development. Care must be used to avoid problems that occur while developing and verifying part programs with these features. The programmer must specify transformations correctly, must verify that geometry is transformed correctly, must verify that cut vectors are transformed correctly, and must understand APT processing in a transformation environment.

Verification of Transformation

Because matrix transformations map geometry or cut vectors from one coordinate system into another, their effect must be visualized to eliminate sources of error in specifying the matrix. For this purpose, matrix forms explicitly revealing the transformation are recommended. These forms include modifiers for rotation, translation, and mirroring. The canonic form should be avoided, if possible, as should computation of matrix elements for a composite matrix representing a succession of transformations. Elementary matrices should be multiplied to yield a composite matrix (remember that matrix multiplication is not commutative).

Correct matrix selection can be verified by transforming known entities (a point, for example, such as SETPT) in short program segments written for this purpose. The REFSYS feature can be verified by noting canonic form parameters printout and, during development, by contouring ON drive surfaces for selected portions of the curve to define its shape.

Program Development

Develop the part program in manageable segments. Upon verification, combine them to yield the whole. Apply transformation techniques only when they are justified and well understood. Properly structure the program to observe the hierarchical ordering

TABLE 7.3 APT SYSTEM PROCESSING

Feature	APT section		
	1	2	3
REFSYS	Entities tagged with matrix for transformation	Entities transformed when referenced	
COPY		COPY bounds established; original cut vectors computed	CL-data copied and transformed
TRACUT		TRACUT bounds established; original cut vectors computed	CL-data transformed

of COPY-TRACUT features (see Section 7.5.2). There are limits on the number of levels of COPY nests and TRACUT overlapping segments.

The partitioning of the APT system functions into Sections is described in Section 1.4. The effects these Sections have on a part program with REFSYS-COPY-TRACUT features are summarized in Table 7.3. Cut vectors are computed in Section 2 only. There is an illusion of motion when cut vectors are generated with COPY-TRACUT operations in Section 3. Tool-to-surface relationships should at all times be consistent with the geometry of the part to be produced even though cut vectors must be suppressed for tool positioning purposes.

Miscellaneous

Emphasize clarity and simplicity in program preparation to highlight and retain program design intent. Apply the techniques and precautions for good programming practices given in prior chapters. Use the PRINT/3 statement liberally for entity parameter verification under transformation. Omit the DNTCUT-CUT statements until positioning paths are verified.

PROBLEMS

7.1. The symmetrical part of Figure P7.1 suggests the XYROT option of the COPY statement. With $R_1 = 0.625$ in. and $R_2 = 1.5$ in., write an APT part program with COPY to take a ¼-in.-diameter tool counterclockwise around the part.

7.2. Rework Problem 4.8 with the XYROT option of the COPY statement.

7.3. Rework Problem 4.9 with the XYROT option of the COPY statement.

7.4. Rework Problem 5.6 with the XYROT option of the COPY statement.

7.5. The symmetrical part of Figure P7.5 suggests the XYROT option of the COPY statement. Write an APT part program with XYROT to take a ¼-in.-diameter tool clockwise around the part.

7.6. Rework Problem 5.7 with the TRANSL option of the COPY statement.

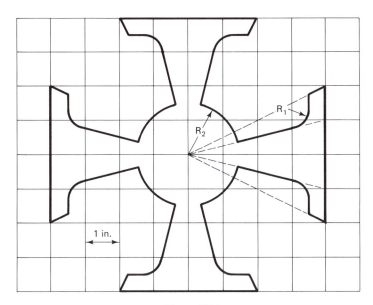

Figure P7.1

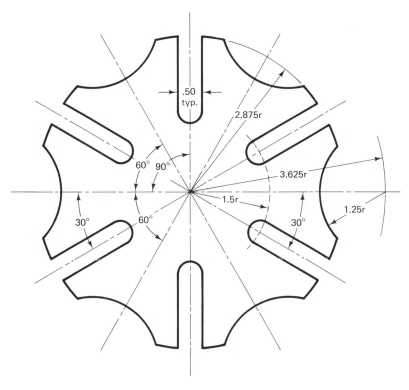

Dimensions in inches

Figure P7.5

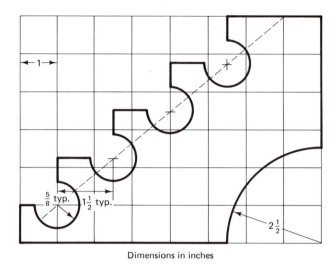

Dimensions in inches

Figure P7.7

7.7. The part of Figure P7.7 suggests the TRANSL option of the COPY statement. Write an APT part program with this option to take a ¼-in.-diameter tool clockwise around the part.

7.8. The part of Figure P7.8 has two portions that are nearly identical. The XYROT option of the COPY statement cannot be used because the portions are of differing lengths. However, if the cut vectors for one portion are rotated and then translated, the cut vectors of the other portion are produced. Write an APT part program to produce the part with the MODIFY option of the COPY statement. It is necessary to define a matrix for this purpose. Take a ½-in.-diameter tool counterclockwise around the part.

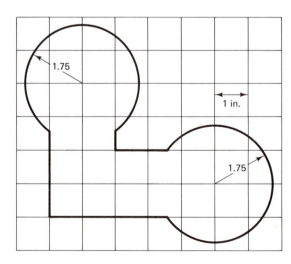

Figure P7.8

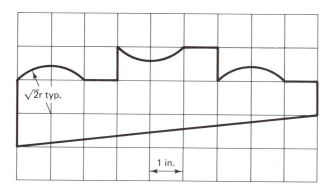

√2r typ.

1 in.

Figure P7.9

7.9. The part of Figure P7.9 has a top edge with three segments whereby the middle segment is a displaced inversion of the segments to either side of it. Let us assume that this part is to be engraved, that is, its outline is to be followed by a tool of zero diameter. In that case, only one of the segments of the top edge need be defined and its resulting cut vectors merely translated and inverted to generate the remaining cut vectors. Write an APT part program with the MODIFY option of the COPY statement to engrave the part by taking a tool clockwise around the part. Define an appropriate matrix for the COPY statement.

7.10. The part of Figure P7.10 shows a symmetrical arm that has been repeated at 45° increments. Write an APT part program with nested COPY statements to produce the part by taking a ¼-in.-diameter tool clockwise around the part.

7.11. The part of Figure P7.11 shows repetition of a segment which itself contains repetition of geometry, thus implying nested COPY statements. Write an APT part program with nested COPY statements to produce the part by taking a 0.2-in.-diameter tool clockwise around the part.

7.12. Rework Problem 4.10 with the TRACUT and COPY statements to produce the part.

7.13. Rework Problem 5.9 with the TRACUT and COPY statements to produce the part.

7.14. The part of Figure P7.14 shows circular cutouts spaced along the part by a constant multiplying (scaling) factor $k = 1.5$. The first cutout has values of $r_1 = 3$ in., $r_2 = 3\frac{1}{2}$ in., and $r_3 = \frac{1}{4}$ in. After scaling, these values are modified for the second cutout to kr_1, kr_2, and kr_3. Similarly, for the third cutout the corresponding values become k^2r_1, k^2r_2, and k^2r_3. Thus the part could be produced by referencing a suitable matrix with the MODIFY option of the COPY statement. However, problems with tool offset will arise in the copies and tolerance error buildup will occur with the modified cut vectors. If the tool diameter is reduced to zero, the tool offset problem will disappear. If we accept the tolerance error buildup in the copies, we can then use the COPY statement. Write an APT part program to produce the part by taking a zero-diameter tool clockwise around the part. Note the effects of scaling on the produced part.

7.15. The sides of the part of Figure P7.15 contain irregular cutouts derived from simple common geometry, thus suggesting the TRACUT feature. Define the matrices for the TRACUT statements and write an APT part program to take a ¼-in.-diameter tool counterclockwise around the part.

7.16. Rework Problem 5.8 with the TRACUT and COPY statements to produce the part.

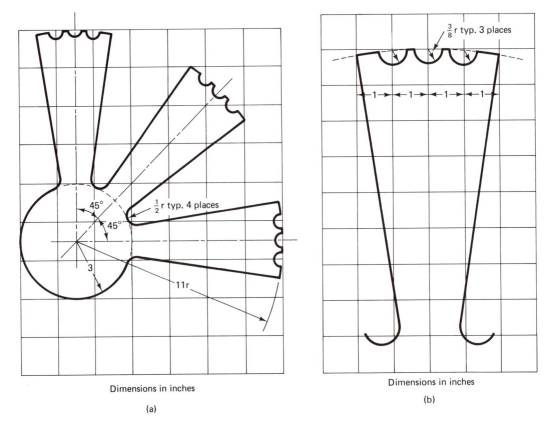

Dimensions in inches

(a)

Figure P7.10a

Dimensions in inches

(b)

Figure P7.10b

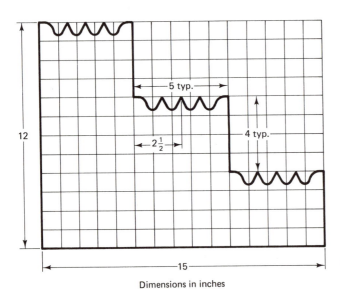

Figure P7.11a

Dimensions in inches

(a)

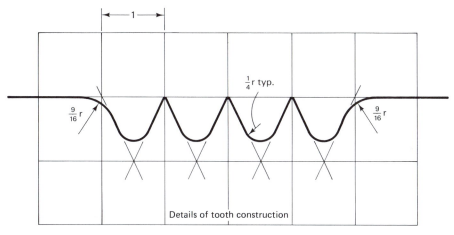

Details of tooth construction

Dimensions in inches

(b)

Figure P7.11b

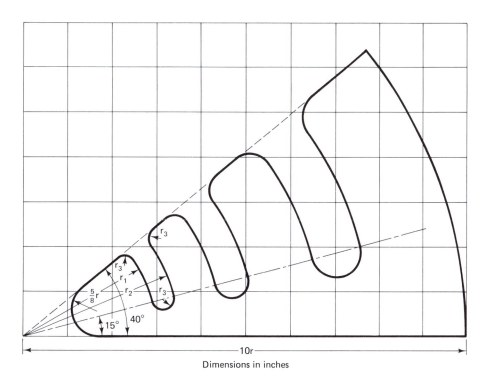

Dimensions in inches

Figure P7.14

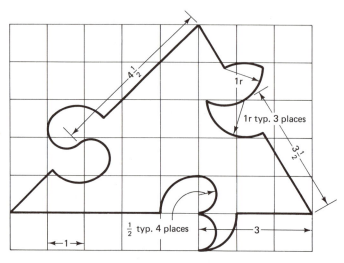

Dimensions in inches

Figure P7.15

7.17. The part of Figure P7.17 shows a complex arrangement of rather simple geometric elements. Since few different geometric elements are involved, it is appropriate to define them and, in conjunction with TRACUT statements, use them with the proper motion commands to produce the part. Write an APT part program using this technique to produce the part. Take a ¼-in.-diameter tool counterclockwise around the part.

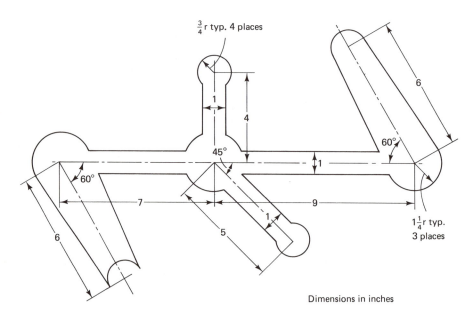

Dimensions in inches

Figure P7.17

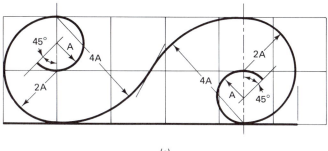

(a)

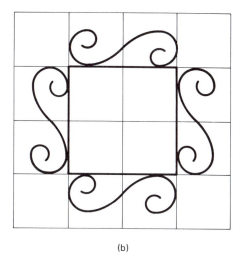

(b)

Figure P7.18

7.18. The design of Figure P7.18(a) is formed from tangent circles and a straight line. Four such designs rotated and translated form the composite design of Figure P7.18(b). For A = 1 inch, write an APT part program to engrave, or draw, this composite design.

7.19. Front, top, and side views of a three-dimensional object appear in Figure P7.19(a). A trimetric projection of this object looks like Figure P7.19(b). The separate views must each be transformed to give the illusion of looking at a three-dimensional object. For A = 10 inches, write an APT part program to engrave, or draw, the trimetric projection. Let one angle ϕ = 30° rotate the object clockwise when viewed from the top, then tilt the object toward the viewer through the angle ψ = 25°.

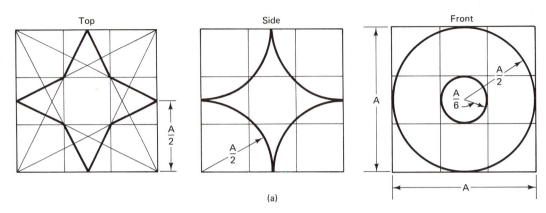

(a)

Figure P7.19a

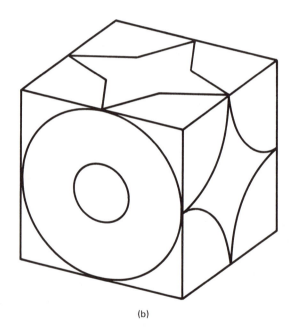

(b)

Figure P7.19b

CHAPTER 8

Planes and Part Surface Specification

The discussion thus far emphasized the need for maintaining the tool within tolerance of the drive surface only because the part surface was assumed to be the *xy*-plane and was ignored. Now we will allow the part surface to be a plane parallel with or inclined to the *xy*-plane, but not perpendicular to it. As a result, we will discuss the tool-to-part-surface relationship for the general case. We will have to maintain the tool within tolerance of the part surface also. Part surfaces of arbitrary shape will be admitted in Chapter 13.

8.1 PLANE DEFINITIONS

A **plane** is a surface that contains all points of a straight line that connects any two points in the surface. In APT, planes may be used as part surfaces, drive surfaces, or check surfaces. They may be parallel with the *xy*-plane or inclined to it. Planes perpendicular to the *xy*-plane are inherently defined by a line definition.

The following development leads to the definition of a plane in terms of the elements of the coordinate system in which it appears. The result [equation (8.7)] is fundamental to APT. One purpose in following a mathematical derivation is to lay more groundwork for the material of Chapter 13.

A plane is shown in Figure 8.1, where the directed length $ON = p$ is defined as a vector normal to the plane with direction angles α, β, and γ (not marked). If we let $P(x_1, y_1, z_1)$ be a point in the plane, then the directed length $OP = \rho$ is a vector from the origin O to the point P with direction angles α_1, β_1, and γ_1. Let θ be the angle between ON and OP. We then have

$$OP \cos \theta = ON$$

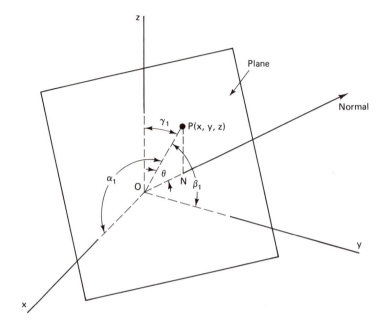

Figure 8.1 **Geometry showing the normal to a plane, *ON*, and a point in the plane, *P(x,y,z)*.**

or

$$\rho \cos \theta = p \tag{8.1}$$

Equation (8.1) says that p is the component of OP in the direction ON. To obtain an expression involving the direction cosines of the normal and the coordinate system variables, we first derive an identity for $\cos \theta$. We draw on the material of Section 7.1.1.

In Figure 8.2(a) we have unit vector **a** at an angle θ to unit vector **b**. The component of **a** along **b** (or the component of **b** along **a**) is given by

$$|\mathbf{a}|\,|\mathbf{b}| \cos \theta = \cos \theta \quad (\text{since } |\mathbf{a}| = |\mathbf{b}| = 1) \tag{8.2}$$

However, **a** and **b** can each be resolved into components along the coordinate axes as shown in Figure 8.2(b) and (c) for **a** and **b**, respectively. Since **a** and **b** are each of unit magnitude, these components are the direction cosines. Now, to get the component of **a** along **b**, we sum the products of each component of **a** with each component of **b** times the cosine of the angle between the associated component pair (there are nine terms in the sum). Since the angle will be either 0° or 90° (nonzero products will occur only for components along the same axis since $\cos 0° = 1$ and $\cos 90° = 0$) the sum of the nonzero component pairs is

$$a_x b_x \cos \theta_x + a_y b_y \cos \theta_y + a_z b_z \cos \theta_z = |\mathbf{a}||\mathbf{b}| \cos \theta$$

$$\cos \alpha_a \cos \alpha_b \cos 0° + \cos \beta_a \cos \beta_b \cos 0° + \cos \gamma_a \cos \gamma_b \cos 0° = \cos \theta$$

$$\cos \alpha_a \cos \alpha_b + \cos \beta_a \cos \beta_b + \cos \gamma_a \cos \gamma_b = \cos \theta \qquad (8.3)$$

Equation (8.3) is the component of **a** along **b** (also known as the vector dot product for unit vectors) and is the desired identity for $\cos \theta$ in equation (8.2). We will substitute into equation (8.1) the identity for $\cos \theta$ from equation (8.3) and relabel the direction angles to conform with the definitions of Figure 8.1 to get

$$\rho(\cos \alpha_1 \cos \alpha + \cos \beta_1 \cos \beta + \cos \gamma_1 \cos \gamma) = p \qquad (8.4)$$

But for any point $P(x, y, z)$,

$$\rho \cos \alpha_1 = x$$

$$\rho \cos \beta_1 = y \qquad (8.5)$$

$$\rho \cos \gamma_1 = z$$

Upon substituting equations (8.5) into (8.4), we get

$$x \cos \alpha + y \cos \beta + z \cos \gamma = p \qquad (8.6)$$

Equation (8.6) is called the **normal form** of the equation of a plane. It is used for writing the equation of a plane when the length and direction of its normal are known. It can be rewritten as

$$ax + by + cz = d \qquad (8.7)$$

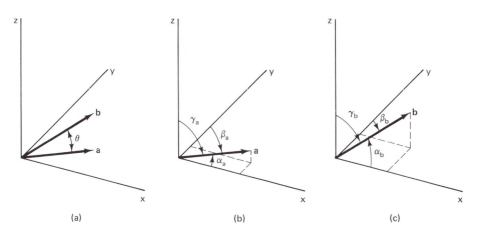

Figure 8.2   Direction angles for two unit vectors a and b.

where a, b, and c are the direction cosines of the unit vector perpendicular to the desired plane and d is the perpendicular distance from the origin to the plane. The coefficients for equation (8.7) are the canonical form entries for definition Plane-1 of Table 8.1 and for LINE (see Table 3.4). If a, b, and c are not direction cosines but instead are direction numbers, then d is the product of the length of the vector whose direction numbers are a, b, and c by the perpendicular distance from the origin to the plane.

The APT general form of plane definition is

[Statement-label)]   Symbolic-name = PLANE/minor-section

The plane is assumed to be infinite in extent. If the statement is labeled and reexecuted, either the canonic form of plane definition must be used or the symbolic name must be subscripted and a different index used for each execution; otherwise, a multiply defined geometric statement error will result. The rules for symbol naming apply here also. The minor section is completed according to any one of several different formats provided for the convenience of the part programmer. A selection of plane definitions is given in Table 8.1. When required in the minor section, scalar values can be numeric literals, scalar variables (subscripted or nonsubscripted), arithmetic expressions, or some mixture of these.

8.2 PART SURFACES AND TOOL-TO-PART-SURFACE POSITIONING

Part, drive, and check surfaces were discussed in Section 2.3.1 and illustrated in Figure 2.8, where part surfaces are planes parallel with the xy-plane. Thus far, it was not necessary to specify a part surface since by default APT assumed it to be the xy-plane. As a result, cut vectors had zero z-coordinate values except for those beginning or ending at a SETPT with a nonzero z-coordinate value. The plane definitions of Section 8.1 now allow us to define part surfaces different from the xy-plane. We were already using planes as drive surfaces; they were called lines. We will still maintain the tool axis parallel with the z-axis. This will not prevent us from writing part programs for an important class of parts in three dimensions.

There are four ways to specify part surfaces: (1) with the PSIS statement, (2) with the two- and three-surface startup GO statements, (3) with the two-surface OFFSET statement, and (4) with the AUTOPS statement. These are discussed in the paragraphs that follow.

8.2.1 The PSIS Statement

A surface may be declared a part surface with the PSIS (part surface is) statement as follows

[Statement-label)]   PSIS/*surface*

where *surface* is a previously defined or nested definition of a geometric entity qualifying

TABLE 8.1 PLANE DEFINITIONS

| 1 | By the coefficients of the plane equation (canonical form): |

PLANE/*a, b, c, d*

The parameters are the coefficients of the plane equation $ax + by + cz = d$ [see equation (8.7)]. When *a, b,* and *c* are direction cosines, *d* is the perpendicular distance from the plane to the origin of the coordinate system. When *a, b,* and *c* are direction numbers, *d* must be the scaled version of the perpendicular distance defined above.

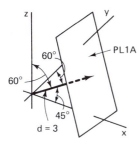

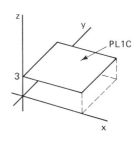

Direction cosines:

$a = \cos(45) = 0.707$
$b = \cos(60) = 0.5$
$c = \cos(60) = 0.5$

PL1A = PLANE/.707, .5, .5, 3

Direction numbers: $a = -4$
$b = -3$
$c = -2$

$p = \sqrt{a^2 + b^2 + c^2} = 5.385$
$d = p\sqrt{a^2 + b^2 + c^2} = 29$

PL1B = PLANE/−4, −3, −2, 29

PL1C = PLANE/0, 0, 1, 3

| 2 | By three noncollinear points: |

PLANE/*point1, point2, point3*

The defined plane passes through the three noncollinear points (not all on the same line) *point1, point2, point3.*

PL2 = PLANE/P1, P2, P3

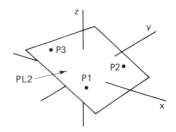

| 3 | Through a point and parallel to another plane: |

PLANE/*point*, PARLEL, *plane*

The defined plane passes through *point* and is parallel to *plane*.

PL3B = PLANE/P1, PARLEL, PL3A

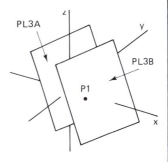

TABLE 8.1 (CONTINUED)

| 4 | Parallel to but offset from another plane: |

$$\text{PLANE/PARLEL, } plane, \begin{Bmatrix} \text{XLARGE} \\ \text{XSMALL} \\ \text{YLARGE} \\ \text{YSMALL} \\ \text{ZLARGE} \\ \text{ZSMALL} \end{Bmatrix}, d$$

The defined plane is parallel to *plane* but offset from it a perpendicular distance *d*. The modifier is chosen by observing the axis intercepts as both planes cross the x-, y-, or z-axis and selecting that modifier which represents the relative intercept value of the desired plane with respect to the given plane.

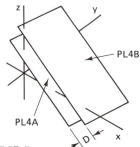

PL4B = PLANE/PARLEL, PL4A, ZLARGE, D

or

PL4B = PLANE/PARLEL, PL4A, XLARGE, D

| 5 | Through a point and perpendicular to a vector: |

PLANE/*point*, PERPTO, *vector*

The defined plane passes through *point* and is positioned such that *vector* is normal to the plane.

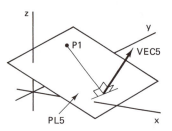

PL5 = PLANE/P1, PERPTO, VEC5

| 6 | Through two points and perpendicular to another plane: |

PLANE/PERPTO, *plane*, *point1*, *point2*

The defined plane passes through *point1* and *point2* and is positioned to be perpendicular to *plane*. A line through the two points must not be perpendicular to *plane*. One, or both, of the points may be on the intersection of the two planes.

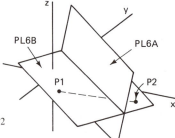

PL6B = PLANE/PERPTO, PL6A, P1, P2

TABLE 8.1 (CONTINUED)

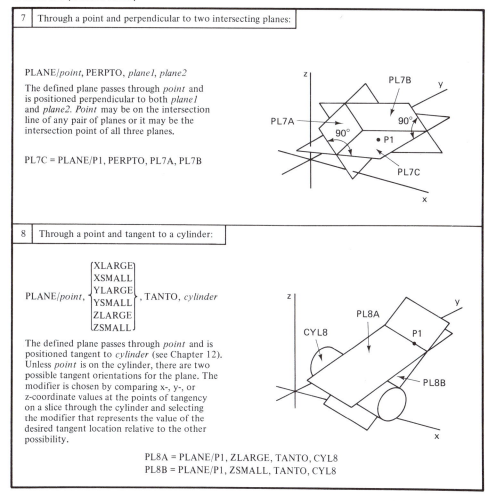

| 7 | Through a point and perpendicular to two intersecting planes: |

PLANE/*point*, PERPTO, *plane1*, *plane2*

The defined plane passes through *point* and is positioned perpendicular to both *plane1* and *plane2*. *Point* may be on the intersection line of any pair of planes or it may be the intersection point of all three planes.

PL7C = PLANE/P1, PERPTO, PL7A, PL7B

| 8 | Through a point and tangent to a cylinder: |

$$PLANE/\textit{point}, \begin{Bmatrix} XLARGE \\ XSMALL \\ YLARGE \\ YSMALL \\ ZLARGE \\ ZSMALL \end{Bmatrix}, TANTO, \textit{cylinder}$$

The defined plane passes through *point* and is positioned tangent to *cylinder* (see Chapter 12). Unless *point* is on the cylinder, there are two possible tangent orientations for the plane. The modifier is chosen by comparing x-, y-, or z-coordinate values at the points of tangency on a slice through the cylinder and selecting the modifier that represents the value of the desired tangent location relative to the other possibility.

PL8A = PLANE/P1, ZLARGE, TANTO, CYL8
PL8B = PLANE/P1, ZSMALL, TANTO, CYL8

as a surface, such as a plane or the entities described in Chapter 12. This is a Section 2 executable statement. It is modal and remains in effect until countermanded by another PSIS statement, by a part surface declared in a multisurface startup statement, or by the AUTOPS statement. There is no limit on the number of PSIS statements permitted in a part program. This statement is illustrated in the examples of Section 8.3.

8.2.2 The Multisurface GO Startup Statement

The GO startup command defined in Section 2.3.3 is restated here:

$$
\text{GO/}\begin{Bmatrix} \text{TO} \\ \text{ON} \\ \text{PAST} \end{Bmatrix}\text{,}\textit{drive-surf}\left[\text{,}\begin{Bmatrix} \text{TO} \\ \text{ON} \\ \text{PAST} \end{Bmatrix}\text{,}\textit{part-surf}\left[\text{,}\begin{Bmatrix} \text{TO} \\ \text{ON} \\ \text{PAST} \\ \text{TANTO} \end{Bmatrix}\text{,}\textit{check-surf}\right]\right]
$$

Until now, this statement was used as a one-surface startup statement where *part-surf* defaulted to the *xy*-plane with a default modifier of TO. In a three-dimensional context *part-surf* must be specified, in which case the tool height (and in some cases the location of the longitudinal center of the tool) must be considered. Unless the tool height is specified, it defaults to 5 units (see Section 2.2). To visualize the final tool position under a multisurface GO startup command, and to select the proper set of modifiers, we must remember that the tool height (top) may influence tool positioning, especially when PAST is the *part-surf* modifier. Figure 8.3 shows initial and final tool positions when the initial position is remote from *part-surf*. Tool positioning with the startup command is also sensitive to the initial location of the tool relative to *part-surf*. If the tool initially intersects *part-surf*, a condition called gouging, the location of the longitudinal center of the tool relative to *part-surf* is used to determine the outcome. Possible initial tool locations are shown in Figure 8.4, where for tool-positioning purposes, locations A and B are considered equivalent and locations C and D are considered equivalent. Cutter height is also significant for tool positioning purposes when the tool is near the surfaces introduced in Chapter 12.

The INDIRP-INDIRV statements can be used to establish a sense of direction for three-dimensional parts in a manner analogous to that discussed in Section 2.3.3. From the two-surface startup GO statement we can generalize to the three-surface startup GO statement. The third surface, *check-surf*, allows further control over the final tool position. An example is shown in Figure 8.5.

8.2.3 The Two-Surface OFFSET Statement

The part surface may be specified in the two-surface OFFSET statement. This statement was defined in Section 2.3.3 and illustrated there in a one-surface startup context. The statement is restated here.

$$
\text{OFFSET/}\begin{Bmatrix} \text{TO} \\ \text{ON} \\ \text{PAST} \end{Bmatrix}\text{,}\textit{drive-surf}\left[\text{,}\begin{Bmatrix} \text{TO} \\ \text{ON} \\ \text{PAST} \end{Bmatrix}\text{,}\textit{part-surf}\right]
$$

As a one-surface startup statement, the part surface in effect when this command is executed is used to help locate the final tool position. As a two-surface startup statement, *part-surf* becomes the new part surface for locating the final tool position.

The INDIRV or INDIRP statement must precede the OFFSET statement. When

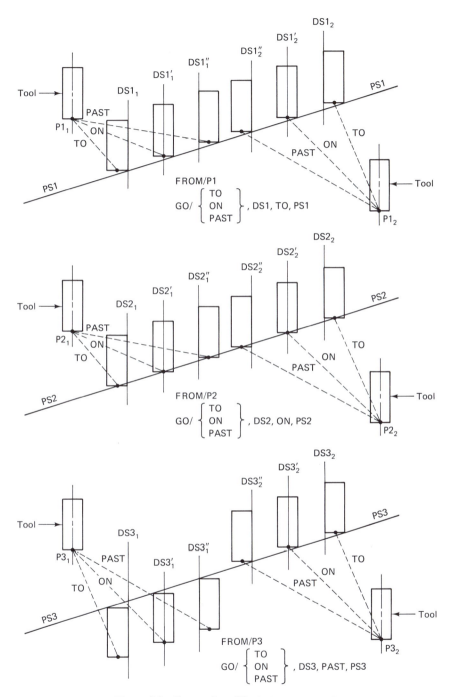

Figure 8.3   Two-surface GO startup movements.

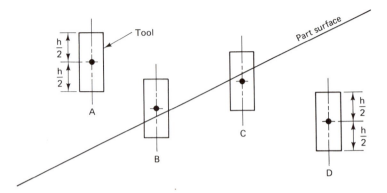

Figure 8.4   Tool locations remote from and intersecting the part surface.

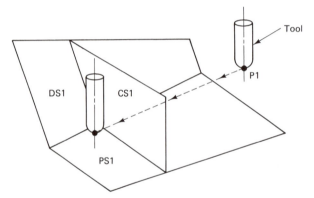

FROM/P1
GO/TO, DS1, TO, PS1, PAST, CS1

**Figure 8.5   Geometry for a three-sur-
face GO startup command.**

used in a three-dimensional context, the APT system locates the final tool position from
the OFFSET statement as follows:

1. A normal to *drive-surf* is computed at the point where the vector defined by the
   INDIRV or INDIRP statement intersects *drive-surf*.
2. The tool is moved along the normal until it is positioned as required by the drive
   surface positional modifier (TO, ON, or PAST).
3. The tool is then moved along its longitudinal axis until it is positioned as required
   by the part surface positional modifier (TO, ON, or PAST).

Coordinates of the final tool position only are output to the CL-file. The height of
the tool may be used to determine the final tool position in a manner similar to that

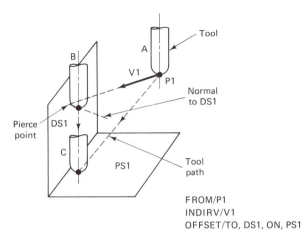

FROM/P1
INDIRV/V1
OFFSET/TO, DS1, ON, PS1

Figure 8.6   Tool positioning geometry for the two-surface OFFSET statement.

described in Section 8.2.2. Figure 8.6 illustrates tool movement as described above. The APT system computes the tool position at location B according to the vector of the INDIRV statement and the drive surface TO modifier of the OFFSET statement. It then computes the final tool position by projecting along the longitudinal axis to location C (ON the new part surface).

8.2.4 The AUTOPS Statement

The AUTOPS (automatic part surface) statement is a single-word modal command that establishes a new part surface parallel with the xy-plane at a z-coordinate intercept derived from the tool endpoint z-coordinate at the end of the preceding motion command. It allows dynamic definition of an implicit part surface that remains in effect until countermanded by any of the methods of Sections 8.2.1 through 8.2.3.

In addition to its application as described here, the AUTOPS statement may be used in point-to-point applications (see Section 9.2.4) and with multiaxis tool orientations (see Section 14.2.4). This statement must be used with care because of the invisible nature of the defined part surface. Such definitions are contrary to good programming practice. While contouring it is wise at all times to constrain the tool within tolerance of known surfaces to avoid unexpected and unwanted tool movements.

The results of applying the AUTOPS statement are shown in the two examples of Figure 8.7. In Figure 8.7(a) the AUTOPS statement is used before a one-surface GO to establish a part surface for subsequent contouring. The part surface is derived from the z-coordinate of P1. In Figure 8.7(b) it is used to establish a new part surface after contouring to C1 along PL2 as the part surface. We then take the tool around C1 on the implicit part surface and check on PL1, where the tool is no longer within tolerance of PL2.

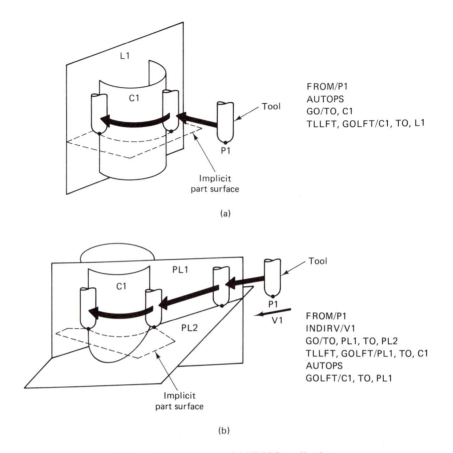

FROM/P1
AUTOPS
GO/TO, C1
TLLFT, GOLFT/C1, TO, L1

(a)

FROM/P1
INDIRV/V1
GO/TO, PL1, TO, PL2
TLLFT, GOLFT/PL1, TO, C1
AUTOPS
GOLFT/C1, TO, PL1

(b)

Figure 8.7   Examples of AUTOPS application.

8.2.5 The Complete General Motion Command

Once the tool is positioned with a startup command, tool-to-surface relationships must be maintained thereafter during execution of a general motion command. The general motion command was described in Section 2.3.2, where it was discussed in a two-dimensional contouring context. For motion in three dimensions the command is as follows:

$$
\begin{bmatrix} TLLFT, \\ TLON, \\ TLRGT, \end{bmatrix}
\begin{bmatrix} TLONPS, \\ TLOFPS, \end{bmatrix}
\begin{Bmatrix} GOFWD \\ GOBACK \\ GORGT \\ GOLFT \\ GOUP \\ GODOWN \end{Bmatrix}
/drive\text{-}surf,
\begin{Bmatrix} TO \\ ON \\ PAST \\ TANTO \\ PSTAN \end{Bmatrix}
,[n,INTOF,]check\text{-}surf
$$

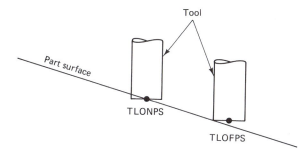

Figure 8.8   Interpretation of the tool-to-part surface modifier.

Additional modifiers are added to constrain the tool with respect to the part surface. The tool-to-part-surface modifier, TLONPS (tool on part surface) or TLOFPS (tool off part surface) should be specified for the first motion command following startup and, thereafter, whenever the tool-to-part-surface relationship changes. Interpretation of the modifier is shown in Figure 8.8.

The modifiers GOUP and GODOWN establish a sense of direction relative to the longitudinal axis of the tool. GOUP is used when the dominant direction of movement is from the bottom of the tool toward its top; GODOWN is the reverse of GOUP. The volume within which neither modifier should be used assumes the form shown in Figure 8.9. This volume is generated by revolving two lines about the tool axis, each line 2° from a normal to the axis. Figure 8.9 also shows the direction modifier suffix relative to the direction of the last tool movement. Except for GOUP and GODOWN, which are

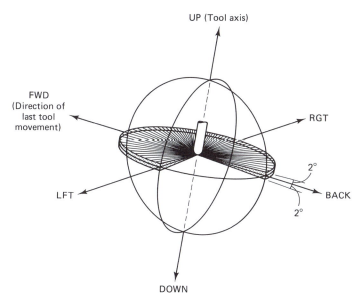

Figure 8.9   Direction modifiers relative to the last tool movement.

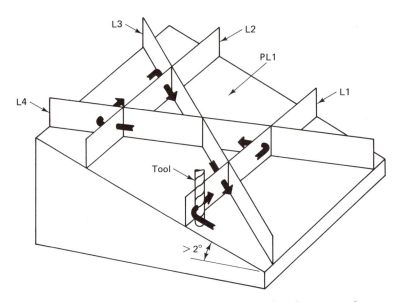

Figure 8.10   Part geometry to illustrate the general motion command.

measured relative to the tool axis, the direction may be explicitly given with the INDIRV or INDIRP command.

   If the sphere of Figure 8.9 were solid, admissible directions for GORGT, GOLFT, GOFWD, and GOBACK would lie within the solid volumes remaining after $\pm 2°$ wedges were removed from along axes in line with or normal to the direction of last tool movement. The regions shown in Figure 2.9 correspond to the solid volumes just described. Some of the modifiers just described appear in the following two equivalent part program segments that command the tool along the path shown in Figure 8.10.

GO/TO,L1,TO,PL1	GO/TO,L1,TO,PL1
TLRGT,TLOFPS,GORGT/L1,TO,L4	TLRGT,TLOFPS,GORGT/L1,TO,L4
TLLFT,GOUP/L4,PAST,L2	TLLFT,GOLFT/L4,PAST,L2
GORGT/L2,TO,L3	GORGT/L2,TO,L3
TLRGT,GODOWN/L3,PAST,L1	TLRGT,GORGT/L3,PAST,L1

   The tool-to-part-surface tolerance may be specified as described in Section 2.4. It becomes important for part surfaces of arbitrary shape as described in Chapter 12. PSTAN is illustrated in Section 13.4.

8.2.6 Resolving Surface Directions: The SRFVCT Statement

Instances arise where it is difficult to determine the TO or PAST side of one or more surfaces to be specified in a GO startup command. It is especially difficult to determine the sides when the tool is in an ON condition relative to one or more of the surfaces. To

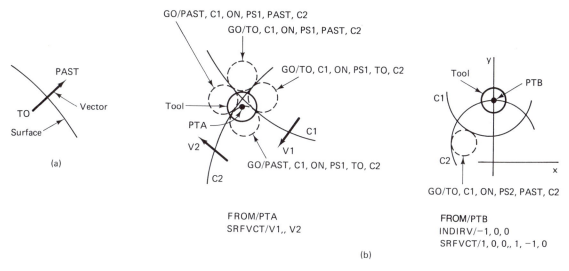

Figure 8.11   SRFVCT illustrated for drive and check surfaces.

resolve uncertainty in selecting the modifier(s), the part programmer may use the SRFVCT (surface vector) statement to designate the TO and PAST sides. The format of this statement is

SRFVCT/[*ds-vect*] [,[*ps-vect*] [,*cs-vect*]]

where *ds-vect*, *ps-vect*, and *cs-vect* are vectors that apply to the drive, part, and check surfaces, respectively, of the immediately following GO command. Each vector determines the TO-PAST sides as shown in Figure 8.11(a). Each vector can be given by its components or by a symbol. As many vectors as there are surfaces in the GO command may be specified, but not more. Vectors may be omitted when not needed, but leading commas must be retained to denote omitted vector positions. Trailing commas need not be included. Following are valid SRFVCT statements with surface associations identified:

SRFVCT/VEC1,VEC2	ds, ps
SRFVCT/,VECT	ps
SRFVCT/,V1,V5	ps, cs
SRFVCT/VTA,,VTB	ds, cs
SRFVCT/,,VCT	cs
SRFVCT/0,1,0	ds
SRFVCT/VA,1,0,0	ds, ps
SRFVCT/,,2,1,0	cs
SRFVCT/VX,VY,VZ,,VEC	ds(scalar variables), cs
SRFVCT/ − 1,2,0,1,1, − 1,VX,VY,VZ	ds, ps, cs(scalar variables)

Geometric examples of the SRFVCT statement are shown in Figure 8.11(b). The SRFVCT statement does not eliminate the need for the INDIRV and INDIRP statements.

8.3 EXAMPLES

The first example contours a part that requires several part surface declarations. For this example, all part surfaces are parallel with the *xy*-plane. The second example contours on a canted plane. Because of the sloping part surface, the tool will leave a scallop. Computations for directing the tool to maintain an acceptable scallop height are included for tool motion parallel with the gradient of the plane and for motion normal to the gradient of the plane. The third example shows how *z*-axis movement may be obtained with multilevel transformations.

Example 8.1: Part Surface Declaration for Contouring

Figure 8.12 shows a part that requires several part surface declarations for contour machining. All part surfaces are parallel with the *xy*-plane and are labeled in the figure. The coordinate axes and symbolically labeled curves are shown in Figure 8.13. Since some facing is required, a flat-end 1-in.-diameter tool is selected. A suitable part program follows.

SETPT	= POINT/6.5, −4,2.5		100
LX	= LINE/XAXIS		110
LY	= LINE/YAXIS		120
C1	= CIRCLE/0,0,1.5		130
L1	= LINE/XAXIS, −2		140
L2	= LINE/XAXIS,1.5		150
L2A	= LINE/XAXIS,.75		160
L3	= LINE/YAXIS,3.5		170
L3A	= LINE/YAXIS,4.25		180
L4	= LINE/YAXIS,5		190
C2	= CIRCLE/2.5,1.5,1.5		200
PL1	= PLANE/0,0,1,0		210
PL2	= PLANE/PARLEL,PL1,ZLARGE,.375		220
PL3	= PLANE/PARLEL,PL1,ZLARGE,.75		230
PL4	= PLANE/PARLEL,PL1,ZLARGE,1.125		240
PL5	= PLANE/PARLEL,PL1,ZLARGE,2		250
	FROM/SETPT		260
	GO/TO,L3A,TO,PL4,TO,L1	$$ MACHINE ON PL4	270
	TLRGT,TLOFPS,GORGT/L3A,ON,LX		280
	TLON,GOLFT/LX,TO,L3		290
	TLLFT,GOLFT/L3,PAST,L1		300
	GORGT/L1,PAST,LY		310
	INDIRV/0,1,0		320
	PSIS/PL3		330
	GO/PAST,C1	$$ MACHINE ON PL3	340
	TLLFT,TLOFPS,GORGT/C1,ON,LX		350
	TLON,GOLFT/LX,ON,LY		360
	GOLFT/LY,TO,C1		370
	TLRGT,GORGT/C1,PAST,LY		380
	INDIRV/0,1,0		390
	PSIS/PL2		400
	GO/PAST,LX	$$ MACHINE ON PL2	410

TLLFT,TLOFPS,GORGT/LX,ON,L4	420
TLON,GOLFT/L4,PAST,L2A	430
TLRGT,GOLFT/L2A,PAST,LY	440
INDIRV/0,1,0	450
PSIS/PL1	460
GO/PAST,L2 $$ MACHINE ON PL1	470
TLLFT,TLOFPS,GORGT/L2,PAST,C2	480
GORGT/C2,PAST,2,INTOF,L2	490
GORGT/L2,PAST,L4	500
GORGT/L4,PAST,L1	510
GORGT/L1,PAST,LY	520
GORGT/LY,ON,L2	530
INDIRV/0,1,0	540
GO/PAST,L2,PAST,PL5	550
GOTO/SETPT	560

Simple geometric definitions suffice. L2A and L3A are used for the facing operations since the 1-in.-diameter tool cannot accomplish this in one pass over the surface. The APT line and circle definitions, being infinite in the $\pm z$ direction, are used as drive and check surfaces regardless of the part surface in effect. The bottom of the part is assumed on the xy-plane (PL1), while PL2, PL3, and PL4 are required for the various surfaces parallel with the xy-plane. PL5 is used to bring the tool clear of the part prior to returning it to SETPT. The top of the part does not get machined and is therefore not labeled.

Example 8.2: Contouring on a Canted Plane

Figure 8.14(a) shows a part with a canted plane. We will contour on this plane. Figure 8.14(b) shows the part placed on the coordinate axes and the pertinent surfaces symbolically

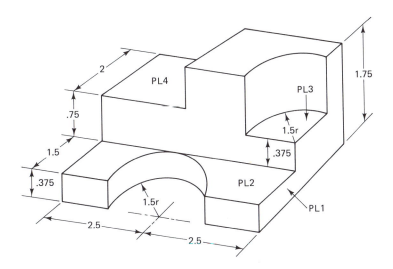

Figure 8.12   Geometry of Example 8.1.

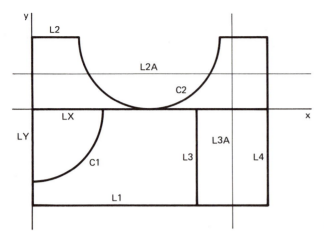

Figure 8.13 Symbolically labeled top view of the part shown in Figure 8.12.

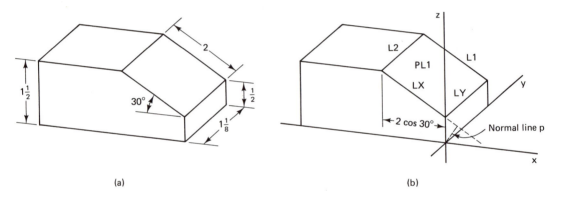

(a)

(b)

Figure 8.14 The part for Example 8.2.

labeled. The normal line p from the origin to the plane makes an angle of 30° with the z-axis. By similar triangles, we have the relationship

$$\frac{p}{1/2} = \frac{2 \cos 30°}{2}$$

from which $p = \frac{1}{2} \cos 30°$. The direction cosines are

$$\alpha = \cos 60° \qquad \beta = \cos 90° = 0 \qquad \gamma = \cos 30°$$

The definition of the plane then becomes

$$PL1 = PLANE/COSF(60),0,COSF(30),(.5*COSF(30))$$

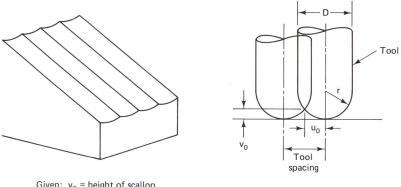

Given: v_0 = height of scallop

Then: tool spacing = $2u_0 = 2\sqrt{r^2 - (r - v_0)^2} = 2\sqrt{2rv_0 - v_0^2} = 2\sqrt{v_0(D - v_0)}$

(a)

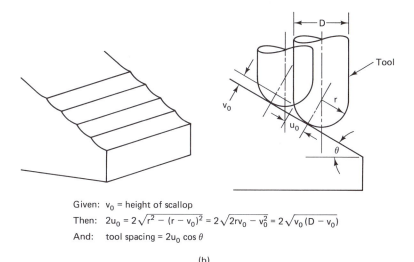

Given: v_0 = height of scallop

Then: $2u_0 = 2\sqrt{r^2 - (r - v_0)^2} = 2\sqrt{2rv_0 - v_0^2} = 2\sqrt{v_0(D - v_0)}$

And: tool spacing = $2u_0 \cos\theta$

(b)

Figure 8.15   Surface finish when machining (a) along the slope and (b) across the slope.

Machining on the canted plane can be performed either along the slope (parallel with the gradient) or across the slope (normal to the gradient) as shown in Figure 8.15. In either case, a scallop will result. Tool spacing for successive cuts with a round-end tool can be calculated as shown in Figure 8.15. The number of successive cuts can either be precomputed and inserted into the part program as a numeric literal or the decision when to stop machining can be made within the part program. The pattern for successive cuts will appear as either of those shown in Figure 8.16, depending on the part width and tool spacing.

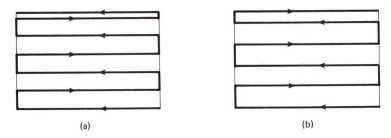

(a)   (b)

Figure 8.16 Successive cutting patterns.

A part program to machine along the slope as shown in Figure 8.15(a) follows.

	D = ...	$$ TOOL DIAMETER	100
	CUTTER/D,(D/2)		110
	V0 = ...	$$ SCALLOP HEIGHT	120
	TS1 = 2*SQRTF(V0*(D − V0))	$$ TOOL SPACING	130
	FROM/SETPT		140
	GO/ON,LY,TO,PL1		150
	TLON,TLOFPS,GORGT/LY,ON,LX		160
	GOLFT/LX,ON,L2		170
	LOOPST		180
	DSP = 0		190
CAL1)	DSP = DSP + TS1		200
	IF (DSP − 1.125) CAL2, CAL4, CAL4		210
CAL2)	LTEMP = LINE/CANON,0,1,0, DSP		220
	GORGT/L2,ON,LTEMP		230
	GORGT/LTEMP,ON,LY		240
	DSP = DSP + TS1		250
	IF (DSP − 1.125) CAL3, CAL5, CAL5		260
CAL3)	LTEMP = LINE/CANON,0,1,0, DSP		270
	GOLFT/LY,ON,LTEMP		280
	GOLFT/LTEMP,ON,L2		290
	JUMPTO/CAL1		300
CAL4)	GORGT/L2,ON,L1		310
	GORGT/L1,ON,LY		320
	JUMPTO/CAL6		330
CAL5)	GOLFT/LY,ON,L1		340
	GOLFT/L1,ON,L2		350
CAL6)	LOOPND		360
	GOTO/SETPT		370

A part program to machine across the slope as shown in Figure 8.15(b) follows.

	D = ...	$$ TOOL DIAMETER	100
	CUTTER/D,(D/2)		110
	V0 = ...	$$ SCALLOP HEIGHT	120
	THETA = ...	$$ ANGLE OF SLOPE	130

	D2 = 2*COSF(THETA)	140
	TS2 = 2*SQRTF(V0*(D − V0))*COSF (THETA)	150
	FROM/SETPT $$ ABOVE LINE IS TOOL	160
	GO/ON,LY,TO,PL1 $$ SPACING	170
	TLON,TLOFPS,GORGT/LY,ON,L1	180
	LOOPST	190
	DSP = 0	200
CAC1)	DSP = DSP + TS2	210
	IF (DSP − D2) CAC2, CAC4, CAC4	220
CAC2)	LTEMP = LINE/CANON, − 1,0,0,DSP	230
	GOLFT/L1,ON,LTEMP	240
	GOLFT/LTEMP,ON,LX	250
	DSP = DSP + TS2	260
	IF (DSP − D2) CAC3, CAC5, CAC5	270
CAC3)	LTEMP = LINE/CANON, − 1,0,0,DSP	280
	GORGT/LX,ON,LTEMP	290
	GORGT/LTEMP,ON,L1	300
	JUMPTO/CAC1	310
CAC4)	GOLFT/L1,ON,L2	320
	GOLFT/L2,ON,LX	330
	JUMPTO/CAC6	340
CAC5)	GORGT/LX,ON,L2	350
	GORGT/L2,ON,L1	360
CAC6)	LOOPND	370
	GOTO/SETPT	380

Example 8.3: Transforming Coordinates for z-Axis Contouring

Figure 8.17 shows a part with a cylindrically shaped recess. It is proposed to machine this part with a round-end tool across the axis of the cylinder. This would, of course, leave scallops as shown. Since we have not yet introduced cylinders as part surfaces, two machining

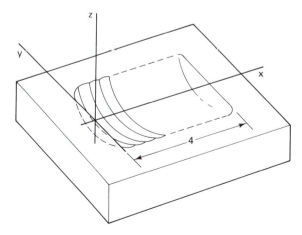

Figure 8.17   Part geometry for Example 8.3.

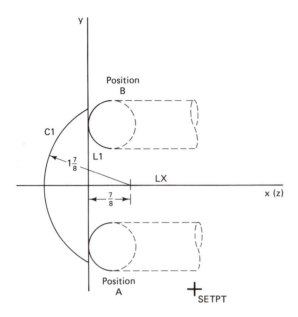

Figure 8.18 Symbolically labeled geometry for Example 8.3.

alternatives are possible. First, we could machine parallel with the axis of the cylinder by changing the part surface and moving the cutter over a small amount for successive passes. This would produce scallops in a different direction than shown in Figure 8.17. Second, we could machine according to equivalent geometry and transform the cut vectors as required. The latter approach will be used.

A cross section of the recess would appear as shown in Figure 8.18, where the z-axis now becomes the x-axis and the y-axis remains unchanged. Line L1 conforms with the top surface of the part and C1 is the bottom of the recess shown in Figure 8.17. A round-end tool, when directed along C1 in the xy-plane, can be used to produce cut vectors for use in the yz-plane when the tool is oriented as shown by its dashed outline in Figure 8.18. The approach will be to direct the tool to position A, then along C1 to position B. While at position B we will displace it to a new part surface selected to keep the scallop height to an acceptable value, then direct the tool along C1 back to position A. The cut vectors from this sequence of tool movements will be rotated with the TRACUT feature, then translated a specified number of times to produce the recess. This recess must be translated in the $+x$-direction to allow for the tool radius. The multilevel transformation part program follows.

```
       CUTTER/1                                    100
SETPT = POINT/2, -2,2                               110
LX    = LINE/XAXIS                                  120
C1    = CIRCLE/.875,0,1.875                         130
L1    = LINE/YAXIS                                  140
PL1   = PLANE/0,0,1,0                               150
PL2   = PLANE/PARLEL,PL1,ZSMALL,.2                  160
```

```
        FROM/SETPT                                              170
        TRACUT/LAST,(MATRIX/TRANSL,.5,0,0)                      180
            INDEX/44                                            190
                TRACUT/(MATRIX/ZXROT, - 90)                     200
                    INDIRV/ - 2,1,0                             210
                    GO/PAST,C1,ON,PL1,TO,L1                     220
                    TLRGT,TLONPS,GOFWD/C1,PAST,LX               230
                    GOFWD/C1,PAST,L1                            240
                    INDIRV/0,0, - 1                             250
                    GO/TO,C1,ON,PL2,TO,L1                       260
                    INDIRV/ - 1, - 1,0                          270
                    TLLFT,TLONPS,GOFWD/C1,PAST,LX               280
                    GOFWD/C1,PAST,L1                            290
                TRACUT/NOMORE                                   300
            COPY/44,TRANSL,.4,0,0,7                             310
        TRACUT/LAST,NOMORE                                      320
        GOTO/SETPT                                              330
```

8.4 AVOIDING PROBLEMS WITH PLANES AND PART SURFACES

A third dimension for constrained tool control slightly complicates part program debugging. The part programmer must make a conscious effort to maintain tool-to-part-surface relationships as demanded by the part geometry. Since, for now, we are restricting part surfaces to planes, our job is made easier. We now review potential sources of difficulty for this class of problems.

Plane Definitions

The location of every plane in the part program should be verified. From the canonical form, printed with the PRINT/3 statement, we can determine the unit normal vector to the plane and the distance of the plane from the origin. It is usually sufficient to visualize the plane orientation from the unit normal vector and to estimate its distance.

Both sides of an algebraic equation may be multiplied by -1 without changing the solution to the equation. This also applies to the plane equation (8.7). The left-hand side of the equation represents the unit normal vector. When multiplied by -1, its direction is reversed. This indicates that the plane is on the side of the origin opposite to that desired. The correction is made by making the distance negative (multiplying the right-hand side by -1). This negative condition may, at times, be observed in the canonical-form printout. The true location of the plane should be visualized by compensating for this negative condition. Also, plane parameters extracted via the OBTAIN statement may be the negative of those desired. It is wise to check for a positive distance before using the extracted parameters. If the distance is negative, the parameters should be multiplied by -1.

The Part Surface

When changing from one part surface to another, the tool must be within tolerance of the new part surface before another contouring command is issued. If it is not, a Section 2 "cutter is out of tolerance with the part surface" diagnostic is printed. The diagnostic can be avoided only if the two part surfaces meet at the check surface of the previous command, allowing for a tool-on or tool-off condition relative to the part surfaces. Otherwise, it is necessary to issue a new startup command.

Tool Positioning

The desired part surface must be included in the OFFSET and startup GO statements, else the tool will be positioned TO the default *xy*-plane. The OFFSET statement requires that the direction derived from its associated INDIRV or INDIRP statement intersect the new drive surface. If it does not, a Section 2 error results. Thereafter, the tool is positioned as described in Section 8.2.3. This means that the tool longitudinal axis must pierce the part surface. As long as part surfaces are planes, this is not a problem unless the part surface is parallel with the tool axis. Ambiguous tool positions can be resolved with the INDIRV-INDIRP, and SRFVCT statements.

Tool positioning with the startup GO statement is sensitive to tool height when the PAST modifier is applied to the part surface. Possible outcomes are shown in Figure 8.3. These were verified for the IBM APT-AC system but tool positioning different from that shown may occur with other APT implementations and should be checked with a part program written for this purpose.

The direction modifier of the general motion command should be chosen to describe most naturally the desired tool movement. This is usually the dominant direction (see Figure 8.9). Unless describing tool directions for looping constructs, unnatural or contrary directions should be avoided. The INDIRV-INDIRP statements are normally used for contrary directions. INDIRV is used for this purpose in the examples of Section 5.5. The tool-to-drive-surface and tool-to-part-surface modifiers should always be specified for the first motion command following startup and, thereafter, whenever the tool-to-surface relationship changes.

PROBLEMS

Because point-to-point commands have not yet been introduced, the GOTO/SETPT and contouring commands only are to be used in the following problems. DNTCUT-CUT statements should be considered while positioning the tool to avoid otherwise awkward sequences of tool motion, especially when changing part surfaces that do not intersect.

8.1 The object of Figure P8.1 is to be machined from 1⅜-in.-thick material. Write the APT part program to machine the part with a 1-in.-diameter flat-end tool. Allow for small overlap of cutting motions.

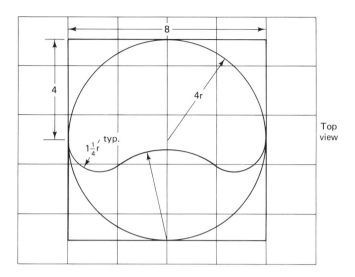

4r

$1\frac{1}{4}$r / typ.

8

4

Top
view

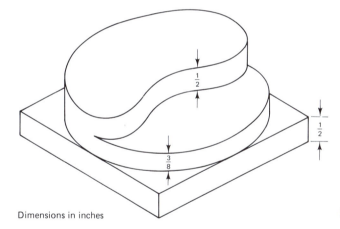

$\frac{1}{2}$

$\frac{1}{2}$

$\frac{3}{8}$

Dimensions in inches

Figure P8.1

8.2. The object of Figure P8.2 is to be machined from 2-in.-thick material. The rounded corners of the recesses as shown in the top view would be produced by a ½-in.-diameter flat-end tool. Write the APT part program to machine the part with this tool. Part symmetry suggests the COPY feature. Allow for small overlap of cutting motions.

8.3 The object of Figure P8.3 is to be machined from 1-in.-thick material. Write the APT part program to machine the part with a ½-in.-diameter flat-end tool. Allow for small overlap of cutting motions.

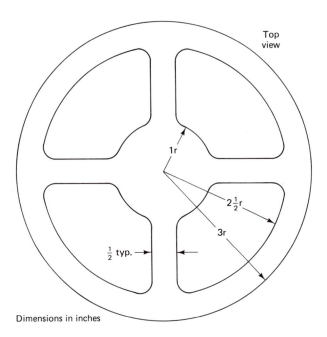

Top
view

1r

$2\frac{1}{2}$r

3r

$\frac{1}{2}$ typ.

Dimensions in inches

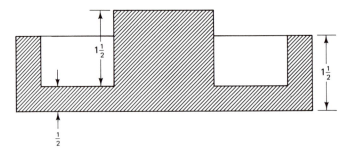

$1\frac{1}{2}$

$1\frac{1}{2}$

$\frac{1}{2}$

Figure P8.2

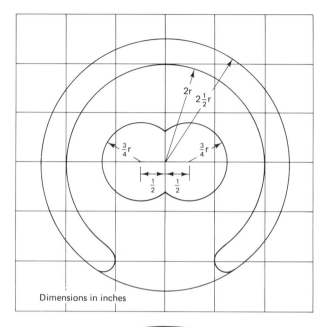

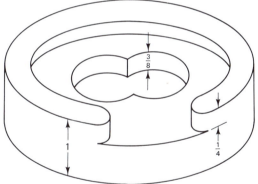

Figure P8.3

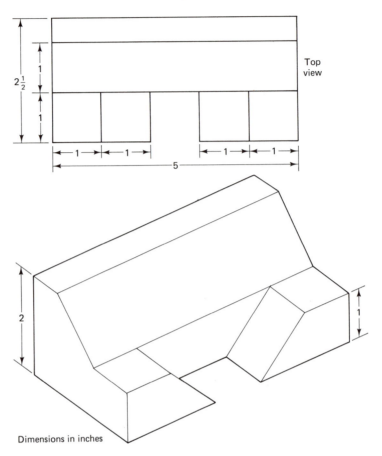

Top view

Dimensions in inches

Figure P8.4

8.4 The object of Figure P8.4 is to be machined from 2¼-in.-thick material. Write the APT part program to machine the part with a ¾-in.-diameter flat-end tool, except on the sloping surfaces, which are to be machined with a ⅜-in.-diameter tool. Cut up and down along the sloping surfaces, leaving a scallop height (perpendicular distance from the sloping surface to the highest point of uncut material between adjacent passes of the tool) of 0.03 in.

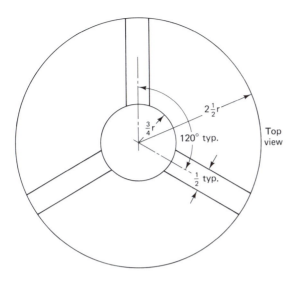

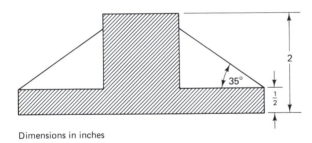

Dimensions in inches   **Figure P8.5**

8.5 The object of Figure P8.5 is to be machined from 2-in.-thick material. Write the APT part program to machine the part with a ½-in.-diameter flat-end tool. Cut up and down along the flat sloping surfaces, leaving a scallop height of no more than 0.03 in. Part symmetry suggests the COPY feature. Allow for small overlap of cutting motions.

8.6 We will call the object of Figure P8.6 a welded mount. It consists of plates welded together along their intersecting edges and along their intersection with the base. Simulate welding with a cutter diameter of zero. No machining is required. Write the APT part program to guide the welder along the intersecting edges.

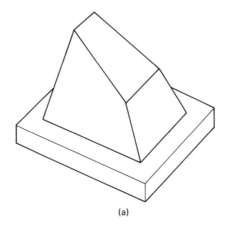

(a)

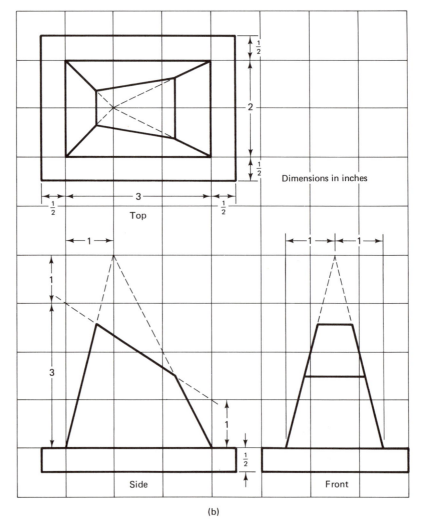

Dimensions in inches

$\frac{1}{2}$

2

$\frac{1}{2}$

3

$\frac{1}{2}$

Top

1

1

3

1

$\frac{1}{2}$

Side

1

1

Front

(b)

Figure P8.6 Welded mount.

282

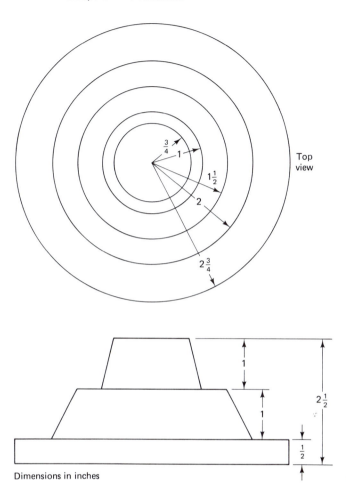

Top
view

Dimensions in inches

Figure P8.7

8.7 The object of Figure P8.7 is to be machined from 2½-in.-thick material. Write the APT part program to machine the part with a 1-in.-diameter flat-end tool. The sloping circular surfaces are to be machined in a circular fashion leaving a scallop height of no more than 0.05 in.

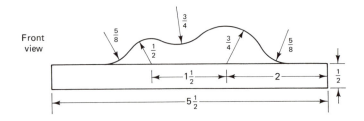

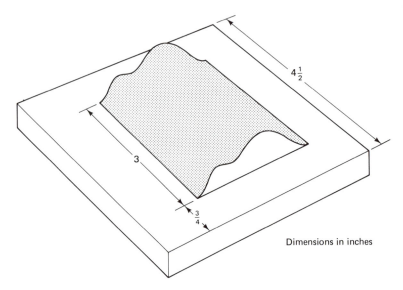

Figure P8.8

8.8 Figure P8.8 shows an object whose top surface cannot be machined conveniently using only planes as a part surface. However, by guiding the tool along the profile of the side view, and with TRACUT and COPY statements for translation and repetition, the object can be machined. If a 1-in.-diameter round-end tool were used with passes separated by 0.3 in., a satisfactorily textured surface would result. Write the APT part program to machine the part in this manner. Assume that in the actual machining operation the 1-in.-diameter round-end tool will be used for the irregular top surface and that a 1-in. flat-end tool will be used for the planar top surface and sides. To indicate that a tool change will occur, take the tool back to set point and begin again with what would normally be the other tool.

CHAPTER 9

Point-to-Point
Part Programming

Some machines operate such that the tool path between points in three-dimensional space is not determined by constraining surfaces during tool movement. The objective is to position the tool on an imaginary point. This type of machine control is referred to as point-to-point control. As a result, part programming considerations differ from those required for continuous motion control. Special tool motion commands and point pattern definitions are available for the point-to-point programming concept. These commands and point pattern definitions are described in this chapter.

9.1 THE APT POINT-TO-POINT PART PROGRAMMING CONCEPT

The point-to-point terminology is derived from machine control action. It connotes the positioning of a tool from one point in three-dimensional space to another point in three-dimensional space. The tool path between points depends solely on the mechanism of control for each machine. The tool may be controlled simultaneously along each dimension as it is commanded from one point to another, it may be controlled in a sequential manner along each axis individually as it is commanded from one point to another, or it may be controlled in a manner representing some combination of the two extremes above. Figure 9.1 shows possible tool paths for performing the same point-to-point operations on differently controlled machines. Perhaps drilling or tapping operations would be performed on the workpiece.

While processing contouring commands, the APT system computes cut vectors to direct the tool in the prescribed manner to produce a part. The cut vectors are computed and written on the CL-file as absolute coordinate values; that is, only one coordinate

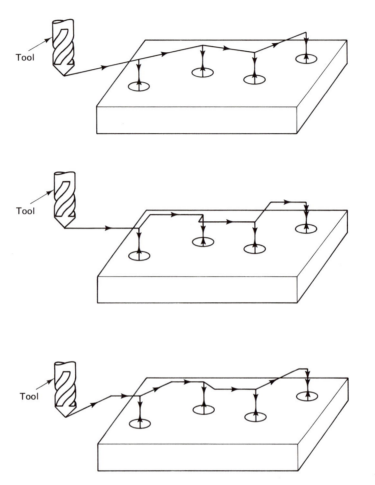

Figure 9.1 Possible tool paths resulting from different point-to-point machine control systems.

system is used and the coordinate values of each cut vector are with respect to the origin of this coordinate system. Whether or not the machine responds to absolute values or incremental values is of no consequence, the APT system always produces absolute values. Point-to-point machines may require absolute or incremental values for tool positioning, in which case it is a function of the postprocessor to convert the absolute values to incremental values or transformed absolute values for use by a given point-to-point controlled machine.

Concerns of the part programmer when writing a program for a point-to-point controlled machine are (1) to maintain clearance for obstructions as the tool is positioned from one point to another, and (2) to generate a set of cut vectors to produce the part.

Presumably, the part programmer knows how the machine operates and avoids writing commands that cause the tool to interfere with part projections as it is being positioned. Because cut vectors are written to the CL-file as absolute coordinate values regardless of whether they were computed from contouring or point-to-point commands makes the APT system transparent to any machine, with respect to the type of programming. If curves are not involved, which result in cut vectors being spaced closely together and not likely to be programmed in a point-to-point manner, it is impossible to distinguish between cut vectors produced from point-to-point commands or while contouring along straight-line segments. Thus a part program for a point-to-point application may consist solely of point-to-point commands or some combination of point-to-point and contouring commands.

The three-dimensional capability of APT is inherent in the point-to-point programming concept. Furthermore, the point-to-point commands are useful in contouring applications, as will be shown in examples to follow.

9.2 POINT-TO-POINT PART PROGRAMMING STATEMENTS

The APT motion commands not relying on part, drive, and check surfaces are FROM, GOTO, and GODLTA. These commands, together with the ZSURF statement, are used for point-to-point programming applications. Each command writes absolute coordinate values on the CL-file to cause three-dimensional movement of the tool to position the tool-end on a point. The CUTTER statement is not needed for point-to-point programming. However, it may be needed for a postprocessor.

9.2.1 The FROM and GOTO Commands

The FROM and GOTO commands were introduced in Sections 2.3.3 and 2.3.4, respectively. They are now discussed in the context of point-to-point programming. For completeness, they are repeated here.

$$\text{FROM}/\left\{\begin{array}{l} \textit{x-coord,y-coord}[,\textit{z-coord}] \\ \textit{point} \end{array}\right\}$$

$$\text{GOTO}/\left\{\begin{array}{l} \textit{x-coord,y-coord}[,\textit{z-coord}] \\ \textit{point} \end{array}\right\}$$

Each statement writes on the CL-file the coordinate values of the point referenced. A zero value is assigned to the z-coordinate when $\textit{z-coord}$ is omitted, except when used in conjunction with the ZSURF statement as described in Section 9.2.3. The formats for both commands are expanded in Chapter 13 for multiaxis programming.

The FROM command establishes a reference point for the tool end. When first used in the program it causes no movement of the tool. Later FROM statements usually cause tool movement from the current tool location to the point given in the command. The GOTO command causes the tool to move from its current location to the point

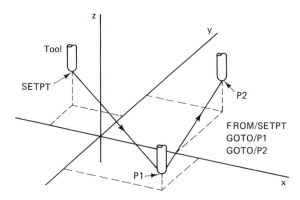

Figure 9.2 Tool positioning with FROM and GOTO commands.

specified. For both commands, the point specified refers to the tool endpoint. The commands may be labeled as needed for program flow purposes.

Tool movement with these commands is illustrated in Figure 9.2. There is no provision for tool offset when using point-to-point commands. Depending on the machine control action, the actual path of the tool may or may not follow the linear paths connecting the points as shown in the figure.

9.2.2 The GODLTA Command

The GODLTA command directs the tool from its current location to its new location with a set of incremental distances measured along the three principal axes. The part programmer does not furnish absolute coordinate values in the command. Following is the format of the command:

$$\text{GODLTA/} \begin{Bmatrix} dx, dy, dz \\ dtlax \\ vector \end{Bmatrix}$$

The incremental values dx, dy, and dz are added to the x-, y-, and z-axis coordinates, respectively, of the current tool location to yield the new tool location in absolute coordinate values. The incremental distances can be positive or negative constants or scalar variables. Negative values denote movement in the negative axis direction. The new tool location is written to the CL-file. Figure 9.3 shows the new tool position derived from this form of the GODLTA command. The command may be labeled as needed for program flow purposes. An equivalent move is obtained from *vector*, a symbol for a vector definition as given in Section 12.4. In this case, the components of *vector* are the incremental distances.

The single value *dtlax* corresponds to an incremental distance along the tool axis. When the value is positive, the direction denoted is for tool withdrawal. When negative the direction is for tool insertion. Until we discuss arbitrary tool axis orientation in Chapter 13, the tool axis is maintained parallel with the z-axis. Therefore, withdrawal and insertion corresponds to the $+z$ and $-z$ directions, respectively. Each form of the GODLTA

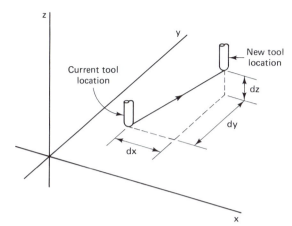

Figure 9.3   Tool positioning with the GODLTA command.

command is illustrated in Figure 9.4. In the interest of program clarity and self-documentation, we will not again use GODLTA/*dtlax* until Chapter 13.

9.2.3 ZSURF

Programming with point-to-point commands is simplified when points have all three coordinate values. All coordinate values can usually be computed for points derived from simple part geometry, but nontrivial part geometries may not allow coordinate values to be easily determined.

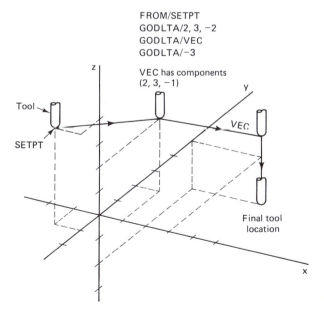

FROM/SETPT
GODLTA/2, 3, −2
GODLTA/VEC
GODLTA/−3

VEC has components
(2, 3, −1)

Figure 9.4   Tool positioning with GODLTA commands.

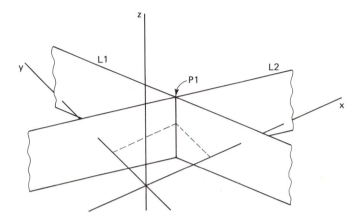

Figure 9.5   Point defined at the intersection of two lines.

So far, only the Point-1 definition format allows us to specify the z-coordinate value. All other point definition formats of Table 3.1 could yield an infinite number of z-coordinate values because APT treats curves as surfaces. We see this in Figure 9.5, where point P1 is defined at the intersection of lines L1 and L2. Since defined lines in APT are infinite planes parallel with the z-axis, their intersection is an infinitely long derived line also parallel with the z-axis. The x and y coordinates of the point are uniquely determined but the z-coordinate could be any of an infinite number of values. Unless something is done as described below, by default the z-coordinate is assigned the value zero because such points are defined by the projection of the lines onto the xy-plane. The z-coordinates of any points used to define lines are irrelevant when determining the z-coordinate of the point of their intersection. The comments above do not apply to the point definition formats of Section 12.6.1, where the z-coordinate value is uniquely determined.

Point z-coordinate values may be implicitly assigned with the ZSURF statement. This statement has the form

$$\text{ZSURF/}\begin{Bmatrix} plane \\ z \end{Bmatrix}$$

where *plane* is the symbol for a previously defined plane or is a nested plane definition and z is a constant or scalar variable. When the z form is used, a plane is defined parallel with the xy-plane with a z-axis intercept of z. *Plane* cannot be parallel with the z-axis. The statement may be labeled.

ZSURF is a modal statement that defines a plane for providing z-coordinate values for subsequent point definitions that do not have explicit z-coordinate assignments. These point definitions, or their equivalent in a motion command, will have their z-coordinate value determined by projecting the point onto *plane*. A ZSURF statement can be counter-

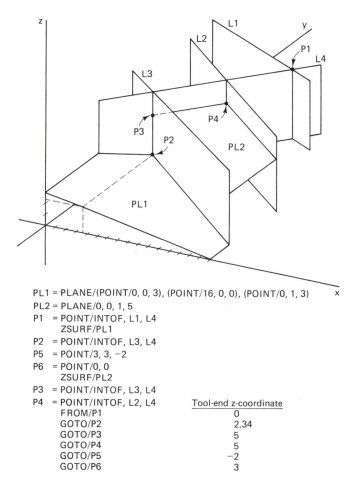

PL1 = PLANE/(POINT/0, 0, 3), (POINT/16, 0, 0), (POINT/0, 1, 3)
PL2 = PLANE/0, 0, 1, 5
P1   = POINT/INTOF, L1, L4
         ZSURF/PL1
P2   = POINT/INTOF, L3, L4
P5   = POINT/3, 3, −2
P6   = POINT/0, 0
         ZSURF/PL2
P3   = POINT/INTOF, L3, L4
P4   = POINT/INTOF, L2, L4   Tool-end z-coordinate
         FROM/P1   0
         GOTO/P2   2.34
         GOTO/P3   5
         GOTO/P4   5
         GOTO/P5   −2
         GOTO/P6   3

Figure 9.6   Assigning point *z*-coordinates with ZSURF statements.

manded only by another ZSURF statement. Points with implicit *z*-coordinates defined prior to the first ZSURF statement of a program will have zero assigned as their *z*-coordinate value. The PSIS statement does not function as a ZSURF statement and the ZSURF statement does not replace a PSIS statement. Figure 9.6 shows the ZSURF statement in action.

9.2.4 More on the AUTOPS Statement

We implicitly defined part surfaces with the AUTOPS statement in Section 8.2.4 for resuming contouring operations. There we constrained the tool within tolerance of the drive and part surfaces at all times. We will now show that new part surfaces may be

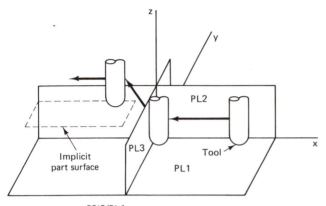

```
PSIS/PL1
INDIRV/−1, 0, 0
TLLFT, GOFWD/PL2, TO, PL3
GODLTA/−2, 0, 2
AUTOPS
INDIRV/−1, 0, 0
TLLFT, GOFWD/PL2, . . .
```

Figure 9.7 AUTOPS with point-to-point command.

defined with the AUTOPS statement after repositioning the tool with point-to-point commands. We may then continue contouring, provided that the tool is within tolerance of the drive surface when resuming tool motion.

Figure 9.7 shows an example where the AUTOPS statement is invoked after a GODLTA statement. We are able to resume contouring because we assume that PL2 is perpendicular to the *xy*-plane. This would not be possible if it were a canted plane because, for the example, the tool would no longer be within tolerance of the drive surface. Not all APT systems may allow contouring to resume in this manner. This example includes a combination of point-to-point and contouring commands. There is no need for AUTOPS with only point-to-point commands.

9.3 EXAMPLES

The first example is a straightforward application of point-to-point commands to drill a pattern of holes. The second example defines subscripted point and line symbols and drills holes within a nested loop structure.

Example 9.1: Hole Drilling with Point-to-Point Commands

Figure 9.8 shows the hole pattern for a drilling operation. The holes on concentric circles exhibit angular symmetry and will be stagger drilled by drilling at P1 and P2 within the range of a COPY structure. The three holes in each corner exhibit rotational and mirror symmetry about the center of the part. Drilling at P3, P4, and P5 within the range of a nested COPY structure will complete the part. A part program to perform the hole drilling operation follows.

```
SETPT  =  POINT/0, −4,2                                          100
C1     =  CIRCLE/0,0,2.25                                        110
C2     =  CIRCLE/0,0,1.25                                        120
L1     =  LINE/YAXIS                                             130
L2     =  LINE/(POINT/0,0),ATANGL, −60                           140
          ZSURF/.75                                              150
P1     =  POINT/YSMALL,INTOF,L1,C1                               160
P2     =  POINT/YSMALL,INTOF,L2,C2                               170
P3     =  POINT/ − 3.5, − 2.5                                    180
P4     =  POINT/ − 3, − 2                                        190
P5     =  POINT/ − 2.5, − 2.5                                    200
          FROM/SETPT                                             210
          INDEX/10                                               220
             GOTO/P1                                             230
                GODLTA/0,0, − 1                                  240
                GODLTA/0,0,1                                     250
             GOTO/P2                                             260
                GODLTA/0,0, − 1                                  270
                GODLTA/0,0,1                                     280
          COPY/10,XYROT,60,5                                     290
          INDEX/20                                               300
             GOTO/P3                                             310
                GODLTA/0,0, − 1                                  320
                GODLTA/0,0,1                                     330
             GOTO/P4                                             340
                GODLTA/0,0, − 1                                  350
                GODLTA/0,0,1                                     360
             GOTO/P5                                             370
                GODLTA/0,0, − 1                                  380
                GODLTA/0,0,1                                     390
          COPY/20,MODIFY,(MATRIX/MIRROR,YZPLAN),1                400
          COPY/20,XYROT,180,1                                    410
          GOTO/SETPT                                             420
```

The workpiece is assumed to be 0.5 in. thick and on the xy-plane. The ZSURF statement, line 150, is used to define the points in lines 160 through 200 whose z-coordinate is above the top part surface.

The various GOTO statements position the drill endpoint on the coordinates of the points referenced. Since their z-coordinates were set at 0.75 with the ZSURF statement, the drill is now properly positioned to begin the drilling operation. Actual drilling occurs with the GODLTA statements, which take the drill down 1 in. and then retract it. This allows for the drill penetrating the workpiece and for clearing it when retracted. Feed rates, which must ultimately be specified, are discussed in Chapter 10 and are ignored here. The diameter of the cutter (here drill) is also ignored.

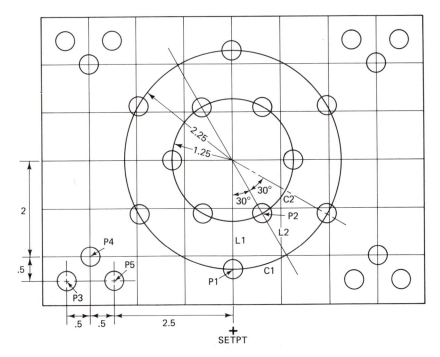

Figure 9.8   Part geometry for Example 9.1.

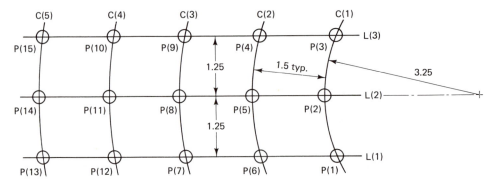

Figure 9.9   Part geometry for Example 9.2.

Example 9.2: Hole Drilling with Subscripted Symbols

Figure 9.9 shows a part with holes to be drilled on a pattern formed by the intersection of concentric circles and parallel lines. The smallest circle has the radius shown, the others each have a radius 1.5 in. greater than its immediately inner circle. The centers of the circles are located on the middle of the parallel lines shown. The lines, circles, and points are labeled with subscripted symbols. A part program to perform the hole drilling operation follows.

```
          RESERV/L,3,P,15                                      100
SETPT  =  POINT/−1,−2,2                                        110
C1     =  CIRCLE/CENTER,(POINT/0,0),RADIUS,1.75               120
          ZSURF/.75                                            130
L(1)   =  LINE/XAXIS,−1.25                                     140
L(2)   =  LINE/XAXIS                                           150
L(3)   =  LINE/XAXIS,1.25                                      160
          LOOPST                                               170
          R = 1.75                                             180
          J = 0                                                190
          K = 1                                                200
          M = 1                                                210
          I = 0                                                220
CIR1)  I = I + 1                                               230
          R = R + 1.5                                          240
          C1 = CIRCLE/C1,CANON,,,,,,,R                         250
          K = −K                                               260
          M = M + K                                            270
PNT1)  M = M − K                                               280
          J = J + 1                                            290
          P(J) = POINT/XSMALL,INTOF,L(M),C1                    300
          IF (3*I − J) PNT2, PNT2, PNT1                        310
PNT2)  IF (15 − J) PNT3, PNT3, CIR1                            320
PNT3)  FROM/SETPT                                              330
          J = 0                                                340
HDR1)  J = J + 1                                               350
          GOTO/P(J)                                            360
             GODLTA/0,0,−1                                     370
             GODLTA/0,0,1                                      380
          IF (15 − J) HDR2, HDR2, HDR1                         390
HDR2)  GOTO/SETPT                                              400
          LOOPND                                               410
```

The points are defined above the workpiece (assumed to be 0.5 in. thick) in a manner similar to that for Example 9.1. The ZSURF statement, line 130, is needed for the point definitions of line 300. Figure 9.10 shows a skeleton flowchart for this program.

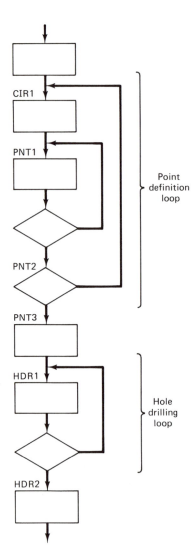

Figure 9.10   Skeleton flowchart for Example 9.2.

9.4 PATTERNS

The point-to-point programming method, although conceptually simple, is not without its implementation drawbacks, among which are the need to define many points and the need to program tool operations at each point. We will address the first drawback in this section by describing the pattern feature but will postpone discussion of the second drawback until Chapters 10 and 11.

The pattern feature is a convenience feature that permits ease of point definition for special cases and simplicity of point specification in motion commands. In essence, we will define a collection of points with a pattern geometric definition statement. There-after, we will reference the pattern by its symbolic name. Tool movement from point to point among the points of the pattern conforms to that described in Section 9.1.

The part programmer must realize that pattern programming features are not stan-dardized to the extent of those features described thus far. Differences among APT implementations must be expected and the reference manual must be consulted. We make no attempt to describe all pattern features nor do we attempt to document a given APT system. We do describe a useful subset of features likely to be found in most systems.

9.4.1 Pattern Definition Formats

A selection of pattern definition formats appears in Table 9.1. A characteristic of all formats is the ordering (numbering) of the points in the pattern to allow any point to be located by its position within the pattern. Except for Pattern-13, the formats allow no exception to forming the structure of the pattern. This follows an "old" pattern definition philosophy whereby exceptions to using all points of the pattern are taken in the motion commands. Pattern-13 follows a "new" philosophy whereby exceptions to the pattern structure are built into the definition and no exceptions are taken in motion commands. The Pattern-13 format is found in the IBM APT-AC system.

The formats in Table 9.1 allow most common point patterns to be described. They may be defined as a linear pattern (in a straight line), on a circular arc, in arbitrary locations, and as a two-dimensional grid (linear only in each dimension). Pattern defi-nitions cannot be transformed with the REFSYS feature, nor can z-coordinates of their points be set with the ZSURF statement. The z-coordinates are determined as described for each format.

9.4.2 Point-to-Point Programming with Patterns

We saw the need for traversing a pattern of points earlier in this chapter. Now, we will traverse the points of the patterns in Section 9.4.1. Surely, nothing is accomplished by merely traversing a pattern of points. Some function must be performed at each point location to make the traversal worthwhile. This function might be, say, the drilling, tapping, or punching of a hole. Because we are delaying the discussion of postprocessor commands until Chapter 10, we are somewhat handicapped in discussing fully the traversal of points in a pattern.

For the moment, we will assume that with each visit to a point of a pattern some function is performed. Later, we will consider that function in detail. Below we will show how that function is invoked. The general form of the command for traversing the points of a pattern is as follows:

$$\text{GOTO}/patt \text{ [,INVERS] [,AVOID,}dz,list] \left[, \begin{Bmatrix} \text{OMIT} \\ \text{RETAIN} \end{Bmatrix} ,list \right]$$

Patt is the name of the pattern being traversed. It will be traversed in the forward direction (as the points are numbered when *patt* is defined) or in the reverse direction (when the

TABLE 9.1 PATTERN DEFINITIONS

| 1 | On a line connecting given endpoints: |

PATERN/LINEAR, *point1, point2, n*

The pattern consists of *n* equally spaced points on a line from *point1* (the first point) to *point2* (the *n*th point). The x, y, and z-coordinates of each point are determined by linearly interpolating along the line between *point1* and *point2*.

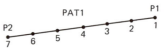

PAT1 = PATERN/LINEAR, P1, P2, 7

| 2 | On a line from a point and with a vector for direction and spacing: |

PATERN/LINEAR, *point, vector, n*

The pattern consists of *n* equally spaced points beginning at *point* (the first point) and extending on a line in the direction of *vector*. Spacing between points is | *vector* |. The coordinates of the first point are given by *point*. Thereafter, the point coordinates are determined from the length and direction of *vector*.

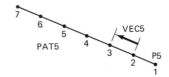

PAT5 = PATERN/LINEAR, P5, VEC5, 7

| 3 | On a line from a point, a vector for direction, and sets of uniform increments: |

PATERN/LINEAR, *point, vector*, INCR, n_1, AT, d_1 [, INCR, n_2, AT, d_2] . . .

The pattern begins at *point* (the first point) and extends on a line in the direction of *vector*. It consists of one or more subpatterns, each defined by the instantiation of an INCR, n_i, AT, d_i phrase. Each phrase specifies the number of increments n_i to be generated along the line at equal distances d_i. The subpatterns are generated in the order in which the phrases appear in the statement. The coordinates of the points are determined in a manner similar to that of Pattern-2. The total number of points in the pattern is $1 + n_1 + n_2 + \cdots + n_k$.

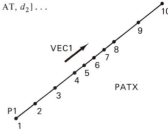

PATX = PATERN/LINEAR, P1, VEC1, INCR, 3, AT, 2, INCR, 4, AT, 1, INCR, 2, AT, 2.5

| 4 | On a line from a point, a vector for direction, and arbitrary increments: |

PATERN/LINEAR, *point, vector*, INCR, $d_1, d_2, \ldots, d_n$

The pattern begins at *point* (the first point) and extends on a line in the direction of *vector*. Successive points are spaced a distance d_i from their preceding point as governed by the left-to-right order of the d_i in the statement. The coordinates of the points are determined in a manner similar to that of Pattern-2. The total number of points in the pattern is $1 + n$, where *n* is determined by the number of d_i in the statement.

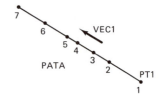

PATA = PATERN/LINEAR, PT1, VEC1, INCR, 3, 1.5, 1.5, 1, 2, 2.5
or
PATA = PATERN/LINEAR, PT1, VEC1, INCR, D(1, THRU, 6)
(inclusive subscripts, with D a scalar array)

298

TABLE 9.1 (CONTINUED)

| 5 | On a circular arc connecting endpoints defined by starting and ending angles: |

PATERN/ARC, *circle*, α, β, $\begin{Bmatrix} \text{CLW} \\ \text{CCLW} \end{Bmatrix}$, *n*

The pattern consists of *n* equally spaced points on *circle* in a clockwise (CLW) or counterclockwise (CCLW) direction from α (the first point) to β (the *n*th point). Angles α and β are in degrees measured from the positive x-axis of a local coordinate system whose origin is at the center of *circle*. The points are given the z-coordinate value of the center of *circle*.

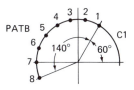

PATB = PATERN/ARC, C1, 60, 200, CCLW, 8

| 6 | On a circular arc, from a starting angle, and arbitrary angular increments: |

PATERN/ARC, *circle*, α, $\begin{Bmatrix} \text{CLW} \\ \text{CCLW} \end{Bmatrix}$, INCR, $d\phi_1$, $d\phi_2$, ..., $d\phi_n$

The pattern consists of points in a clockwise (CLW) or counterclockwise (CCLW) direction on *circle* with the first point at α. Successive points are spaced an angular distance $d\phi_i$ from their preceding point as governed by the left-to-right order of the $d\phi_i$ in the statement. Angles α and the $d\phi_i$ are in degrees, with α measured from the positive x-axis of a local coordinate system whose origin is at the center of *circle*. The points are given the z-coordinate value of the center of *circle*. The total number of points in the pattern is $1 + n$, where *n* is determined by the number of $d\phi_i$ in the statement.

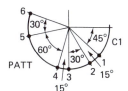

PATT = PATERN/ARC, C1, -45, CLW, INCR, 15, 30, 15, 60, 30

or

PATT = PATERN/ARC, C1, -45, CLW, INCR, PHI(1, THRU, 5)
(inclusive subscripts, with PHI a scalar array)

| 7 | On a circular arc, from a starting angle, and sets of uniform increments: |

PATERN/ARC, *circle*, α, $\begin{Bmatrix} \text{CLW} \\ \text{CCLW} \end{Bmatrix}$, INCR, n_1, AT, $d\beta_1$[, INCR, n_2, AT, $d\beta_2$] ...

The pattern consists of points in a clockwise (CLW) or counterclockwise (CCLW) direction on *circle* with the first point at α. It consists of one or more sub-patterns, each defined by the instantiation of an INCR, n_i, AT, $d\beta_i$ phrase. Each phrase specifies the number of increments n_i to be generated along *circle* at equal angular distances $d\beta_i$. The subpatterns are generated in the order in which the phrases appear in the statement. Angles α and the $d\beta_i$ are in degrees with α measured from the positive x-axis of a local coordinate system whose origin is at the center of *circle*. The points are given the z-coordinate value of the center of *circle*. The total number of points in the pattern is $1 + n_1 + n_2 + \cdots + n_k$.

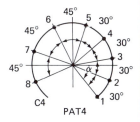

PAT4 = PATERN/ARC, C4, ALF, CCLW, INCR, 4, AT, 30, INCR, 3, AT, 45

TABLE 9.1 (CONTINUED)

| 8 | By points and other patterns at arbitrary locations: |

PATERN/RANDOM, $\left\{\begin{matrix} patt1 \\ point1 \end{matrix}\right\}$ $\left[,\left\{\begin{matrix} patt2 \\ point2 \end{matrix}\right\}\right]$ $\cdots$

The pattern defined is a collection of points (*pointi*) and/or patterns (*patti*). The points being defined are consecutively numbered as *pointi* and/or *patti* as encountered while the list is being processed in a left-to-right manner. The x-, y-, and z-coordinate values are the same as when *pointi* and *patti* were defined.

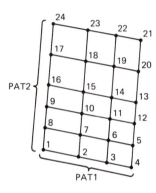

PATR = PATERN/RANDOM, P1, P2, PAT4, P6, PAT2, PAT8, P7

| 9 | By two linear patterns: |

PATERN/PARLEL, *patt1*, *patt2*

The pattern defined is a collection of points derived from linear patterns *patt1* and *patt2*, which will form adjacent sides of a parallelogram shaped grid structure. *patt1* and *patt2* usually must share their first or last points for this definition. Lines parallel to *patt1* are established through the points of *patt2* and lines parallel to *patt2* are established through the points of *patt1*. The intersection points of these grid lines are the points of the pattern being defined. The z-coordinates for points on the grid are the same as for the corresponding points of *patt1*. Call *patt1* grid line 1 and designate the numbering direction for it and all odd-numbered grid lines parallel with it in the same direction; the even-numbered grid lines are numbered in the reverse direction. The points of the pattern being defined are then numbered consecutively in a back-and-forth manner beginning with the first point of *patt1*. The total number of points in the pattern is $n_{patt1} * m_{patt2}$, where n_{patt1} and m_{patt2} are the number of points in *patt1* and *patt2*, respectively.

PATG1 = PATERN/PARLEL, PAT1, PAT2

| 10 | By a linear pattern, a vector for direction, and sets of uniform increments: |

PATERN/PARLEL, *patt*, *vector*, INCR, n_1, AT, d_1 [, INCR, n_2, AT, d_2] $\ldots$

The pattern defined is a collection of points in a grid structure similar to that of Pattern-9. Linear pattern *patt* is replicated parallel to itself in the direction of *vector* a number of times and at the spacing given by the INCR, n_i, AT, d_i phrases. Each phrase specifies the number of replications n_i a distance d_i in the direction of *vector*. Replications occur in the order in which the phrases appear in the statement. The points of the pattern being defined are numbered consecutively in a back-and-forth manner beginning with the first point of *patt* in the manner described for Pattern-9. The coordinates for replicated points are determined from the direction of *vector*. The total number of points in the pattern is $(1 + n_1 + n_2 + \cdots + n_k) * m_{patt}$, where m_{patt} is the number of points in *patt*.

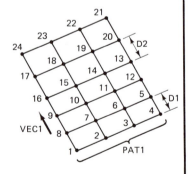

PATG2 = PATERN/PARLEL, PAT1, VEC1, INCR, 3, AT, D1, INCR, 2, AT, D2

TABLE 9.1 (CONTINUED)

| 11 | By a linear pattern, a vector for direction, and arbitrary increments: |

PATERN/PARLEL, *patt*, *vector*, INCR, $d_1, d_2, \ldots, d_n$

The pattern defined is a collection of points in a grid structure similar to that of Pattern-9. Linear pattern *patt* is replicated parallel to itself in the direction of *vector* at successive spacings of d_i as governed by the left-to-right order of the d_i in the statement. The points of the pattern being defined are numbered consecutively in a back-and-forth manner beginning with the first point of *patt* in the manner described for Pattern-9. The coordinates for replicated points are determined from the direction of *vector*. The total number of points in the pattern is $(1 + n) * m_{patt}$, where n is determined by the number of d_i in the statement and m_{patt} is the number of points in *patt*.

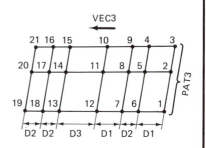

PATG3 = PATERN/PARLEL, PAT3, VEC3, INCR, D1, D2, D1, D3, D2, D2

| 12 | By a linear pattern and a vector for spacing and direction: |

PATERN/PARLEL, *patt*, *vector*, n

The pattern defined is a collection of points in a grid structure similar to that of Pattern-9. Linear pattern *patt* is replicated parallel to itself $n - 1$ times in the direction of *vector* at successive spacings of | *vector* |. The points of the pattern being defined are numbered consecutively in a back-and-forth manner beginning with the first point of *patt* in the manner described for Pattern-9. The coordinates for replicated points are determined from the length and direction of *vector*. The total number of points in the pattern is $n * m_{patt}$, where m_{patt} is the number of points in *patt*.

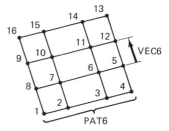

PATG4 = PATERN/PARLEL, PAT6, VEC6, 4

TABLE 9.1 (CONTINUED)

| 13 | By copying a pattern to each point of another pattern: |

PATERN/COPY, *patt1* [, ATTACH, *point*] [, XYROT, α[, INCR, β]] $\left[, \left\{ {SAME \atop UNLIKE} \right\} \right]$, \$

ON, *patt2* $\left[, \left\{ {SAME \atop UNLIKE} \right\} \right] \left[, \left\{ {OMIT \atop RETAIN} \right\} , n_1, n_2, \ldots, n_k \right]$

The pattern defined is formed by translating, perhaps with rotation, *patt1* to each point of *patt2*. It is attached with respect to *point*, an arbitrary reference point that need not be a member of *patt1*. *point* defaults to the first point of *patt1*. The z-coordinate of each point of the copies is determined from the expression

$$z\text{-coord}(patt1) + [z\text{-coord}(patt2) - z\text{-coord}(point)]$$

where, for *patt1* and *patt2*, only one point at a time is considered. The XYROT phrase rotates *patt1* about *point* through α degrees before translation. When the INCR, β phrase is included, *patt1* is rotated α degrees at the first point of *patt2*, then progressively by β degrees for each remaining point of *patt2*. This option is effective at all points of *patt2*, even though some points are bypassed with an OMIT–RETAIN modifier. *patt1* may be copied only at selected points of *patt2* by including the RETAIN modifier followed by a list of n_i, point positions of *patt2* at which copying is to take place. Conversely, copies of *patt1* may be omitted at selected points of *patt2* by including the OMIT modifier followed by a list of n_i, point positions of *patt2* at which copying is not to take place. Points of *patt2* at which copies are not made when OMIT or RETAIN is used are not copied to the pattern being defined. Only one SAME or UNLIKE modifier is allowed per statement; it defaults to SAME for *patt1*. The modifier affects the numbering of points in the pattern being defined as illustrated below.

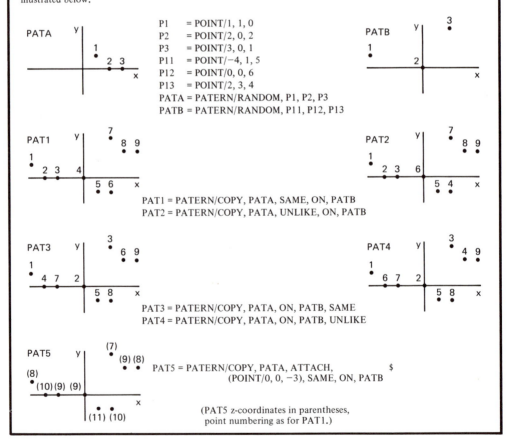

P1 = POINT/1, 1, 0
P2 = POINT/2, 0, 2
P3 = POINT/3, 0, 1
P11 = POINT/−4, 1, 5
P12 = POINT/0, 0, 6
P13 = POINT/2, 3, 4
PATA = PATERN/RANDOM, P1, P2, P3
PATB = PATERN/RANDOM, P11, P12, P13

PAT1 = PATERN/COPY, PATA, SAME, ON, PATB
PAT2 = PATERN/COPY, PATA, UNLIKE, ON, PATB

PAT3 = PATERN/COPY, PATA, ON, PATB, SAME
PAT4 = PATERN/COPY, PATA, ON, PATB, UNLIKE

PAT5 = PATERN/COPY, PATA, ATTACH, \$
(POINT/0, 0, −3), SAME, ON, PATB

(PAT5 z-coordinates in parentheses,
point numbering as for PAT1.)

INVERS modifier is included). Whenever some points of the pattern are not to be visited, they must be designated in *list* following the OMIT modifier. Optionally, it may be more convenient to specify points to be visited, in which case they must be designated in *list* following the RETAIN modifier. The points of *list* should be kept in numerically ascending order. OMIT and RETAIN cannot both be in the same statement.

A distinction must be made between the numbering of points when a pattern is defined and the numbering of points when the pattern is referenced in the GOTO statement. In the latter case, when traversing the pattern in a forward direction, they are referenced by the number assigned to them during pattern definition. When traversing the pattern in the reverse direction (when iNVERS is included), they are assumed to have been renumbered in the reverse direction for that GOTO statement only. The points to be omitted or retained are designated in *list* by their position number in the pattern as just described. Sometimes, the numbering of points for GOTO statement referencing is referred to as the output order. *List* contains integers only, which may be scalar variables. A list of consecutive numbers may be designated with the phrase, *m*,THRU,*n*, where *m* and *n* are starting and ending numbers of the succession. This phrase may appear more than once in *list*.

To avoid obstructions between points, perhaps because of clamps or part protrusions, the tool is retracted between points by the incremental amount *dz* above the pattern point *z*-coordinate value by designating point positions in *list* following the AVOID modifier. Point positions are determined as described for the OMIT and RETAIN modifiers. The points in *list*, which should be kept in numerically ascending order and which may include the *m*,THRU,*n* phrase, designate the starting point for tool retraction. The tool is again protracted by the amount *dz*, but not until it is over the next point of the traversal sequence. This next point is affected by the points of *list* following OMIT or RETAIN. It is recommended that the AVOID option precede the OMIT/RETAIN option.

Once defined, the pattern is never altered. Exceptions to point locations in the pattern are taken with the OMIT or RETAIN lists. This traversal philosophy is analogous to the "old" pattern definition philosophy. The "new" philosophy is represented in the motion commands of the IBM APT-AC system, where exceptions are taken during pattern definition. Pattern motion command statement functioning for a given APT implementation may differ from that described above. The principles presented here will work with most APT systems. The reference manual must be consulted for details of command interpretation.

Numerous examples of pattern GOTO statement preparation and interpretation according to the description above are given in Figure 9.11. For illustrative purposes, PATA could be defined as follows:

    PT1 = POINT/0,0,1
    PT2 = POINT/8,0,9
    PATA = PATERN/LINEAR,PT1,PT2,9

PATA then lies in the *zx*-plane. Alternative GOTO statements are given for parts (a) and (b). The AVOID movement is always parallel with the *z*-axis.

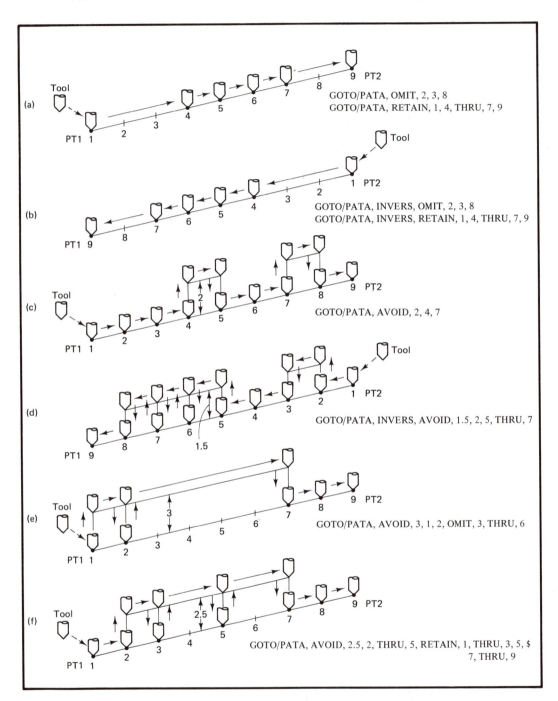

Figure 9.11 Pattern points output with various modifiers in the GOTO command.

```
0008 PPRINT/  USE OMIT AND AVOID.
0009   GOTO/
                    0.0            0.0          1.00000000
0009 CYCLE /        NOMORE
0009   GOTO/
                    0.0            0.0          4.00000000
0009   GOTO/
                    1.00000000     0.0          5.00000000
0009 CYCLE /        ON
0009   GOTO/
                    1.00000000     0.0          2.00000000
0009 CYCLE /        NOMORE
0009   GOTO/
                    1.00000000     0.0          5.00000000
0009   GOTO/
                    6.00000000     0.0         10.00000000
0009 CYCLE /        ON
0009   GOTO/
                    6.00000000     0.0          7.00000000
0009   GOTO/
                    7.00000000     0.0          8.00000000
0009   GOTO/
                    8.00000000     0.0          9.00000000
```

(e)

```
0011 PPRINT/  USE RETAIN AND AVOID.
0012   GOTO/
                    0.0            0.0          1.00000000
0012   GOTO/
                    1.00000000     0.0          2.00000000
0012 CYCLE /        NOMORE
0012   GOTO/
                    1.00000000     0.0          4.50000000
0012   GOTO/
                    2.00000000     0.0          5.50000000
0012 CYCLE /        ON
0012   GOTO/
                    2.00000000     0.0          3.00000000
0012 CYCLE /        NOMORE
0012   GOTO/
                    2.00000000     0.0          5.50000000
0012   GOTO/
                    4.00000000     0.0          7.50000000
0012 CYCLE /        ON
0012   GOTO/
                    4.00000000     0.0          5.00000000
0012 CYCLE /        NOMORE
0012   GOTO/
                    4.00000000     0.0          7.50000000
0012   GOTO/
                    6.00000000     0.0          9.50000000
0012 CYCLE /        ON
0012   GOTO/
                    6.00000000     0.0          7.00000000
0012   GOTO/
                    7.00000000     0.0          8.00000000
0012   GOTO/
                    8.00000000     0.0          9.00000000
```

(f)

Note: The brackets span point pairs between which an AVOID movement is effected.
       CYCLE actions are not performed at these point pairs.

Figure 9.12   CL-printout for the GOTO statements of parts (e) and (f) of Figure 9.11.

Figure 9.12 is the CL-printout for Figure 9.11(e) and (f) when applied to the pattern defined above. The PPRINT statement is described in Section 9.5. The CYCLE statements intermingled with the cut vectors interest us now since they invoke the function desired. The CYCLE statement is a postprocessor statement discussed in Chapter 10. The function it performs depends on its implementation within the postprocessor. This function is turned on or off with the CYCLE/ON and CYCLE/NOMORE commands, respectively, inserted in the CL-file by the APT processor. These are modal commands and are the CYCLE statements of Figure 9.12. The CYCLE/ON command appears before pattern points being output. It could invoke, say, a drilling operation at each of the points following it. The CYCLE/NOMORE command appears before the tool is raised when a pattern point is referenced in an AVOID phrase. No drilling operation would be performed at the points following it.

9.4.3 Examples

The convenience afforded through point pattern definitions and the simplicity of pattern traversal with the GOTO statement and its options is not always the solution to programming this class of problem. This is illustrated in Examples 9.3 and 9.4 which are reworks of Examples 9.1 and 9.2. Example 9.5 exploits the pattern definition and traversal features.

Example 9.3: A Second Look at Example 9.1

The hole pattern symmetry of Figure 9.8 implies the pattern feature. However, in Example 9.1 we stagger drilled the holes on the concentric circles to expedite the machining operations since staggering led to the shortest tool path. We cannot stagger visit points defined as two different circular patterns. Now we will traverse one circular pattern before proceeding to the other pattern. We will define the three holes in the lower left corner of Figure 9.8 as a pattern but will resort to the COPY feature for their replication in the other corners. It is assumed that a CYCLE statement will eventually be inserted that defines the drilling function for each pattern point visited. The modified program is as follows:

SETPT	= POINT/0, −4,2	100
C1	= CIRCLE/0,0,2.25	110
C2	= CIRCLE/0,0,1.25	120
	ZSURF/.75	130
PAT1	= PATERN/ARC,C1, −90,CCLW,INCR,5,AT,60	140
PAT2	= PATERN/ARC,C2, −60,CCLW,INCR,5,AT,60	150
P3	= POINT/ −3.5, −2.5	160
P4	= POINT/ −3, −2	170
P5	= POINT/ −2.5, −2.5	180
PAT3	= PATERN/RANDOM,P3,P4,P5	190
	FROM/SETPT	200
	GOTO/PAT2	210
	GOTO/PAT1	220
	INDEX/20	230
	GOTO/PAT3	240
	COPY/20,MODIFY,(MATRIX/MIRROR,YZPLAN),1	250
	COPY/20,XYROT,180,1	260
	GOTO/SETPT	270

We reduced just slightly the number of point definition statements and we eliminated some motion statements. However, we had to compromise on the order in which we visited the points. This compromise is adverse to the goals of a production operation.

Example 9.4: A Second Look at Example 9.2

The hole pattern symmetry of Figure 9.9 is adverse to direct pattern application because the varying radius for each arc combined with lines that are parallel does not fit pattern definition formats. Angular relationships cannot easily be defined with existing pattern definition formats. If we insist on using a pattern, we will retain the subscripted point definitions of Example 9.2 and place the points in a pattern with the RANDOM format. The statements of lines 340 through 410 will be as follows:

```
PNT3)   FROM/SETPT
        PATT = PATERN/RANDOM,P(1,THRU,15)
        GOTO/PATT
        GOTO/SETPT
        LOOPND
```

The loop for outputting points is eliminated. It is assumed that a CYCLE statement will be inserted in the program.

Example 9.5: A Part with Several Hole Patterns

The holes in the part of Figure 9.13 can be adequately described as point patterns. From the dimension of the hole spacings we can define patterns with the linear, arc, grid, and copy formats. The barriers along the edges will require the AVOID option in the GOTO statement. Because of two part surface heights, care must be used to prevent rapid tool motion into the part. Some symbols are marked on the part of Figure 9.13(b). ORIG is the coordinate system origin for pattern definition. A part program for the hole patterns follows.

```
SETPT     = POINT/6.5,4,3                                          100
ORIG      = POINT/0,0                                              110
          ZSURF/.375                                              120
P1        = POINT/6.5,2                                            130
VEC1      = VECTOR/ − 1,0,0                                        140
PAT1      = PATERN/LINEAR,P1,VEC1,INCR,4,AT,1.5                    150
VEC2      = VECTOR/0, − 4,0                                        160
PAT2      = PATERN/PARLEL,PAT1,VEC2,2                              170
ALF       = ATAN2F(2, − .5)                                        180
C1        = CIRCLE/.75,0,2.25                                      190
L1        = LINE/ORIG,ATANGL,ALF,XAXIS                             200
PC1       = POINT/YLARGE,INTOF,L1,C1                               210
BETA      = ANGLF(C1,PC1)                                          220
PAT3      = PATERN/ARC,C1,BETA, − BETA,CCLW,5                      230
C2        = CIRCLE/0,0,2.5                                         240
PAT4      = PATERN/ARC,C2, − 60,CLW,INCR,4,AT,15                   250
PAT5      = PATERN/COPY,PAT4,ATTACH,(POINT/0,0, − .125),   $       260
            XYROT, − 30,INCR,30,SAME,ON,PAT2,              $       270
            RETAIN,2,8                                             280
          FROM/SETPT                                              290
PPRINT    FROM UPPER RIGHT,GO LEFT NEAR EDGE (PAT2).              300
          GOTO/PAT2,AVOID,.5,1,THRU,4,RETAIN,1,THRU,5            310
PPRINT    GO AROUND CIRCULAR EDGE (PAT3).                        320
          GOTO/PAT3                                              330
```

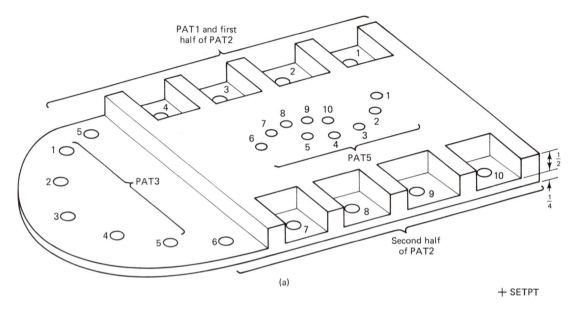

(a)

+ SETPT

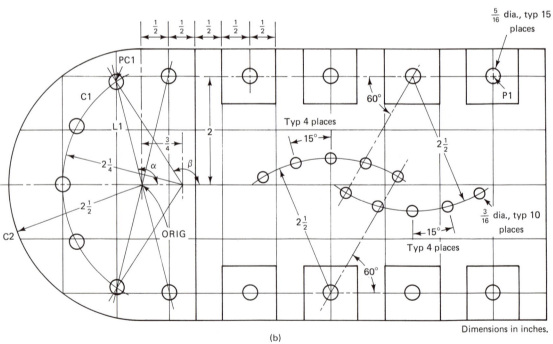

(b)

Dimensions in inches.

Figure 9.13   (a) Part geometry for Example 9.5; (b) top view of Figure 9.13(a).

```
PPRINT   GO RIGHT NEAR FRONT EDGE (PAT2).                        340
         GOTO/PAT2,AVOID,.5,6,THRU,10,RETAIN,6,THRU,10          350
PPRINT   VISIT POINTS ON UPPER SURFACE (PAT5).                  360
         GOTO/PAT5                                              370
         GOTO/SETPT                                             380
```

The tool-end is kept ⅛ in. above the surfaces. The z-coordinates for points of PAT1, PAT2, and PAT3 are assigned via the ZSURF statement. The z-coordinates for points of PAT5 are determined from PAT4 (via the ZSURF statement and C2), from PAT2, and from the ATTACH point (see the computational expression for Pattern-13 in Table 9.1). The ANGLF function of line 220 is used to determine the limits for the arc of PAT3.

The order in which the points are visited is marked in Figure 9.13(a) for each pattern. The AVOID option is used with PAT2 to clear the barriers in a rapid movement. The PPRINT statements (see factor 7 in Section 9.5) allow annotation of the CL-printout for pattern traversal identification.

9.5 SUGGESTIONS AND REMINDERS

Point-to-point programming, with its absence of constraining surfaces, does not eliminate the need to carefully position the tool for sequences of machine operation. Also, good programming practices must not be ignored to realize the benefits discussed earlier. Important factors associated with point-to-point programming are as follows.

1. Use the GODLTA command with care to avoid error buildup. Increments of motion should be based on numerical values containing few fractional digits. Inaccurate numerical conversion and approximations should be avoided.

2. Employ short sequences of cascaded GODLTA commands. This avoids having to change several such commands when dimensional changes are made to a command placed early in the sequence. Consider using variables in commands for development parts to accommodate dimensional changes.

3. Carefully check pattern definitions with the formats of the reference manual. Uncertainty of interpretation should be resolved by programming test patterns before proceeding with part program planning.

4. The order in which the points of a pattern are numbered when defined is important. When the order is in doubt, it can be determined from the printout of a GOTO/*patt* statement. Pattern point coordinates should also be verified at this time. Draw a pattern number diagram for complex patterns.

5. The points output from a GOTO/*patt* statement should be in the pattern definition order unless part structural factors clearly dictate the need for the INVERS modifier without adversely affecting part program understanding.

6. Keep the pattern point numbers in numerical order for AVOID and OMIT-RETAIN modifiers in a GOTO/*patt* statement to promote readability and to enhance statement understanding.

7. Printouts for patterns usually do not include the symbolic name of the pattern, thus making it difficult to easily identify specific points. Use the PPRINT statement to

insert identifying comments in strategic locations of the CL-printout. Characters following the PPRINT (PPRINT must be in columns 1 through 6 of the part program) will appear in the CL-file and in the CL-printout (see Figure 9.12).

8. Review, for possible application, the Point-13 definition of Table 3.1. It defines for external reference a given point of a pattern.

9. Review, for possible application, the NUMF and ANGLF functions of Table 4.2. The value of NUMF is the number of points of a pattern. ANGLF is useful when defining patterns on a circular arc (see Example 9.5).

PROBLEMS

9.1 Figure P9.1 shows a ½-in.-thick part bounded by two lines and two circles. Many holes are drilled in the part. Write an APT part program to contour the part boundary and to simulate the drilling of the holes with point-to-point commands.

9.2 Rework Problem 8.2 with point-to-point commands for positioning the tool onto the part surfaces before resuming contouring.

9.3 Repeat Problem 9.2 but for Figure P8.3.

9.4 Figure P9.4 shows a ½-in.-thick plate with equal-sized holes drilled in it. Write an APT part program with point-to-point commands to simulate the drilling of the holes in a shortest-path sequence.

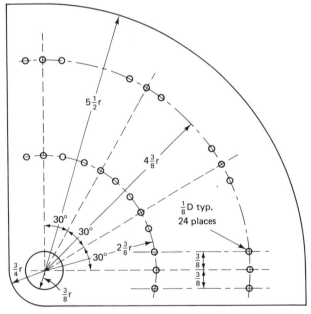

Dimensions in inches

Figure P9.1

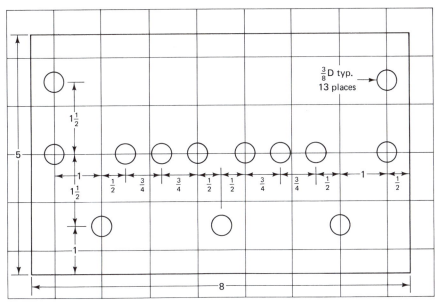

Dimensions in inches

Figure P9.4

9.5 The points on the curve of Figure P9.5 are locations at which holes are to be drilled. They are on equal abscissa value increments on the curve whose equation is given (θ varies over the single cycle shown). With A = 3, write the APT part program with point-to-point commands to drill the holes in the sequence shown in ½-in.-thick material.

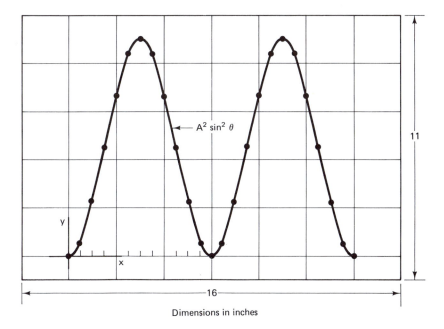

Dimensions in inches

Figure P9.5

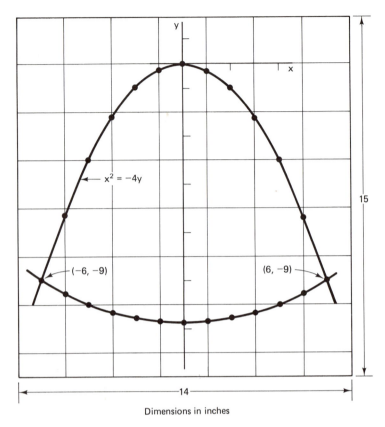

Dimensions in inches

Figure P9.6

9.6 Figure P9.6 shows points on a circular arc and on a downward opening parabola at which holes are to be drilled. Write the APT part program to drill the holes in ½-in.-thick material. The holes are spaced at equal abscissa value increments on each curve.

9.7 Figure P9.7 shows a stepped block with equal-sized holes drilled on each step. Write the APT part program to drill the holes to the depth shown in a shortest-path sequence.

9.8 Figure P9.8 shows a part with equal-sized holes drilled on circular arcs. Write the APT part program to drill the holes to the depth shown in a shortest-path sequence.

9.9 The welded mount of Problem 8.6 contained only straight-line edges along which welding took place. The outcome from contouring is a set of cut vectors representing the line endpoints. These could just as well be obtained with point-to-point commands. Rework that problem using point-to-point commands.

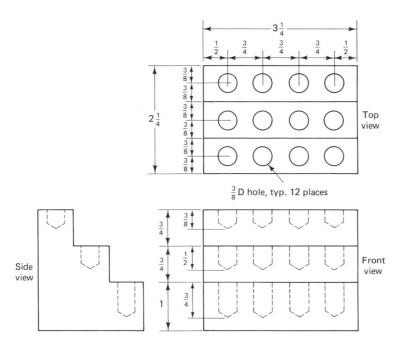

Top view

$\frac{3}{8}$ D hole, typ. 12 places

Side view

Front view

Dimensions in inches

Figure P9.7

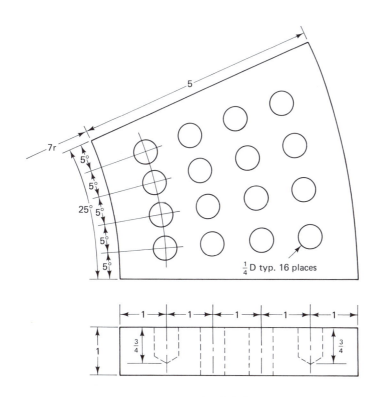

$\frac{1}{4}$ D typ. 16 places

Dimensions in inches

Figure P9.8

CHAPTER **10**

Postprocessor Commands

Part programming concepts introduced thus far had the principal objective of obtaining a set of cut vectors to produce the part. However, the cut vectors must be combined with special commands that control machine functions, such as the feed rate, coolant, auxiliary functions, and so on, that collectively result in automatic control of the machine. Many of these special commands are not part of the APT system but are unique to some machines and therefore cannot be processed by the APT system. Another computer program is needed to process these special commands in conjunction with the cut vectors computed by the APT part program. This additional computer program is referred to as a **postprocessor** program. Its use with the APT system is briefly discussed in this chapter.

10.1 THE ROLE OF THE POSTPROCESSOR

The APT system processes the part program to produce the CL-file, which contains the cut vectors, a copy of the postprocessor commands, and other information, such as tolerances and the tool description. The CL-file is created at the time APT Section 3 processing is finished, which is upon completion of the general part program solution. At this time the requirement for the APT system has been fulfilled and a file in a standardized format is available for postprocessing during which a machine control tape is produced.

Figure 10.1 shows the sequence of computer processing required to obtain a machine control tape (either a paper tape or a magnetic tape or, for a DNC system, another file). The machine control tape contains coded information which will be used for starting and stopping the machine, controlling the coolant, spindle speeds, and movement of the

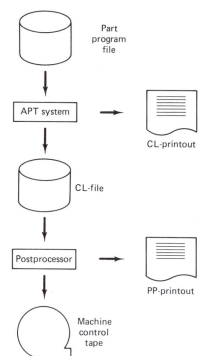

Figure 10.1   Sequence of computer processing to obtain a machine control tape.

slides, changing tools, and other mechanical operations for proper machine control. The coded information on the machine control tape must be compatible with the coding scheme recognized by the controller for the particular machine to be controlled. More specifically, the controller is the electrical interface between the machine control tape and the machine tool.

According to Figure 10.1, the CL-file is written by the APT system. Conversion from the coding and recording format of the CL-file to the coding and recording format of the machine control tape are functions of the postprocessing program. Although the postprocessing program, also called a **postprocessor**, is not part of the APT system, it is automatically executed in Section 4 of the APT system when appropriate postprocessor commands are included in the part program. The postprocessor is not a machine. It is a computer program, usually written in a combination of the FORTRAN and computer system assembly languages, that is unique for a given machine-tool/controller combination or for a family of machine tools and controllers with nearly identical characteristics. The diversity of machine tool control requirements demands a large number of postprocessors, nearly one for each machine-tool/controller combination. The part programmer usually is not concerned with the writing of the postprocessor but requires its availability. The documentation for a postprocessor provides information about the commands available for controlling a specific machine tool. These commands are called **postprocessor commands** and are included in the APT part program.

The function of the postprocessor is complicated because of differences between the desired machine tool response and the actual machine tool control provided by the machine-tool/controller combination. This complexity may arise because (1) the simplicity of a given machine-tool/controller combination requires that much of the control action be determined within the postprocessor and that additional control commands be issued to effect proper control, (2) a given machine-tool/controller combination is so sophisticated in its operation that the simple cut vector sequence determined by the APT system must be augmented with postprocessor commands so that the machine tool capability is fully utilized, or (3) some other combination of circumstances. As a result, the demands placed on the postprocessor are varied and its function is expanded to include more than the simple coding and format conversion requirement. A particular postprocessor will perform some, or all, of the following functions:

1. Read in data from the CL-file, which is usually a magnetic tape or a magnetic disk
2. Convert the cut vectors from the part programmers coordinate system to the machine tool coordinate system. This may require, individually or in combination, coordinate system translation and rotation.
3. Perform computations for nonlinear motions such as result from multiaxis tool movement and circular and parabolic interpolation
4. Convert coordinate values for use with absolute or incremental systems
5. Determine whether or not machine tool limitations have been exceeded. This requires comparing the machine tool feedrates, spindle speeds, travel distances, machine clearances, and availability of auxiliary functions with those specified in the CL-file
6. Establish tool position and velocity components to maintain acceptable machine tool dynamic characteristics with respect to percent of overshoot and undershoot, acceleration and deceleration rates, dwell times for tape reading and control system settling, and so on
7. Prepare a machine control tape which has the correct character coding and recording format. It is usually recorded on paper tape or magnetic tape
8. Prepare an edited printout of the information on the machine control tape, together with warning, computational, and logical error diagnostics

10.2 SOME POSTPROCESSOR COMMANDS

Postprocessor commands are reserved words inserted in logically appropriate places of the part program to convey to the postprocessor some predetermined meaning concerning machine control action. The action of the postprocessor is to interpret the command and insert into the machine control tape the proper machine control equivalent or to establish some modal condition within the postprocessor to remain in effect during subsequent processing of the CL-file by the postprocessor. The general format of the postprocessor command is

Major word/minor section

where a single major word (command) is always used but a minor section may not be required. When the major word does not require a minor section it is used without the slash. The minor section consists of keywords, parameters, and modifiers used individually or in combination, depending on the scope of control provided by the major word. The major word, keywords, and modifiers are all reserved words of the APT system.

Except for TRACUT, INDEX, and COPY, which are classified as postprocessor commands, all other postprocessor commands and modifiers are not considered logically or computationally by the APT system but are checked against the reserved word list. The collection of postprocesor commands and modifiers included within a particular APT system is dependent on the requirements of the postprocessors that have been implemented for that system. Different postprocessors may require different postprocessor commands for identical machine control functions or the same postprocessor command may be interpreted differently by different postprocessors. Thus the meaning assigned to a particular postprocessor command can only be determined by reference to the appropriate machine postprocessor documents which explain their use. However, the meaning of certain postprocessor commands has received rather uniform recognition, thus permitting the number of postprocessor words in the reserved word list to be kept small.

The selection of postprocessor commands given in Table 10.1 is often found in an APT system. Their interpretation is generally recognized as given here, but this must be verified for a particular postprocessor before they are used. Nearly all postprocessor commands initiate a modal condition. Additional postprocessor words are contained in the reserved word list in the Appendix.

10.3 USING POSTPROCESSOR COMMANDS

There is normally only one input to the postprocessor, the CL-file. Therefore, postprocessor commands must be inserted in the APT part program in a logically correct sequence so as to properly control the machine.

When APT processing for a part program is finished, postprocessing (Section 4) automatically begins if the part program contains a MACHIN statement and if the postprocessor specified in the MACHIN statement has been incorporated into the APT system. Computer processing by the APT system and the postprocessor will appear to be continuous. The MACHIN statement may be placed anywhere in the part program but will usually be found either at the beginning after PARTNO or at the end before FINI. Nearly all other postprocessor commands must be combined with the APT motion commands such that the machine tool will be automatically controlled to produce the part.

Except for the feed rate specification, all postprocessor commands must be constructed and interpreted exactly as stated in the postprocessor documentation. The post-

TABLE 10.1   POSTPROCESSOR COMMANDS

AUXFUN/n, $\begin{Bmatrix} \text{ON} \\ \text{OFF} \end{Bmatrix}$	This modal command turns on or off the auxiliary function identified by the number n. The specific meaning of the auxiliary function depends on the controller being used. The number n may be a numeric literal or scalar variable.
CLEARP/XYPLAN,z	This modal command establishes the z-axis intercept of a plane parallel with the xy-plane. It is used in conjunction with the RETRCT statement as a tool clearance plane for noncutting tool movements. The intercept z is in the same units as used throughout the part program and may be a numeric literal or scalar variable.
COOLNT/$\begin{Bmatrix} \text{FLOOD} \\ \text{MIST} \\ \text{OFF} \end{Bmatrix}$	This modal command turns on the flood or mist coolant, or turns off the coolant.
CYCLE/$\begin{Bmatrix} \text{OFF} \\ \text{ON} \\ \text{NOMORE} \\ \text{BORE} \\ \text{DRILL} \\ \text{FACE} \\ \text{TAP} \end{Bmatrix}$, ...	This modal command initiates a canned cycle positioning operation. Modifiers following the slash select the canned cycle operation, specify a z-axis plunge distance, select a feed rate, and so on.
DELAY/t	This command causes the machine tool controller to delay any further motion of the machine tool for t seconds. The machine tool will dwell at its current location with zero motion.
END	This single word command signals the end of a logical section of the part program. It causes the controller, tape reader, and machine tool to stop, and it turns off all auxiliary functions. The startup procedure is reset and the postprocessor program is prepared for a new part program.
FEDRAT/f, $\begin{Bmatrix} \text{IPM} \\ \text{IPR} \end{Bmatrix}$	This modal command sets the machine tool feed rate to f inches per minute (IPM) or inches per revolution (IPR).
LOADTL/n	This command causes the spindle to move to the tool change position and reload the spindle with tool number n. A TOOLNO statement must have been programmed prior to this command.
MACHIN/$codeword,n$	This modal command identifies to the postprocessor the machine controller (*codeword*) to be used with a particular machine number (n). The result is that a machine control tape will be made from the CL-tape for that combination of controller and machine tool.

TABLE 10.1   POSTPROCESSOR COMMANDS (CONTINUED)

OPSTOP	This single word command means operator stop. It is a planned stop at the tool operator's option and commands the machine tool to stop only if the operator has pushed the OPSTOP button on the controller.
RAPID	This single word command initiates the rapid traverse mode of machine tool operation. It is usually in effect only during the one motion command that follows it.
RETRCT	This command causes the spindle to retract in the rapid mode to the plane given in the CLEARP statement.
REWIND/n	This command initiates the automatic rewinding of the machine control tape to the identification mark whose value is n. The mark must have been written by the TMARK command. The value of n may be given as a numeric literal or scalar variable.
SEQNO/n	This command assigns sequence number n to the next machine control tape block to be written. It is used to assign non-sequential numbers to the tape blocks but blocks are numbered sequentially from this point until another SEQNO command is encountered. The number n may be a numeric literal or a scalar variable.
$\text{SPINDL/}\left\{\begin{array}{c}\text{OFF}\\ rpm\end{array}\right\},\left\{\begin{array}{c}\text{CCLW}\\ \text{CLW}\end{array}\right\},\left\{\begin{array}{c}\text{HIGH}\\ \text{LOW}\\ n\end{array}\right\}$	This modal command sets the spindle speed to the rpm specified in a clockwise (CLW) or counter-clockwise (CCLW) direction and in the spindle speed range designated as HIGH, LOW, or by the integer n. The spindle stays on until a SPINDL/OFF, STOP, or END command is given.
STOP	This single word command stops the tape reader and machine tool in a normal manner but leaves the controller on. Subsequent tool movement is from the last tool location just prior to the STOP command. This command permits changing of clamps, inspection of the cutter, or some other operation requiring operator intervention.
TMARK/n	This command initiates the writing of an identification mark whose value is n. The mark is written on the machine control tape at the time this command is executed. The tape can later be rewound to this point by referencing the value of the mark in a REWIND command. The value of n may be given as a numeric literal or scalar variable.
TOOLNO/n,q	This modal command specifies a tool number to be used with a LOADTL command. It defines tool number n, which has an effective tool length of q inches (or a length consistent with the unit of measurement in use).

processor feed rate command is FEDRAT but an optional feed rate specification format may be used. This optional specification is constructed by adding to any APT general motion statement (FROM, OFFSET, GO, GOTO, GODLTA, GOFWD, GORGT, GOLFT, GOBACK, GOUP, or GODOWN) a feed rate specification (numeric literal or scalar variable) which is separated from the minor section of the motion command by a comma. An example of this optional format is

    GOFWD/L3,PAST,L6,8

where a feed rate of 8 in./min will be in effect until countermanded by another feed rate specification. The feed rate specification is modal regardless of the manner in which it is specified.

Postprocessor commands are not considered logically or computationally by the APT processor but do appear on the CL-printout in a location consistent with their appearance in the part program. This aids the part programmer in verifying the part program without running the machine tool.

The part program of Figure 10.2 applies to the part shown in Figure 10.3. It uses an assortment of postprocessor commands for a Sheldon vertical machining center. The CL-printout appears in Figure 10.4 and the postprocessor printout (PP-printout) is in Figure 10.5. The part geometry is devised to illustrate various APT features in a postprocessing context.

We assume that the blank is mounted on buttons on the $z = 0$ plane and we will ignore clamping for clarity. The PSIS statement (ISN 27) locates the part surface just below the bottom surface of the blank to leave clear cut edges. We command the tool from SETPT and ignore other tool adjustments for clarity. Loose tolerances are chosen to reduce the output for compactness of the figures.

The MACHIN statement selects the Sheldon postprocessor. Comments via PPRINT statements identify the three principal parts of the machining operation. These comments are carried forward to the CL-printout and PP-printout. The tool is specified and loaded by the automatic tool changer with the TOOLNO and LOADTL commands. The spindle is caused to rotate 2500 rpm in a clockwise direction in a flood coolant. The milling tool is taken clockwise around the periphery and counterclockwise around the hole. The feed rate is illustrated with the commands RAPID and FEDRAT and with trailing parameters on contouring and point-to-point commands (ISNs 32, 36, and 45).

The z-coordinate values for the pattern points were established above the blank with the ZSURF statement (ISN 11) via definitions for P1 and C2. P1 also allows for tool positioning prior to ramping into the hole (ISN 40) and for retraction from the hole (ISN 45). The hole pattern definition allows the CYCLE command to be invoked (ISN 54) and, because it is modal, it must also be disabled (ISN 56). The END statement stops the operation.

The CL-printout of Figure 10.4 shows the ISN numbers carried forward from the part program listing of Figure 10.2. The CL-file record numbers are listed under the heading REC. These records are read by the postprocessor as it processes the CL-file. The numbers in the CARD column are carried forward from the part program of Figure

```
ISN 00001 PARTNO  POSTPROCESSOR COMMANDS EXAMPLE.
ISN 00002         CLPRNT
ISN 00003         INTOL/.0005,.01
ISN 00004         OUTTOL/.0005,.01
ISN 00005         CUTTER/.5
ISN 00006 SETPT = POINT/-1,-.75,1.5
ISN 00007 LX    = LINE/XAXIS
ISN 00008 LY    = LINE/YAXIS
ISN 00009 L1    = LINE/XAXIS,2.5
ISN 00010 L2    = LINE/(POINT/5,0),ATANGL,-55
ISN 00011         ZSURF/.375
ISN 00012 P1    = POINT/2,1.25
ISN 00013 C1    = CIRCLE/CENTER,P1,RADIUS,.75
ISN 00014 L3    = LINE/P1,ATANGL,-20
ISN 00015 C2    = CIRCLE/CENTER,P1,RADIUS,1.125
ISN 00016 PAT1  = PATERN/ARC,C2,-135,CCLW,INCR,3.AT,90
ISN 00017 PL1   = PLANE/0,0,1,-.031
ISN 00018         MACHIN/SHELDN,0
ISN 00019 PPRINT***********************************
ISN 00020 PPRINT*         MILL PERIPHERY          *
ISN 00021 PPRINT*         LOAD 1/2 INCH MILL      *
ISN 00022 PPRINT*********************************} *******
ISN 00023         TOOLNO/6,0
ISN 00024         LOADTL/6
ISN 00025         SPINDL/2500,CLW
ISN 00026         COOLNT/FLOOD
ISN 00027          PSIS/PL1
ISN 00028          FROM/SETPT
ISN 00029          INDIRV/1,0,0
ISN 00030         RAPID
ISN 00031          GO/TO,LY,TO,PL1
ISN 00032          TLLFT,GOLFT/LY,PAST,L1,25
ISN 00033          GORGT/L1,PAST,L2
ISN 00034          GORGT/L2,PAST,LX
ISN 00035          GORGT/LX,PAST,LY
ISN 00036          GODLTA/0,0,.375,75
ISN 00037 PPRINT***********************************
ISN 00038 PPRINT*         MILL HOLE               *
ISN 00039 PPRINT***********************************
ISN 00040          GOTO/P1
ISN 00041         FEDRAT/25
ISN 00042          INDIRV/0,-1,0
ISN 00043          GO/TO,C1,TO,PL1
ISN 00044          TLLFT,GOLFT/C1,TO,3,INTOF,L3
ISN 00045          GOTO/P1,75
ISN 00046         SPINDL/OFF
ISN 00047 PPRINT***********************************
ISN 00048 PPRINT*         DRILL 4 HOLES           *
ISN 00049 PPRINT*         LOAD 11 (.191) DRILL    *
ISN 00050 PPRINT***********************************
ISN 00051         TOOLNO/13,0
ISN 00052         LOADTL/13
ISN 00053         SPINDL/1500,CLW
ISN 00054         CYCLE/DRILL,.5,20,IPM,0
ISN 00055          GOTO/PAT1
ISN 00056         CYCLE/NOMORE
ISN 00057         RAPID
ISN 00058          GOTO/SETPT
ISN 00059         COOLNT/OFF
ISN 00060         SPINDL/OFF
ISN 00061         END
ISN 00062         FINI
```

Figure 10.2   APT part program to produce the part of Figure 10.3.

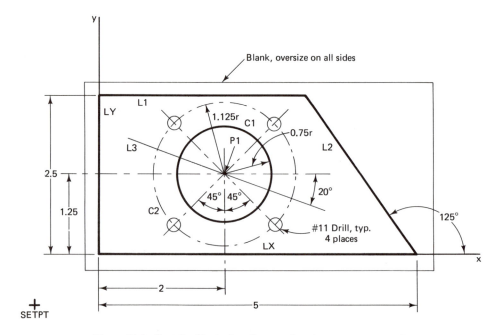

Figure 10.3 Part for illustrating the use of postprocessor commands.

10.2 (they were omitted in the figure). These numbers are added when the part program is created with the editor program.

The PP-printout, an edited version of the machine control tape, is produced by the postprocessor (see Figure 10.5). The sequence numbers of the machine control tape blocks appear in the left column. Some control over the generation of these numbers is usually provided by the SEQNO postprocessor command. The CARD numbers of Figure 10.4 are carried forward and listed in the column headed INCARD. The M and G machine code interpretation is left to the reader. However, inspection of the PP-printout reveals the drilling depth as generated by the CYCLE command.

Some statistical printout may be expected from a postprocessor. In this case, the cumulative cutting time (1.2 min) and the machine control tape length (6.8 ft) are computed.

```
ISN                                                                                        LABEL  REC M    CARD
0001 PARTNO/  POSTPROCESSOR COMMANDS EXAMPLE.                                                      00002 00000100
0003  INTOL/      0.00050000        0.01000000                                                     00004 00000300
0004  OUTTOL/     0.00050000        0.01000000                                                     00006 00000400
0005  CUTTER/     0.50000000                                                                       00008 00000500
0018  MACHIN/        SHELDN         0.0                                                            00010 00001800
0019  PPRINT/*******************************************                                           00012 00001900
0020  PPRINT/*        MILL PERIPHERY            *                                                  00014 00002000
0021  PPRINT/*        LOAD 1/2 INCH MILL        *                                                  00016 00002100
0022  PPRINT/*******************************************                                           00018 00002200
0023  TOOLNO/     6.00000000        0.0                                                            00020 00002300
0024  LOADTL/     6.00000000                                                                       00022 00002400
0025  SPINDL/  2500.00000000              CLW                                                      00024 00002500
0026  COOLNT/        FLOOD                                                                         00026 00002600
0027                                                                                               00027 00002700
0028   FROM/         SETPT                                                                         00029 00002800
               -1.00000000       -0.75000000        1.50000000
0029                                                                                               00030 00002900
0030  RAPID                                                                                        00032 00003000
0031   GOTO/         LY                                                                            00034 00003100
               -0.25000000       -0.75000000       -0.03100000
0032  FEDRAT/    25.00000000                                                                       00036 00003200
0032   GOTO/         LY                                                                            00037 00003200
               -0.25000000        2.75000000       -0.03100000
.0033  GOTO/         L1                                                                            00039 00003300
                3.37962292        2.75000000       -0.03100000
0034   GOTO/         L2                                                                            00041 00003400
                5.48024553       -0.25000000       -0.03100000
0035   GOTO/         LX                                                                            00043 00003500
               -0.25000000       -0.25000000       -0.03100000
0036  FEDRAT/    75.00000000                                                                       00045 00003600
0036   GOTO/                                                                                       00046 00003600
               -0.25000000       -0.25000000        0.34400000
0037  PPRINT/*******************************************                                           00048 00003700
0038  PPRINT/*        MILL HOLE                 *                                                  00050 00003800
0039  PPRINT/*******************************************                                           00052 00003900
0040   GOTO/         P1                                                                            00054 00004000
                2.00000000        1.25000000        0.37500000
0041  FEDRAT/    25.00000000                                                                       00056 00004100
0042                                                                                               00057 00004200
0043   GOTO/         C1                                                                            00059 00004300
                2.00000000        0.75000000       -0.03100000
0044  SURFACE        C1                          CIRCLE        DS(IMP-TO)                           00061 00004400
                2.00000000        1.25000000        0.0
                0.0               0.0               1.00000000        0.75000000
0044   GOTO/         C1                                                                            00062 00004400
                2.13530149        0.75827497       -0.03100000
                2.36781303        0.89671035       -0.03100000
                2.49677414        1.13460740       -0.03100000
                2.48587839        1.40499094       -0.03100000
                2.33819326        1.63173986       -0.03100000
                2.09529655        1.75101753       -0.03100000
                1.82557100        1.72924370       -0.03100000
                1.60493246        1.57254835       -0.03100000
                1.49355168        1.32504594       -0.03100000
                1.52816828        1.05641584       -0.03100000
                1.69361970        0.84228550       -0.03100000
                1.94532639        0.74293906       -0.03100000
                2.21242534        0.78634552       -0.03100000
                2.32139380        0.86697778       -0.03100000
0045  FEDRAT/    75.00000000                                                                       00064 00004500
0045   GOTO/         P1                                                                            00065 00004500
                2.00000000        1.25000000        0.37500000
0046  SPINDL/        OFF                                                                           00067 00004600
0047  PPRINT/*******************************************                                           00069 00004700
0048  PPRINT/*        DRILL 4 HOLES             *                                                  00071 00004800
0049  PPRINT/*        LOAD 11 (.191) DRILL      *                                                  00073 00004900
0050  PPRINT/*******************************************                                           00075 00005000
0051  TOOLNO/    13.00000000        0.0                                                            00077 00005100
0052  LOADTL/    13.00000000                                                                       00079 00005200
0053  SPINDL/  1300.00000000              CLW                                                      00081 00005300
0054  CYCLE /         DRILL        0.50000000       20.00000000            IPM                      00083 00005400
                0.0
0055   GOTO/                                                                                       00085 00005500
                1.20450487        0.45450487        0.37500000
0055   GOTO/                                                                                       00086 00005500
                2.79549513        0.45450487        0.37500000
0055   GOTO/                                                                                       00087 00005500
                2.79549513        2.04549513        0.37500000
0055   GOTO/                                                                                       00088 00005500
                1.20450487        2.04549513        0.37500000
0056  CYCLE /        NOMORE                                                                        00090 00005600
0057  RAPID                                                                                        00092 00005700
0058   GOTO/         SETPT                                                                         00094 00005800
               -1.00000000       -0.75000000        1.50000000
0059  COOLNT/        OFF                                                                           00096 00005900
0060  SPINDL/        OFF                                                                           00098 00006000
0061    END                                                                                        00100 00006100
0062 ***** FINI *****                                                                              00102 00006200
....END OF SECTION 3....
                                SECTION 3 ELAPSED CPU TIME IN MIN/SEC IS 0( 00/00.0206
```

Figure 10.4   CL-printout for Figure 10.2.

```
SHELDON 2040VC A. B.  POSTPROCESSOR COORDINATE LISTING     DATE: FEB-12-85   TIME: 07:45:56        PAGE  1
PARTNO         POSTPROCESSOR COMMANDS EXAMPLE.
*****************************************************************) ***********************************************
                                                                              INCARD        ELPSD C.T.
I$     FIRST REWIND STOP (IBM)
(TN,POSTPROC)
$      SECOND REWIND STOP
G17$G70$G90$
N0002                                                    S(120     M03    00000100          0.0
PPRINT    **************************************    00001900
PPRINT    *         MILL PERIPHERY          *       00002000
PPRINT    *       LOAD 1/2 INCH MILL        *       00002100
PPRINT    **************************************    00002200
N0003                                                    T0606 M06    00002400          0.0
N0004                                                    S:500     M03    00002500          0.0
N0005                                                              M08    00002600          0.0
N0006 G01 X-1.      Y-.75      Z1.5                 F300.                 00002800          0.0000
N0007     X-.25              Z-.031                                      00003100          0.0000
N0008 G94                                           F25.                 00003200          0.0000
N0009 G01           Y2.75                                               00003200          0.1400
N0010     X3.3796                                                        00003300          0.2852
N0011     X5.4502    Y-.25                                               00003400          0.4317
N0012     X-.25                                                          00003500          0.6609
N0013                                               F75.                 00003600          0.6609
N0014               Z.344                                               00003600          0.6659
PPRINT    **************************************    00003700
PPRINT    *         MILL HOLE               *       00003800
PPRINT    **************************************    00003900
N0015     X2.       Y1.25      Z.375                                    00004000          0.7020
N0016                                               F25.                 00004100          0.7020
N0017               Y.75       Z-.031                                   00004300          0.7277
N0018 G03 X2.5      Y1.25                 I0.0     J.5                   00004400          0.7602
N0019     X2.       Y1.75                 I.5      J0.0                  00004400          0.7927
N0020     X1.5      Y1.25                 I0.0     J.5                   00004400          0.8252
N0021     X2.       Y.75                  I.5      J0.0                  00004400          0.8576
N0022     X2.3214   Y.8669                I0.0     J.5                   00004400          0.8631
N0023                                               F75.                 00004500          0.8631
N0024 G01 X2.       Y1.25      Z.375                                    00004500          0.8716
N0025                                                              M05    00004600          0.8716
PPRINT    **************************************    00004700
PPRINT    *         DRILL 4 HOLES           *       00004800
PPRINT    *       LOAD 11 (.191) DRILL      *       00004900
PPRINT    **************************************    00005000
N0026                                                    T1313 M06    00005200          0.8716
N0027                                                    S:500     M03    00005300          0.8716
N0028 G81 X1.2045   Y.4545     Z-.125     R.375    F20.                 00005500          1.0972
N0029     X2.7955                                                        00005500          1.1292
N0030               Y2.0455                                             00005500          1.1611
N0031     X1.2045                                                        00005500          1.1931
N0032 G80                                                               00005600          1.1931
N0033 G01 X-1.      Y-.75      Z1.5                 F300.                 00005600          1.2056
N0034                                                              M09    00005900          1.2056
N0035                                                              M05    00006000          1.2056
N0036                                                              M30    00006100          1.2056

   0 POSTPROCESSOR DIAGNOSTICS IN ABOVE PROGRAM

CURRENT CUMULATIVE CUTTING TIME IS    1.2 MINUTES

CURRENT TAPE LENGTH IS    6.8 FEET
```

Figure 10.5   PP-printout for Figure 10.4.

PROBLEMS

Write APT part programs for the following problems with suitable postprocessor commands to effect part production. Preferably, the parts should be machined. Depending on the material from which the part is made, select suitable feed rates, coolant control, tool rotation direction, and so on. The material should be placed on spacing buttons and held with clamps placed in strategic locations. Provide for tool changes and operator messages. Use PPRINT during program checkout.

10.1. Rework Problem 9.1 for tool changes and clamp relocation.

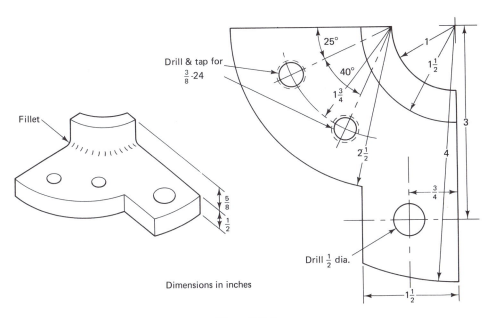

Figure P10.2

10.2. Figure P10.2 shows a part with tapped and untapped holes and with a fillet formed by the tool curvature. Select appropriate tool shapes and sizes and write the APT part program to produce the part. Allow for clamp relocation.

10.3. Figure P10.3 shows a part with three bosses. All top surfaces are to be machined, with boss fillets formed by the tool curvature. Select appropriate tool shapes and sizes and write the APT part program to produce the part. Allow for clamp relocation.

10.4. Rework Problem 9.5 for machining with postprocessor commands. Select a suitable drill size. If patterns are used, invoke the drilling operations with the CYCLE commands, if available.

10.5. Rework Problem 9.6 as described in Problem 10.4.

10.6. Rework Problem 9.7 as described in Problem 10.4.

10.7. Rework Problem 9.8 as described in Problem 10.4.

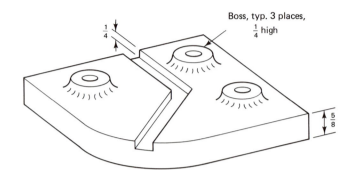

Dimensions in inches

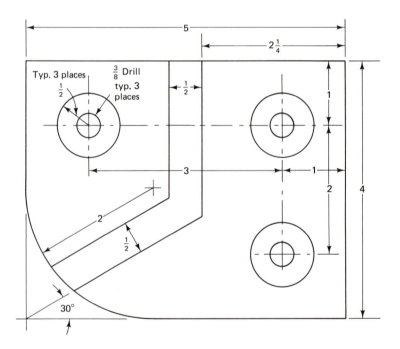

Figure P10.3

10.8. Rework Problem 7.18 with postprocessor commands to control the engraving, or drafting, tool.

10.9. Rework Problem 7.19 as described in Problem 10.8.

CHAPTER 11

Macros

We have seen that some APT features are for the convenience of the part programmer and contribute little else to the APT language. In this chapter we introduce the macro, which at first might seem to offer nothing new and so should be called a convenience feature. But we will show that it is an important concept, it should deserve special consideration, and we will encourage its use. The material of this chapter includes a discussion of the concept of a macro, its definition and its use, and of a convenience feature for performing a pocket operation.

11.1 THE CONCEPT OF A MACRO

Situations arise in which it would be very convenient to the part programmer if certain part programming statements were collected together and defined in a manner such that they could be referenced as a group with a symbolic name, the idea being that a group of statements be written once, assigned a symbolic name, and then referred to as a group by name whenever required elsewhere in the part program. This implies that certain statement sequences occur often in a part program, either identically or with great similarity. The APT feature that permits the above to be realized is known as the **macro** feature.

Macro is the term used for a group of part program statements that have been combined according to established rules so that they may be collectively referenced by name. The macro is thus a part program module which possesses the characteristic for independently defining a part program segment. The use of a macro is suggested by the following situations.

1. A sequence of identical statements must be executed more than once in a part program but does not occur in the repetitious manner that permits the use of the COPY feature.

2. A part program contains sequences of statements that differ slightly, perhaps only in the use of different variables, tool or tolerance specifications, tool-to-surface relationships, feed rates, some drive and check surfaces, and so on.

3. A sequence of similar statements is required in different part programs and would normally be separately prepared for each part program.

4. A part program is so complex and extensive that it would be advantageous to partition it into logical sections for individual preparation and checkout, perhaps even by different part programmers.

We note that the macro concept is not constrained by the limits of a single part program.

Almost any APT statement may be included within a macro, but macros are generally designed for motion commands, arithmetic statements, and similar executable statements. Thus, in order for a macro to perform a function, the flow of control in the part program must eventually transfer from the main part program to the macro and then back again to the main part program after the function of the macro has been completed. The linkage mechanism between the main part program and the macro is inherent in the APT statements that define the macro and which permit its use within the main part program.

Several part program structures that utilize macros are shown in skeleton form in Figure 11.1. Figure 11.1(a) shows a part program structure with one macro. The flow of control is from the main part program to the macro and return for a total of three times, with part program execution ending in the main part program. We say that the macro is *called* by the main part program. In this example, the main part program is the calling program while the macro is the called program. This structure obviously uses a macro to reduce the amount of part program preparation. Whenever the statements of a macro must be executed, flow of control passes to the macro and returns to a point in the main part program immediately following the statement that caused control to pass to the macro. Sequential main part program statement execution continues until the macro is again required. Thus the macro serves to contain a group of statements that appear to be inserted into the main part program and reexecuted as many times as needed to satisfy the part program needs.

Figure 11.1(b) shows a part program structure that includes two macros, each executed more than once. This structure implies that macros are used to reduce the amount of part program preparation. Some of the macros may have been prepared for another part program but found suitable for use in this part program also.

Figure 11.1(c) shows a part program structure that includes two levels of macros. The main part program calls macro A, which in turn calls macro B. This diagram shows nested macros. An APT processor generally imposes limits on the number of levels of macro nesting. Figure 11.1(d) shows a more complicated macro requirement.

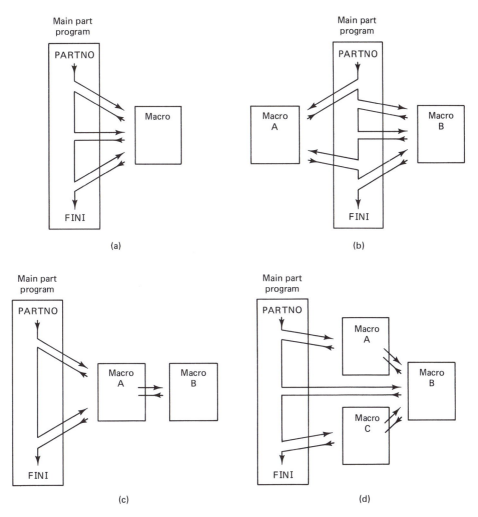

Figure 11.1   Skeleton part program structures with macros.

11.2 DEFINITION AND USE OF SIMPLE MACROS

A macro must first be defined before it can be called in a part program. The APT language includes statements that bound those contained within the macro, and thus become part of the macro definition, and which permit the calling of macros. The bounding statements include the macro identification statement and the macro termination statement. The CALL statement is required to use a macro.

The basic macro structure is as follows:

name = MACRO/*list*

$$\left. \begin{matrix} \vdots \\ \vdots \end{matrix} \right\} \quad \text{Body of the macro}$$

TERMAC

The first macro statement is the macro identification statement. The macro name is *name*, formed according to the usual APT rules for selecting symbolic names. The reserved word MACRO must always appear in the identification statement. The *list* following MACRO is described in Section 11.3 and may not be required for a given macro. When it is not required, the slash may be omitted. The APT statements contained within the macro body follow the macro identification statement. The last macro statement, the termination statement, must contain only the reserved word TERMAC with possibly a statement label. The logic of the macro statements must be such that the flow of control eventually leads to the TERMAC statement. This can be achieved through sequential flow into TERMAC or with a transfer statement or unconditional jump to the TERMAC statement. Each macro to be defined is constructed in exactly this same manner. The MACRO statement, the macro body within the bounding statements, and the TERMAC statement collectively form the macro.

To be useful, the macro must be called into execution. This is done with the CALL statement. Its format is as follows:

CALL/*name,list*

Name is the symbolic name used in the macro definition. Nothing is placed in *list* after *name* in the CALL statement unless macro *name* was defined with dummy variables in its list, in which case it is completed as described in Section 11.3.

The selection of statements for the macro body is governed by the following factors.

1. The function that the macro will perform. This may require motion commands, computing sequences, geometric definitions, postprocessor commands, or other APT statements.

2. The restrictions placed on the type of statements that may be included in a macro definition. LOOPST, LOOPND, FINI, MACRO, and TERMAC may not be contained within the macro. This implies that a macro definition may not include the definition of another macro. However, calls to other macros are permitted. Also, loops are permitted even though LOOPST and LOOPND are not.

3. The normal limitations placed on the use of a given APT statement. For example, since the macro is executed each time it is called, a named geometric definition will be doubly defined if the macro is executed twice unless the name is a subscripted symbol and its index value has not already been used for that geometric definition or it is being defined with the CANON feature. Similarly, the maximum number of unnamed geometric definitions allowed may be exceeded if the macro contains such geometric definitions and if it is called sufficiently often.

Whether or not a macro should be incorporated within a given part program is determined by the requirements of the part to be programmed and if the use of a macro is suggested as outlined in Section 11.1. Inasmuch as a macro is executed each time it is called, a new set of cut vectors will be computed by processing the motion commands each time. This differs from COPY, which applies a transformation to cut vectors that have already been computed. The macro is not to be considered a substitute for COPY. COPY is preferred over a macro whenever it accomplishes the same objective.

The following examples illustrate the preparation and use of **simple macros** (those which do not contain dummy variables) in part programs. Example 11.1 uses a macro to simplify a part program, while Example 11.2 illustrates the application of a computing macro and introduces the notion of global elements.

Example 11.1: Using a Macro to Simplify a Part Program

Consider the point-to-point problem of Example 9.1 where a pattern of holes was drilled. There a series of point-to-point commands in conjunction with the COPY feature positioned the drill and drilled the holes. If we omit the drilling commands from the part program motion statements but indicate their need, we obtain the following series of commands for lines 210 through 420 of that part program:

```
FROM/SETPT
INDEX/10
    GOTO/P1
        drill hole
    GOTO/P2
        drill hole
COPY/10,XYROT,60,5
INDEX/20
    GOTO/P3
        drill hole
    GOTO/P4
        drill hole
    GOTO/P5
        drill hole
COPY/20,MODIFY,(MATRIX/MIRROR,YZPLAN),1
COPY/20,XYROT,180,1
GOTO/SETPT
```

Observe that drilling operations occur at five locations in the part program. Since these operations are identical and occur repeatedly, we will replace 10 GODLTA statements with five CALL statements plus the statements necessary to define the macro. Perhaps a saving in statement preparation will result but other advantages to using a macro must be considered. Among these are (1) part program readability enhancement since fewer motion statements tend to reduce the cluttered appearance of the part program, (2) increased understanding of a part program if the macro name is suggestive of the macro function, and (3) possibly reduced part program debugging time when a macro is used repeatedly since fewer part program statements must be prepared and their effect individually verified.

In this example the macro will not save many statements but we will proceed with its preparation anyway to illustrate the use of a simple macro. The arguments above benefited from hindsight since the problem has already been solved in Example 9.1.

Since a macro is executed each time it is called, we must be sure that its statements are ordered properly for cut vector computation. In this example we require only that a hole be drilled and that the drill then be retracted. The sequence of commands is obvious. We will use different feed rates for drilling (2.5 in./min) and drill retraction (75 in./min). Postprocessor commands are allowed in macros. Thus our macro is as follows:

```
DRIL = MACRO
        GODLTA/0,0, − 1,2.5
        GODLTA/0,0,1,75
      TERMAC
```

The macro name DRIL is suggestive of the application and the flow of control is sequential into TERMAC. DRILL is a reserved word and so cannot be used as a macro name. The part program motion statements now are as follows:

```
FROM/SETPT
INDEX/10
    GOTO/P1
    CALL/DRIL
    GOTO/P2
    CALL/DRIL
COPY/10,XYROT,60,5
INDEX/20
    GOTO/P3
    CALL/DRIL
    GOTO/P4
    CALL/DRIL
    GOTO/P5
    CALL/DRIL
COPY/20,MODIFY,(MATRIX/MIRROR,YZPLAN),1
COPY/20,XYROT,180,1
GOTO/SETPT
```

The complete part program must be structured so that the macro is defined before it is called. In this example it would be placed just before line 210, the FROM/SETPT statement. Point-to-point motion commands were used in the macro but contouring commands also are permitted within a macro, as will be demonstrated in Section 11.3.

Example 11.2: Adapting a Computing Sequence to Macro Form

The part program of Example 4.5 computed within a loop the engraving tick marks for a logarithmically calibrated dial. We now modify that part program to compute tick marks within a macro. Aside from improving program readability, this macro illustrates the notion of global variables. Part program logic follows the flowchart of Figure 4.12. The statements beginning with line 210 of the part program given for Example 4.5 are replaced as follows:

```
$$= = = = = = = = = = = = = = = = = = = = = = = = = =        210
LXMAC = MACRO                                               220
```

	J = J + 1	230
	IF (J − 1) LMK1, LMK1, LMK2	240
LMK1)	NM = LOGF(M)	250
	JUMPTO/LMK3	260
LMK2)	N = N + .5	270
	ANG = − LOGF(N)/NM*THETA − 90	280
	LX = LINE/LX,CANON,COSF(ANG),SINF(ANG)	290
LMK3)	TERMAC	300
$$= =		310
	K = 1	320
	J = 0	330
	N = 1	340
	CALL/LXMAC	350
	JEND = 2*(M − 1)	360
	FROM/SETPT	370
	INDIRP/PX	380
	GO/ON,COUT	390
	TLON,GOBACK/LX,ON,CLONG	400
	GOBACK/LX,ON,COUT	410
LA1)	LOOPST	420
	K = − K	430
	CALL/LXMAC	440
	GORGT/COUT,ON,LX	450
	IF (K) LA2, LA2, LA3	460
LA2)	GORGT/LX,ON,CSHORT	470
	JUMPTO/LA4	480
LA3)	GORGT/LX,ON,CLONG	490
LA4)	GOBACK/LX,ON,COUT	500
	IF (JEND − J) LA5, LA1, LA1	510
LA5)	LOOPND	520
	GOTO/SETPT	530

The macro must be defined before it is referenced. Thus it is placed before its first call in line 350. All computations associated with the logarithm are contained within the macro. Because line 250 is executed only once, J must be incremented before the IF statement of line 240 to thereafter bypass line 250. Consequently, J is initialized to zero in line 330. The macro illustrates that symbol names as well as scalar variables may be **global** (available to both the macro and its calling program by name). Statement labels are **local** (known only to) to the macro and may be duplicated elsewhere without conflict. However, this practice is discouraged to enhance readability.

11.3 MACRO DEFINITION WITH DUMMY VARIABLES

We have shown above that the macro is useful in avoiding the preparation of repeatedly occurring part program statements. Because many part program segments are similar, perhaps differing only in the choice of some motion commands, symbolic names, scalar variables, or other elements, the utility of the macro would be enhanced with the provision for changing elements within the macro to suit some part program requirement. We are saying that each time the macro is called, a general-purpose macro could be made specific by assigning values to elements declared as variable. Since a macro is not executed at the time the definition is processed by the APT system but only when it is called, it becomes possible to substitute for macro elements later when the proper values will be available. This extension of the macro concept over that presented in Section 11.2 requires the declaration of dummy variables in the MACRO statement and their value assignment in the CALL statement.

Dummy variables are the variables in *list* whose symbols appear after the slash in the MACRO statement. These dummy variable symbols must also appear in statements contained within the macro and are the means by which changes are effected within the macro. Each dummy variable must have a value when the macro is executed. This may be a **normal value** (a value assigned to the dummy variable in *list* as part of its declaration in the MACRO statement) or an **assigned value** (a value assigned to the dummy variable in *list* in the CALL statement). When a macro is called, the current value of each dummy variable, either assigned or normal, is used to replace that dummy variable symbol in the statements contained within the macro and then the macro is executed. Thus, by changing the values of the dummy variables, as many different, but similar, part program segments may be created as desired.

The format of the MACRO statement containing dummy variables without normal values is

$$name = \text{MACRO}/d_1, d_2, \ldots$$

where the d_i are dummy variable symbols that also appear in statements within the macro. Dummy variable names are formed according to the usual APT rules for constructing nonsubscripted symbolic names. The format of the MACRO statement containing only dummy variables with normal values is

$$name = \text{MACRO}/d_1 = n_1, d_2 = n_2, \ldots$$

where the d_i are dummy variable symbols and the n_i are the normal values assigned to the respective dummy variables. The following are allowable dummy variable normal values.

1. Any reserved word except LOOPST, LOOPND, RESERV, FINI, MACRO, TERMAC, or CALL
2. The symbolic name of a scalar variable, subscripted or nonsubscripted, or the name of an array
3. The symbolic name of a geometric entity
4. The symbolic name of another macro
5. Numeric literals

Dummy variables with and without normal values may appear in the same MACRO statement. There may be a limit on the number of dummy variables permitted in a MACRO statement by a given APT processor.

The format of the CALL statement that references a macro with dummy variables is

$$\text{CALL}/name, d_1 = v_1, d_2 = v_2, \ldots$$

where the d_i are dummy variable symbols appearing in the MACRO statement referenced by *name* and the v_i are the values to be assigned to the respective dummy variables. Allowable assigned values are the same as permitted for normal values. All dummy variables that do not have normal values assigned in the MACRO statement must have values assigned in the CALL statement. Dummy variables with normal values may have new values assigned in the CALL statement at the option of the part programmer. Thus the format of the CALL statement is the same whether or not dummy variables have normal values assigned in the MACRO statement. When a macro has more than one dummy variable, the order in which the variables are assigned values in the CALL statement does not have to conform with the order they are given in the MACRO statement. The values assigned to dummy variables in a CALL statement apply only during execution of that call and in no other way affect the macro definition or normal value assignments. For all practical purposes, the call to a macro has the same effect as if the statements of the macro were written such that all dummy variable symbols were replaced by their assigned values in the CALL statement or by their normal values if they are not being reassigned.

The notion of a global element as introduced in Example 11.2 also applies to macros with dummy variables. As a matter of review, a global element is a symbolically named APT entity that may be referenced by the same name in both the calling program and in macros for the calling program. A local element is an APT entity that may be referenced by its symbolic name only within the macro in which it appears as local. For example, a line named LA that is local in a macro cannot be referenced by the name LA in the calling program or in any other macro. A macro element is automatically declared as local if its name appears as a dummy variable symbol in a MACRO statement. Macro elements are often made dummy variables with normal values for no other purpose then to restrict them for local use. For scalar variables, the normal value functions as an initial value for looping purposes or it protects the scalar variable from modification while referencing another scalar variable of the same name elsewhere in the part program. The notion of local and global elements is by no means unique to the APT language. They are usually found in other computer programming languages in conjunction with programmer-defined functions.

The following examples illustrate the preparation and use of macros with dummy variables. Example 11.3 contains a dummy variable without a normal value and shows how further simplification is possible for the hole drilling part program reintroduced in Example 11.1. Example 11.4 shows further use of dummy variables, while Example 11.5

is much more extensive and shows how macros can be constructed for general-purpose use and, when combined with TRACUT, can be used in a part program whose main program segment consists almost entirely of calls to macros.

Example 11.3: Hole Drilling with a Macro

The hole drilling operation for Example 9.1 has been simplified with simple macros as shown in Example 11.1. The resulting macro may be further modified to include the point-to-point motion command to direct the drill to the hole. This requires referencing the symbolic name of the hole to which the drill is being directed. Dummy variables simplify the construction of the macro for such references and reduce the main part program segment to a series of macro calls. The macro and motion commands would now appear as follows:

```
$$ = = = = = = = = = = = = = = = = = = = = = = = = = = = = = = = =
DRIL = MACRO/PT
        GOTO/PT
        GODLTA/0,0, − 1,2.5
        GODLTA/0,0,1,75
        TERMAC
$$ = = = = = = = = = = = = = = = = = = = = = = = = = = = = = = = =
        FROM/SETPT
        INDEX/10
            CALL/DRIL,PT = P1
            CALL/DRIL,PT = P2
        COPY/10,XYROT,60,5
        INDEX/20
            CALL/DRIL,PT = P3
            CALL/DRIL,PT = P4
            CALL/DRIL,PT = P5
        COPY/20,MODIFY,(MATRIX/MIRROR,YZPLAN),1
        COPY/20,XYROT,180,1
        GOTO/SETPT
```

The change places the GOTO command within the macro. The dummy variable PT must be assigned a value when the macro is called. This is done in the CALL statement by following the symbol PT with an equals sign and the symbolic name of the point to which the drill is being directed.

Example 11.4: The Dial Calibration Program Revisited

The macro for computing calibration tick marks for the logarithmically calibrated dial in Example 11.2 depended on global variables for transferring entity values to and from the macro. We now rewrite that part program for simplicity in understanding and for generality of application. Recall that once the first long tick mark is made, alternate short and long tick marks are made as calibration marks are engraved. This suggests that two consecutive calls to a macro should be sufficient to produce the desired tick mark pattern, provided that proper check surfaces are identified. These check surfaces can be made dummy variables for contouring commands in the macro. With these changes, the part program segment of Example 11.2 is as follows:

```
$$= = = = = = = = = = = = = = = = = = = = = = = = = =    210
LXMAC = MACRO/CI,CO,THETA = 180                           220
          N = N + .5                                      230
          ANG = −LOGF(N)/NM*THETA − 90                    240
          LX = LINE/LX,CANON,COSF(ANG),SINF(ANG)          250
          GORGT/CO,ON,LX                                  260
          GORGT/LX,ON,CI                                  270
          GOBACK/LX,ON,CO                                 280
          TERMAC                                          290
$$= = = = = = = = = = = = = = = = = = = = = = = = = =    300
          M = 5                                           310
          J = 1                                           320
          N = 1                                           330
          NM = LOGF(M)                                    340
          FROM/SETPT                                      350
          INDIRP/PX                                       360
          GO/ON,COUT                                      370
          TLON,GOBACK/LX,ON,CLONG                         380
          GOBACK/LX,ON,COUT                               390
LA1)      LOOPST                                          400
          J = J + 1                                       410
          CALL/LXMAC,CI = CSHORT,CO = COUT                420
          CALL/LXMAC,CI = CLONG,CO = COUT                 430
          IF (M − J) LA5, LA5, LA1                        440
LA5)      LOOPND                                          450
          GOTO/SETPT                                      460
```

The dummy variables in the contouring commands of the macro are assigned values in lines 420 and 430. This allows flexibility in contouring without additional looping and testing. Economy of statements is also achieved by computing the constant LOGF(M) outside the macro (thus eliminating the IF statement of the previous solution) and by testing against M in the IF statement (thus eliminating the computation of JEND). The zero branching label also had to be changed. Variable M was placed below the macro to emphasize its relationship to the computation. The macro is made somewhat flexible by assigning a normal value to THETA, the angular distribution of the calibration marks. The distribution could be changed by assigning THETA a different value in the macro CALL statements. Of course, THETA is no longer assigned a value near the beginning of the part program as for the previous solutions.

Example 11.5: A Modular Part Program

This example APT program will direct a numerically controlled drafting machine to produce the schematic drawing of the low-pass π filter of Figure 11.2. Computer graphics equipment rather than an NC machine would now be used for this purpose, but the example easily demonstrates the preparation of a modular program so will be used here anyway.

The schematic contains the electrical symbols for an inductor, capacitor, terminal, and connection point connected together with straight lines that represent electrical wiring. We will write separate macros for an inductor, capacitor, and circle (which will be used for

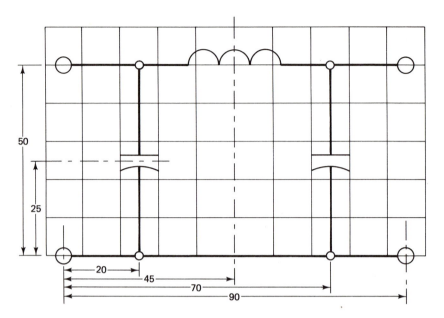

Figure 11.2   Schematic drawing of the low-pass π filter for Example 11.5.

both the terminal and connection points, of radius 2 and 1 unit, respectively) and translate each symbol to its proper location with the TRACUT feature. Therefore, translation point coordinates must be passed to each symbol macro. The unit of length is an arbitrarily chosen integer that could represent graph paper grid mark spacing (0.1 in. or 1 mm) for convenient hand plotting for verification purposes.

We postulate that the drafting machine requires x and y values for coordinate drive signals. The z values are ignored. Instead, pen-up and pen-down motion will be effected with the PENUP and PENDWN postprocessor commands, respectively.

Before writing any macro, it is necessary that we establish the philosophy under which it will be used. The macros defined in the other examples were sufficiently simple that we ignored this obligation. The drafting task we face imposes an almost consistent set of conditions for using each macro. We have already mentioned the need for translating point coordinates. Next, we will assume that redundant pen up/down commands will either be ignored by the postprocessor or will require negligible time to perform. This assumption simplifies the logic of our part program. We will require the pen to be down when we call the inductor and capacitor macros and it is to be down when we exit the macros. Similar requirements for the terminal and connection point macro will be discussed below.

The geometry of the inductor, capacitor, terminal, and connection point appear in parts (a) through (c) of Figure 11.3. The translation reference point for each symbol is noted in the figure. A listing of the macros is in Figure 11.4, while the main program listing is in Figure 11.5.

The leftmost semicircle of the inductor is defined as an unnamed nested circle in the contouring command of the inductor macro in Figure 11.4. The cut vectors from this command are copied twice under translation to form the inductor. The pen must be down when the

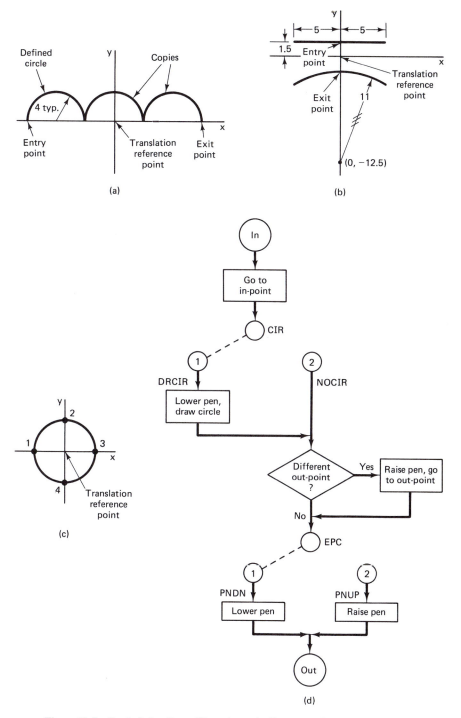

Figure 11.3 Symbols for the π filter schematic diagram and the flowchart for drawing the terminal and connection point circles.

```
§§======    INDUCTOR MACRO   ================================
INDCTR  = MACRO/XT,YT
            TRACUT/(MATRIX/TRANSL,XT,YT,0)      §§ MAKE THREE SEMI-CIRCLES
              GOTO/-12,0,0                      §§ FROM LEFT-TO-RIGHT,
              INDIRV/0,1,0                       §§ LAST TWO WITH COPY.
              INDEX/5
                TLON,GOFWD/(CIRCLE/-8,0,4),TO,(LINE/XAXIS)
              COPY/5,TRANSL,8,0,0,2
            TRACUT/NOMORE
            TERMAC
§§
§§======    CAPACITOR MACRO   ================================
CAPCTR  = MACRO/XT,YT
            TRACUT/(MATRIX/TRANSL,XT,YT,0)
              GOTO/0,1.5,0                  §§ MAKE TOP PLATE,
              GOTO/-5,1.5,0
              GOTO/5,1.5,0
              PENUP
              GOTO/(POINT/YLARGE,INTOF,(LINE/YAXIS,5),(CIRCLE/0,-12.5,11)
              PENDWN
              INDIRV/-1,0,0                §§ NOW MAKE BOTTOM PLATE.
              TLON,GOFWD/(CIRCLE/0,-12.5,11),ON,(LINE/YAXIS,-5)
              PENUP
              GOTO/0,-1.5,0
              PENDWN
            TRACUT/NOMORE
            TERMAC
§§
§§======    GO-TO-POINT MACRO   ======================
GOTOPT  = MACRO/P,R,A,B
            ANG = -(P - 1)*90 + 180      §§    2      POINT NUMBERING
            A = R*COSF(ANG)              §§  1   3    ON CIRCLE WITH
            B = R*SINF(ANG)              §§    4      RADIUS R.
            GOTO/A,B,0
            TERMAC
§§
§§======    CONNECTOR MACRO   ========================
CONN    = MACRO/XT,YT,PIN,POUT,EPC=1,CIR=1,RAD=1
            TRACUT/(MATRIX/TRANSL,XT,YT,0)
              CALL/GOTOPT,P=PIN,R=RAD,A=PXA,B=PYA     §§ POSITION ON
              JUMPTO/(DRCIR,NOCIR), CIR               §§ ENTRY POINT.
DRCIR)        PENDWN
              INDIRV/PYA,PXA,0                        §§ DRAW CIRCLE.
              TLON,GOFWD/(CIRCLE/0,0,RAD),ON,2,INTOF,(LINE/0,0,PXA,PYA)
NOCIR)        IF (PIN - POUT) DIFPT, SAMPT, DIFPT
DIFPT)        PENUP                                  §§ POSITION ON
              CALL/GOTOPT,P=POUT,R=RAD,A=PXA,B=PYA    §§ EXIT POINT.
SAMPT)        JUMPTO/(PNDN,PNUP), EPC
PNDN)         PENDWN
              JUMPTO/OUTC
PNUP)         PENUP
OUTC)       TRACUT/NOMORE
            TERMAC
§§==========================================================
```

Figure 11.4   Macros for the symbols of the schematic diagram of the π filter.

macro is called; it remains down while the macro is executed. The inductor will always be drawn beginning with the leftmost end of the figure.

The top plate of the capacitor is drawn first, as shown in Figure 11.4. It is formed with point-to-point commands while the lower plate is formed while contouring along the circular arc of an unnamed nested circle defined in the contouring command. Several PENUP-PENDWN commands are required to avoid drawing unwanted lines. The pen must be down when the macro is called; it will be down when the macro is exited.

```
SETPT    = POINT/-1,-1
         DRCIR = 1
         NOCIR = 2
         PNDN = 1
         PNUP = 2
         PENUP
         FROM/SETPT
PPRINT*****************************************************************
PPRINT*    DRAW TERMINALS AND CONNECTIONS OF LOWER LINE            *
PPRINT*****************************************************************
         CALL/CONN,XT=0,YT=0,PIN=3,POUT=3,RAD=2
         CALL/CONN,XT=20,YT=0,PIN=1,POUT=3
         CALL/CONN,XT=70,YT=0,PIN=1,POUT=3
         CALL/CONN,XT=90,YT=0,PIN=1,POUT=1,RAD=2,EPC=PNUP
PPRINT*****************************************************************
PPRINT*    DRAW TERM, CONN, & INDUCTOR OF UPPER LINE               *
PPRINT*****************************************************************
         CALL/CONN,XT=0,YT=50,PIN=3,POUT=3,RAD=2
         CALL/CONN,XT=20,YT=50,PIN=1,POUT=3
         CALL/INDCTR,XT=45,YT=50
         CALL/CONN,XT=70,YT=50,PIN=1,POUT=3
         CALL/CONN,XT=90,YT=50,PIN=1,POUT=1,RAD=2,EPC=PNUP
PPRINT*****************************************************************
PPRINT*    DRAW LEFT CAPACITOR                                     *
PPRINT*****************************************************************
         CALL/CONN,XT=20,YT=50,PIN=4,POUT=4,EPC=PNDN,CIR=NOCIR
         CALL/CAPCTR,XT=20,YT=25
         CALL/CONN,XT=20,YT=0,PIN=2,POUT=2,EPC=PNUP,CIR=NOCIR
PPRINT*****************************************************************
PPRINT*    DRAW RIGHT CAPACITOR                                    *
PPRINT*****************************************************************
         CALL/CONN,XT=70,YT=50,PIN=4,POUT=4,EPC=PNDN,CIR=NOCIR
         CALL/CAPCTR,XT=70,YT=25
         CALL/CONN,XT=70,YT=0,PIN=2,POUT=2,EPC=PNUP,CIR=NOCIR
         GOTO/SETPT
```

Figure 11.5   Main program for drawing the π filter schematic diagram.

The macro for the terminal and connection point circles is more difficult to prepare because (1) the pen may be either up or down when the macro is called; (2) it may be required to draw a large or a small circle; (3) the pen may be either up or down when the macro is exited; (4) it may be desired to position the pen at any of the four points shown in Figure 11.3(c) for either entry to or exit from the macro; and (5) it may be desired not to draw a circle but merely to position the pen on a point. This assortment of requirements can be satisfied if when we call the macro, we pass to it the following information.

1. Translation point *x-y* coordinates
2. Entry point
3. Exit point
4. Draw circle flag (yes/no)
5. Exit pen condition (up/down)
6. Circle radius

The drafting requirements will be met if we write the macro called CONN according to the logic of the flowchart in Figure 11.3(d). The flow of control is dependent on the setting of the software switches (symbolically labeled circles) whose normal connections are shown by the dashed lines. It is useful to prepare a separate macro to position the pen on the entry and exit points. This macro, named GOTOPT in Figure 11.4, requires for variable

P the code number of the point (1, 2, 3, or 4) and the variable R, the radius of the circle. It computes the *x-y* coordinates of the point and, in addition to positioning the pen on that point, it returns these coordinate values via variables A and B for use in establishing a vector direction before issuing the circle contouring command in macro CONN shown in Figure 11.4. The entry and exit points are passed to macro CONN via variables PIN and POUT, respectively, when GOTOPT is called from CONN. Since CONN calls GOTOPT, the macro structure resembles that shown in Figure 11.1(c).

Normal values are assigned to dummy variables of macro CONN as follows. The macro will be called more often to draw a circle than to position the pen without drawing. Therefore, variable CIR will be assigned the normal value of 1 so that the program path will take us to the draw circle code. Because connection points normally appear more frequently in a schematic diagram than terminals, the circle radius variable RAD will be assigned the normal value of 1 unit. The pen is more often in a down position after the circle is drawn, so the exit pen condition variable EPC is assigned the normal value of 1 so that the program path will take us to the lower-pen code.

The main program of Figure 11.5 consists principally of macro calls. The pen is raised at SETPT when the part program begins and is returned to SETPT in a raised position at the program end. Four scalar variables are defined at the beginning to allow mnemonics instead of code numbers to be used for the software switch selections of macro CONN. The PPRINT comments explain program action.

11.4 SUMMARY OF IMPORTANT MACRO CHARACTERISTICS AND LIMITATIONS

The macro concept provides a means for assembling statements to collectively exhibit many of the same properties prescribed for the main part program. Each macro may have an individual character but under restrictions not much different from that of the main part program. Since the macro concept is so important, its numerous characteristics and limitations should be carefully distinguished from those of the main part program. The following are important considerations for macro construction.

1. A macro must be defined in a part program before its first reference in a CALL statement.
2. Macro elements that vary with each call to the macro can be represented as macro dummy variables in the MACRO statement. All dummy variables must have a value assigned in the CALL statement or assume a normal value. The dummy variables are local in extent, while all others are global. Dummy variables may not be subscripted symbols.
3. There may be a limit on the number of dummy variables permitted in a MACRO statement by a given APT processor.
4. Geometric definitions within a macro are processed each time the macro is called. To avoid multiply defined symbols, the definitions should (a) have subscripted symbols in which the index differs with each call to the macro, (b) be nested and

unnamed, (c) be defined with the CANON feature, or (d) have dummy variable names in which the symbol differs with each call to the macro.

5. A macro may call another macro, the process being referred to as nesting. There is usually an upper limit on the number of nesting levels permitted by a given APT processor.

6. A macro definition may not be included as part of another macro. Thus MACRO and TERMAC statements must appear only once for a given macro and then only as the identification and termination statements for that macro.

7. Looping and branching may be incorporated within a macro but the LOOPST and LOOPND bounding statements must not be used since these reserved words are among those prohibited within a macro.

8. A RESERV statement must not be included within the macro definition sequence since this would have the effect of reserving a different memory space for the variables each time the macro is called.

9. The FINI statement signifies to the APT system that no more statements for the part program follow. It does not signify the end of a macro and its use for this purpose would prematurely terminate the processing of a part program. Thus its use in a macro is prohibited.

11.5 THE POCKET STATEMENT

The POCKET statement is an APT convenience feature for performing a pocketing operation, that is, removing material from a workpiece so as to form a recess or pocket. Although this feature has limited application, its use can relieve the part programmer of tedious part programming effort. It is included here to suggest what might have to be done were it not available and a macro were to be written to perform this function. Macros are not always as simple as the preceding examples imply.

11.5.1 Description and Use of the POCKET Feature

A frequent machining requirement is the removal of material from the workpiece so as to form a recess or pocket. Although a given pocket may assume only one of an infinite number of possible geometric shapes, there is a recurring requirement for pockets whose sides are straight lines and whose bottom is planar. In APT this restricted class of pocket geometry is limited to that of a convex figure (the interior angle at any vertex must not exceed 180°) composed of no more than 20 sides, each constructed of straight-line segments only. In this chapter we will discuss the POCKET feature as though the tool axis is perpendicular to the xy-plane, but the POCKET feature does not limit tool alignment in this manner. Figure 11.6 shows APT permitted pocket geometries and others that in some way violate POCKET restrictions. A given pocket operation requires a single planar bottom that may be canted. Figure 11.6(e) shows a pocket with a canted bottom. The

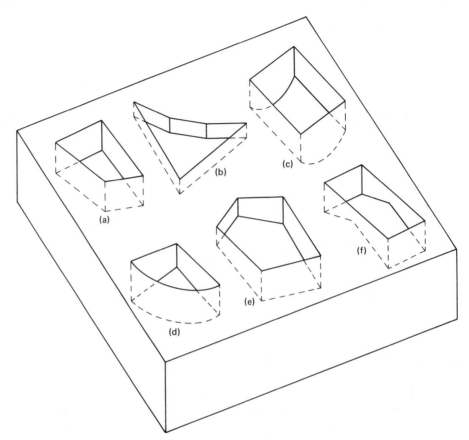

Figure 11.6 Pocket geometries that in APT are: (a) permitted because it is convex and has a planar bottom; (b) not permitted because it is concave; (c) not permitted because it does not have a planar bottom; (d) not permitted because its sides are not all straight-line segments; (e) permitted because it is convex and has a planar bottom, although sloping; (f) not permitted because it does not have a single planar bottom.

pocket of Figure 11.6(f) does not have a single planar bottom and thus cannot be specified in a single pocket operation, although it may be machined with two pocket operations, one for each of the canted bottoms. The pocket of Figure 11.6(b) is concave and thus cannot be specified in a single pocket operation, although it may be machined with three pocket operations by dividing the pocket into three triangular pockets each of which would be convex. Of course, a round tool will produce pockets with round corners, not as shown in the figure.

Cut vectors for a pocket operation are computed by executing the POCKET command formed as described in Section 11.5.2. The pocket is described by defining the corners as points with z-coordinate values on the pocket bottom plane. The first cut vector produced by the pocket command will place the tool on the pocket bottom near the center of the pocket. Since APT assumes that the tool moves in a straight line from one cut vector endpoint to the next, care must be taken to position the tool over the pocket prior to executing the POCKET command so that the tool does not ramp through the pocket wall on its way to the first cut vector endpoint within the pocket. Entry into the pocket will be at a feed rate specified in the POCKET command. Subsequent motion within the pocket results from the cut vectors produced by the POCKET command. The tool is commanded from the first cut vector position within the pocket in ever-increasing concentric polygons, in a clockwise or counterclockwise direction as desired, until a final pass produces the periphery of the pocket. Upon completion of the periphery the pocket operation is finished and part program control passes to the next executable statement following the POCKET command. Upon completion of the pocket operation the tool is within the pocket at one of the corners. The tool must be withdrawn from the pocket with statements following the POCKET command. Care must be taken to avoid ramping through the pocket wall as the tool is withdrawn. It will be necessary to reorient tool direction for contouring commands that follow the pocket operation. If further machining is to be performed within the pocket it will also be necessary to reestablish the part surface since that used for the pocket operation does not automatically become the part surface for subsequent contouring commands.

Figure 11.7 shows tool movement before, during, and after a pocket operation. Although the actual pocket cutting is performed from near the pocket center toward the pocket boundary, computation of the cut vectors is performed from the pocket boundary toward the pocket center. As computation proceeds the cutting polygons become smaller, thus leading to the notion of collapsing polygons as mentioned in Section 11.5.2 during the discussion of the POCKET statement parameters.

11.5.2 Definition Format

The pocket operation is invoked with the following statement:

POCKET/*Re,c,f,F1,F2,F3,Of,p,pt1,pt2,* ...

The parameters of the minor section must be specified according to the following interpretation (numeric values can be supplied as numeric literals or scalar variables).

Re = Effective Cutter Radius

When the pocket part surface (pocket bottom plane) is parallel with the *xy*-plane, and when a flat-end tool is used for pocketing, all points cut by the end of the tool are within tolerance of the pocket part surface. For this tool and pocket bottom geometry the effective cutter radius *Re* is equal to the actual cutter radius as shown in Figure 11.8(a). Actually, *Re* would be selected slightly less than D/2 (where D is the tool diameter of the CUTTER statement in effect while pocketing) for overlap to achieve a smooth cut and to allow for tolerances and tool wear. When the pocket part surface is inclined to the *xy*-plane a flat-end tool cuts as shown in Figure 11.8(b). Now we must overlap the cuts so as never to exceed an allowable scallop height. The scallop height is measured

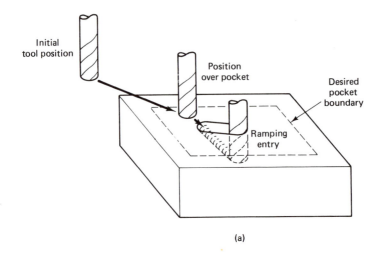

(a)

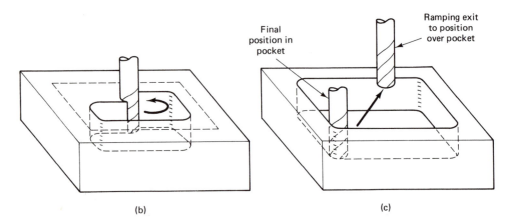

(b)   (c)

Figure 11.7   Pocketing tool positions: (a) entry into pocket; (b) pocketing, counterclockwise direction assumed; (c) removal from pocket.

perpendicular to the pocket part surface. The effective cutter radius Re is now less than D/2 not only because of cutting overlap but because Re is always measured along the pocket part surface.

Figure 11.8(c) shows Re for a fillet-end cutter when the pocket part surface is parallel with the xy-plane. In this case $Re = Re'$. The comparable measurement with an inclined pocket part surface is shown in Figure 11.8(d), where $Re \neq Re'$. The latter measurement applies only for tool motion normal to the pocket bottom gradient (arrow B in Figure 11.9). For tool motion parallel with the pocket bottom gradient (arrow A in

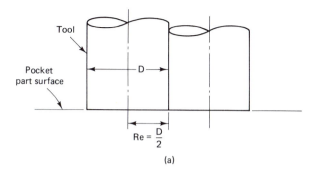

Re = $\frac{D}{2}$

(a)

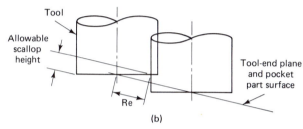

(b)

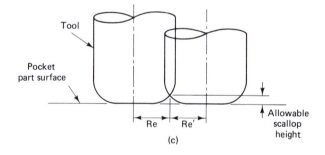

(c)

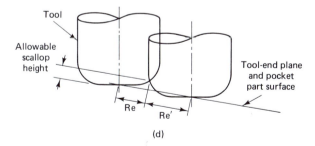

(d)

Figure 11.8   Cross-section geometry for determining *Re* for two different tool shapes, each with two different pocket part surface specifications.

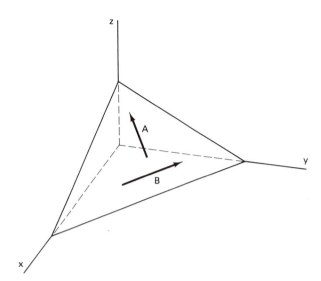

Figure 11.9 Indicating directions along
the gradient (arrow A) and normal to
the gradient (arrow B) on a plane.

Figure 11.9), the tool placement for adjacent cuts and the resulting scallop somewhat resemble that of Figure 11.8(c).

The allowable scallop height should not be confused with the tolerance specification. The most recent INTOL and OUTTOL commands prior to executing the POCKET command are in effect while pocketing and the tool endpoint is maintained on the pocket part surface within this tolerance. On the other hand, Re is used in the calculations to determine whether or not uncut material is left within the pocket as cutting occurs. For this purpose, Re is the effective cutter radius. Depending on the value given to Of, the values for Re, c, and f may be altered so that uncut material is not left within the pocket. Computation of Re is discussed in Example 11.6.

c = Pocketing Cut Offset Factor

Tool offset is the distance separating the tool centerline paths for the same side of adjacent concentric polygons as the pocket is being cut. It is always measured along the pocket part surface. The tool offset must be specified for the polygons interior to the pocket (i.e., all except the peripheral polygon), although it may automatically be re-adjusted by the APT system as certain pocketing requirements arise (see the discussion on the offset override indicator Of). The tool offset for interior polygons is specified with the pocketing cut offset factor c. This factor is determined by dividing the desired tool offset by the true cutter radius, which is one-half the diameter in the CUTTER statement in effect when the POCKET command is executed. The desired tool offset for Figure 11.8(c) would be chosen such that its maximum value is $Re + Re' = 2Re$. Thus, for this figure its maximum value is $c = 2Re/(D/2)$, where D is the tool diameter given in the CUTTER statement. For Figure 11.8(d) we have as a maximum, $c = (Re + Re')/(D/2)$. Physical interpretation of pocketing cut offset is shown in Figure 11.10.

The maximum value for c as given above will cause the allowable scallop height

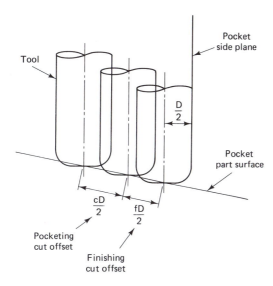

Figure 11.10   Interpretation of offsets.

to be exceeded in the corners of the polygons. Therefore, to compensate for the corner effect c must be chosen such that the following inequality is satisfied:

$$\frac{c\mathrm{D}}{2} \leq Re\left(1 + \sin\frac{\phi}{2}\right)$$

where ϕ is the smallest interior angle of the polygon describing the pocket. This inequality can be derived from the geometric quantities shown in Figure 11.11. Of course, c must not be made so small that unnecessary tool passes are made.

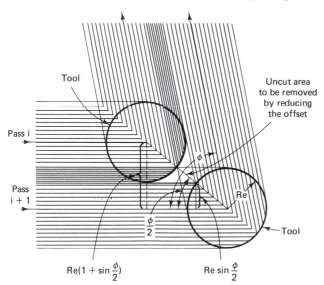

Figure 11.11   Geometry for computing the pocketing cut offset factor inequality.

f = Finishing Cut Offset Factor

The tool offset between the outermost interior polygon and the peripheral cut polygon is specified with the finishing cut offset factor *f*. This factor is determined in a manner analogous to that of the pocketing cut offset factor (see also the discussion on the offset override indicator *Of*). Figure 11.10 shows the finishing cut offset. If no special pocket side finish is desired, *f* is set equal to *c*.

F1 = Pocket Entry Feed Rate

This is the feed rate for the single cut vector that takes the tool from its location immediately preceding the POCKET command to a location near the bottom center of the pocket. It is in the same units as any externally specified feed rate.

F2 = General Pocketing Feed Rate

This is the feed rate for cutting all interior polygons. It does not apply to the finishing cut polygon. Its unit is as described for *F1*.

F3 = Finishing Cut Feed Rate

This is the feed rate for cutting the finishing cut polygon. Since the feed rate specification is modal, this finishing cut feed rate will be in effect for operations following the POCKET command until countermanded by another feed rate specification. Its unit is as described for *F1*.

Of = Offset Override Indicator

The part programmer must specify offset factors *c* and *f*. However, unfortunate combinations of offset factors, tool diameter, tool shape, and pocket geometry may cause uncut portions to remain in the pocket when the pocket operation is completed. This situation could arise from the geometry of Figure 11.12, where a pocket corner is eliminated as a pocket is being reduced, thus leaving uncut material as shown in Figure 11.11. Because pocket computations are made from the outside of the pocket toward the inside, whereas actual cutting occurs from the inside toward the outside, as the computation proceeds a smaller corner angle may develop that does not exist in the geometry of the defined pocket. A new offset, or equivalently new offset factors *c* and *f* must be computed according to the geometry of Figure 11.11 in order to eliminate all material from within the pocket, that is offset factors *c* and *f* must be overridden.

A new offset value may or may not be automatically computed by the APT system, depending on the value of the offset override indicator. The interpretation of the indicator, whose value must be either 0 or 1, is as follows.

0 Offset override feature is in effect: the APT system will test for uncut material and compute new offsets if necessary.

1 Offset override feature is bypassed: all offsets specified by the part programmer will be used while pocketing.

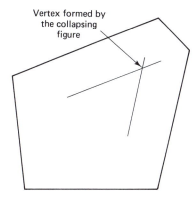

Vertex formed by
the collapsing
figure

**Figure 11.12   Pocket geometry whose
collapsing figure results in the formation
of a vertex with an angular dimension
less than for any corner of the pocket
periphery.**

p = Point-Type Indicator

The geometry of the pocket is determined by the coordinates of the points in the parameter list but subject to interpretation according to the value of the point-type indicator as follows:

0 The specified points define the tool centerline for the pocket finishing cut.

1 The specified points define the vertices produced by the sides of the finished pocket.

2 The points derived by projecting the specified points along the tool axis onto the current part surface define the tool centerline for the pocket finishing cut.

3 The points derived by projecting the specified points along the tool axis onto the current part surface define the vertices produced by the sides of the finished pocket.

When p is 0 or 1, the APT processor creates a plane from the specified points that becomes a temporary part surface for pocketing. Hence these points must all be coplanar. Figure 11.13(a) illustrates a pocket bottom formed in this manner.

When p is 2 or 3, the APT processor uses the current part surface, which must be a plane, as the pocket part surface and projects the specified points along the tool axis onto this surface. Therefore, the specified points need not be coplanar. In Figure 11.13(b) the specified points, defined on the top surface of the part, are projected onto the part surface shown (tool axis assumed to be the z-axis).

When p is 0 or 2, the specified points or their projection onto the pocket part surface becomes the tool-end coordinates during the pocket finishing cut. Therefore, the cutter geometry and diameter determine the actual size of the pocket.

When p is 1 or 3, the specified points or their projection onto the pocket part surface become the vertices of the pocket boundary.

pt1,pt2, . . . = Points Defining the Pocket

These are the point references that determine the geometry of the pocket when interpreted in conjunction with the value for p. The points may be listed by their symbolic

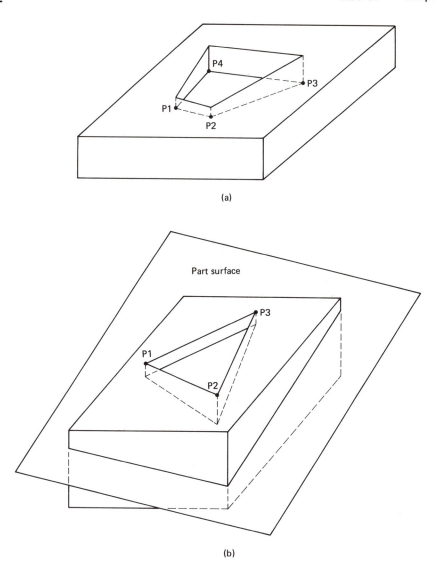

Figure 11.13   Pocket part surface formed by (a) point coordinates and (b) point projection.

names, by their x, y, and z coordinates, or by a combination of these methods. The APT POCKET routine accepts no more than 20 points, there can be no duplicate points in the list, and three or more consecutive points may not be collinear. The ordering of points in the list determines the clockwise or counterclockwise direction of tool movement while pocketing. The point appearing first in the list determines the corner at which the tool will be positioned when the finishing cut is completed. For example, if the points of Figure 11.13(a) are placed in the list in the order P1-P2-P3-P4, pocketing will proceed in a counterclockwise direction with the final tool position at P1. If the points of Figure 11.13(b) are listed in the order P3-P2-P1, pocketing will proceed in a clockwise direction with the final tool position at P3.

11.5.3 Examples

The following examples illustrate the POCKET feature. Example 11.6 provides sample pocket calculations for Re, c, and h for flat-end, ball-end, and fillet-end cutter shapes, Example 11.7 shows how uncut material may be left in the pocket when the offset override feature is bypassed and how this problem may be avoided, and Example 11.8 shows how contouring may be continued in a pocket after the POCKET command has been executed.

Example 11.6: Sample Pocket Calculations

Figure 11.14 shows tool positions for adjacent polygons within the pocket for motion normal to and parallel with the gradient for a sloping pocket bottom. Note that the tool endpoint is shown ON the part surface, which is the pocket bottom, and that the tool cuts below the pocket bottom a distance h'. Formulas are given for calculating the scallop height h, given the offset factor c (except for the fillet-end tool), and for calculating c and Re given the scallop height h. Three special cases are shown for a fillet-end tool cutting normal to the gradient (case 1, right tool radius r passing through left tool endpoint; case 2, right tool radius r passing a distance D/2-r from the left tool axis; case 3, left tool radius r passing a distance r above the right tool endpoint).

The distance h' adds to the scallop height h except for the flat-end tool moving parallel with the gradient, in which case h $\leq$ h', and for some values of c for the ball-end tool similarly oriented. The program part surface should be defined a distance h'sinθ above the part surface shown in the figure to avoid cutting below the desired pocket bottom. It is apparent that the scallop height depends on alignment of the polygon side with respect to the gradient. Tool motion in a direction intermediate to the two cases in the figure yields a scallop height trigonometrically related to values computed from the formulas.

Sample calculations are given in Table 11.1 for D $=$ 1 and θ $=$ 81°. The scallop height for the flat-end and ball-end tools was given as 0.1 (except for the flat-end tool parallel with the gradient, in which case it was equal to h'). Values for c and Re were computed from the formulas of Figure 11.14. The special cases for the fillet-end tool produce values for h, c, and Re.

TABLE 11.1   POCKET PARAMETER COMPUTATIONS

| Parameter | Flat end | | Ball end | | Fillet end | | |
	Normal	Parallel	Normal	Parallel	Normal Case 1	Normal Case 2	Normal Case 3
D	1	1	1	1	1	1	1
r	0	0	0.5	0.5	0.25	0.25	0.25
θ	81°	81°	81°	81°	81°	81°	81°
h	0.1	*	0.1	0.1	0.02995	0.06906	0.19866
h'	0.07822	0.07822	0.00616	0.00616	0.02995	0.02995	0.02995
c	1.2944	2	1.2000	1.2000	1.1280	1.4626	1.9231
Re	0.13753	1	0.22178	0.30000	0	0.24692	0.42857

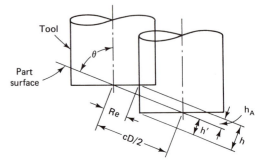

Flat-End

Normal to gradient:

$h' = D/2 \cos \theta$

$h_A = D/2 (c \sin \theta - 1) \cos \theta$

$h = h' + h_A = cD/2 \sin \theta \cos \theta$

$c = \dfrac{2h}{D \sin \theta \cos \theta}$

$Re = D/2 (c \sin \theta - 1) \sin \theta$

Parallel with gradient:

$h' = D/2 \cos \theta$

$h = D/2 (1 - \sqrt{1 - (c/2)^2}) \cos \theta$

$c = \dfrac{4}{D \cos \theta} \sqrt{h (D \cos \theta - h)} \qquad h \le h'$

$Re = cD/4$

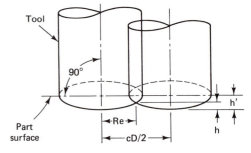

Ball-End

Normal to gradient:

$h' = D/2 (1 - \sin \theta)$

$h = D/2 (1 - \sqrt{1 - (c/2)^2})$

$c = 4/D \sqrt{h (D - h)}$

$Re = D/2 (c/2 - \cos \theta)$

Parallel with gradient:

$h' = D/2 (1 - \sin \theta)$

$h = D/2 (1 - \sqrt{1 - (c/2)^2})$

$c = 4/D \sqrt{h (D - h)}$

$Re = cD/4$

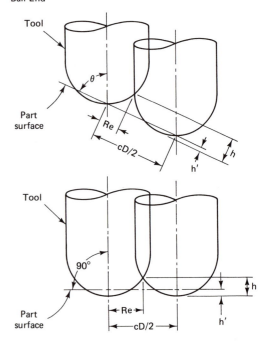

Figure 11.14 Geometry for sample POCKET calculations.

354

Normal to gradient — case 1:

$h = h' = | r (1 - \sin \theta) \sin \theta - D/2 \cos \theta + r \cos \theta (1 + \cos \theta) |$

$c = \left| \dfrac{2}{D \sin \theta} \left(\dfrac{-(D/2 - r) + mr - \sqrt{-2mr - m^2 (D/2)^2 + 2m^2 r (D/2)}}{1 + m^2} \right) \right|$

where $m = -\cot \theta$

$Re = 0$

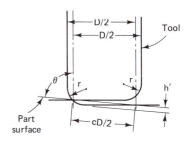

Normal to gradient — case 2:

$h_A = (D/2 - r) \cos \theta$

$h = h' + h_A$

$c = 2/D \, (2(D/2 - r) \sin \theta + \sqrt{-(D - 2r)^2 \cos^2 \theta + r^2} \,)$

$Re = (D/2 - r) \sin \theta$

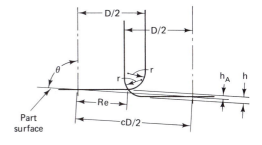

Normal to gradient — case 3:

$h_B = r \sin \theta - D/2 \cos \theta$

$h = h' + h_B$

$c = \dfrac{2}{D \sin \theta} (x' + D/2)$

$Re = x' \sin \theta - y' \cos \theta$

where, $x' = (D/2 - r) \sin^2 \theta - D/2 \cos^2 \theta + \sin \theta \sqrt{- D(D - 2r) \cos^2 \theta + r^2 \sin^2 \theta}$

$y' = -(x' + D/2) \cot \theta + r$

Figure 11.14 (Continued)

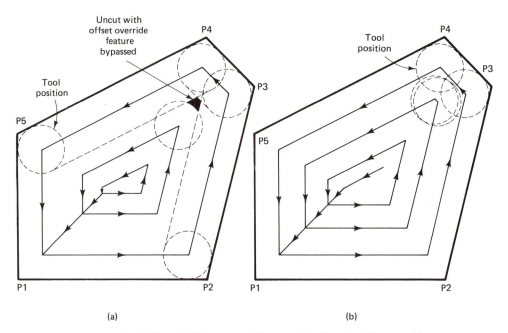

Figure 11.15 (a) Material left uncut within a pocket when the offset override feature is bypassed; (b) the cutter path when the offset override feature is in effect.

Example 11.7: Guarding Against Uncut Material Within a Pocket

We mentioned that uncut material may remain within the pocket if the offset override feature is bypassed because a small angle vertex is formed as the figure is being collapsed (see Figure 11.12). Figure 11.15(a) shows uncut material left in the pocket. The cutter path shown is computed with the following part program:

P1 = POINT/0,0,0	100
P2 = POINT/4,0,0	110
P3 = POINT/5,4,0	120
P4 = POINT/4,5,0	130
P5 = POINT/0,3,0	140
D = 1	150
C = 1.7	160
F = C	170
RE = C*(D/2)/2	180
CUTTER/D	190
POCKET/RE,C,F,F1,F2,F3,1,1,P1,P2,P3,P4,P5	200

The offset override feature guards against leaving uncut material within the pocket, as shown in Figure 11.15(b), where the cutter path was produced with the offset override feature in effect with the following POCKET statement:

POCKET/RE,C,F,F1,F2,F3,0,1,P1,P2,P3,P4,P5

The figure shows that as the pocket was collapsing the APT system automatically adjusted the pocketing cut offset factor to eliminate all material within the pocket.

It is also possible to eliminate all material within the pocket with the offset override feature bypassed by specifying as the first pocket point a vertex that is located in the vicinity of the expected internal small angle vertex. For the pocket geometry of Figure 11.15 this could be either P3 or P4, which result in the cutter paths of Figure 11.16(a) and (b), respectively. The corresponding POCKET statements are

POCKET/RE,C,F,F1,F2,F3,1,1,P3,P4,P5,P1,P2

or

POCKET/RE,C,F,F1,F2,F3,1,1,P4,P5,P1,P2,P3

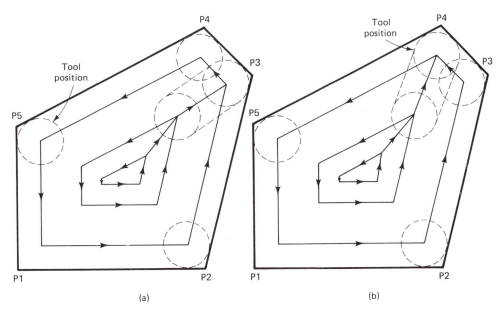

(a)   (b)

Figure 11.16   Complete pocket coverage with the offset override feature bypassed by specifying (a) P3 as the first vertex, and (b) P4 as the first vertex.

Example 11.8: Contouring Within a Pocket

Figure 11.17(a) shows a part with two pockets. The part is symmetrical, suggesting use of the COPY feature. The POCKET statement will be used for a pentagon within the pocket

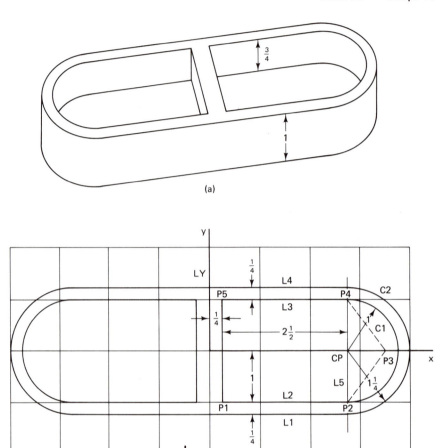

Figure 11.17   Part geometry for Example 11.8.

and contouring will complete the curved portion. A part program to produce the part is as follows:

D = .5		100
CUTTER/D		110
SETPT = POINT/− .5, − 2,2		120
L1	= LINE/XAXIS, − 1.25	130
L2	= LINE/XAXIS, − 1	140
L3	= LINE/XAXIS,1	150
L4	= LINE/XAXIS,1.25	160
LY	= LINE/YAXIS	170

CP	= POINT/2.75,0	180
L5	= LINE/CP,ATANGL,90	190
C1	= CIRCLE/CENTER,CP,RADIUS,1	200
C2	= CIRCLE/CENTER,CP,RADIUS,1.25	210
PL1	= PLANE/0,0,1,0	220
PL2	= PLANE/0,0,1,.25	230
	ZSURF/PL2	240
P1	= POINT/.25, − 1	250
P2	= POINT/INTOF,L2,L5	260
P3	= POINT/3.5,0	270
P4	= POINT/INTOF,L3,L5	280
P5	= POINT/.25,1	290
	PSIS/PL1	300
	FROM/SETPT	310
	GO/TO,L1,TO,PL1,TO,LY	320
	INDIRV/1,0,0	330
	TLRGT	340
	INDEX/7	350
	GOFWD/L1,TANTO,C2	360
	GOFWD/C2,TANTO,L4	370
	GOFWD/L4,ON,LY	380
	COPY/7,XYROT,180,1	390
	GOTO/0, − 1.5,1.25	400
	C = 1.7	410
	F = C	420
	RE = C*(D/2)/2	430
	INDEX/9	440
	GODLTA/2,1.25,0	450
	POCKET/RE,C,F,35,30,20,0,1,P2,P3,P4,P5,P1	460
	DNTCUT	470
	GODLTA/ − .25,.25,0	480
	PSIS/PL2	490
	GO/TO,L2,TO,PL2,TO,L5	500
	TLLFT,GOFWD/L2,TANTO,C1	510
	CUT	520
	GOFWD/C1,PAST,2,INTOF,L5	530
	GODLTA/0, − .25,1	540
	COPY/9,XYROT,180,1	550
	GOTO/SETPT	560

Points for the sides of the pocket are defined in lines 250 through 290. Pocketing is in a counterclockwise direction with the offset override feature in effect. The tool is at vertex P2 when pocketing is finished. Statements 470 through 520 position the tool for contouring within the pocket with the command of line 530. The feedrate values of the pocket command for this example have no machining significance. They are used to identify the cut vectors output by the command.

PROBLEMS

11.1. Figure P11.1 shows a part with some symmetry. Write an APT macro for the contouring motion along the bottom edge. Then use it for machining the bottom and top edges in an APT part program. Take a 0.2-in.-diameter tool counterclockwise around the part.

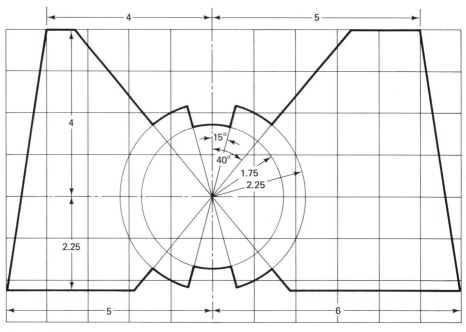

Dimensions in inches

Figure P11.1

11.2. Repeat Problem 9.7 but with macros for cutting the steps and for drilling the holes.

11.3. Repeat Problem 9.8 but with a macro for drilling the holes.

11.4. Figure P11.4 shows coordinate axes engraved with a linear abscissa scale and a logarithmic ordinate scale. Write an APT part program to engrave the axes. Calibrate the linear scale with five major tick marks, each spaced 1 in. apart and each further subdivided with small tick marks into five equal parts. Calibrate the logarithmic scale with three decades, major tick marks each 2.5 in. apart and further subdivided with small tick marks into nine logarithmically spaced parts. Make the major tick marks 0.2 in. long, the subdivision tick marks 0.1 in. long.

11.5. Three portions of Figure P11.5 are curved identically. This part could be programmed with the COPY feature. Instead, write an APT part program to produce the part with a macro for contouring along the three portions. Take a 0.2-in.-diameter cutter counterclockwise around the part.

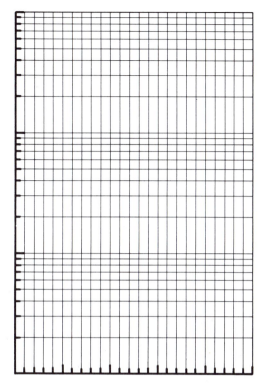

Figure P11.4

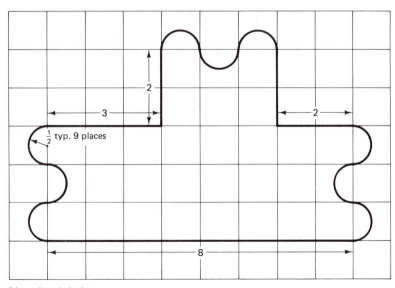

2

3

2

$\frac{1}{2}$ typ. 9 places

8

Dimensions in inches

Figure P11.5

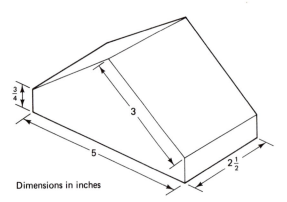

Dimensions in inches   **Figure P11.6**

11.6. Write an APT macro to contour along a sloping surface. Use this macro twice for the sloping surfaces of Figure P11.6. Select an appropriate tool size and scallop height.

11.7. Separate APT macros could be written to engrave each of the 10 decimal digits. The macros could then be called, under suitable transformation, to engrave a particular sequence of digits. Define the geometry and write macros for the digits 6, 8, and 9. The size of the digits should be controlled by height and width parameters initialized at the beginning of the part program. Then, as a special case, write an APT part program to engrave the sequence of digits 98689.

11.8. Figure P11.8(a) shows a part with two recesses, each of which could be machined with a selection of POCKET commands. Details of the boundaries of the recesses are given in Figure P11.8(b) (estimate critical features to the nearest ¼ inch). Select a suitable tool size and write an APT part program to produce the part with POCKET commands.

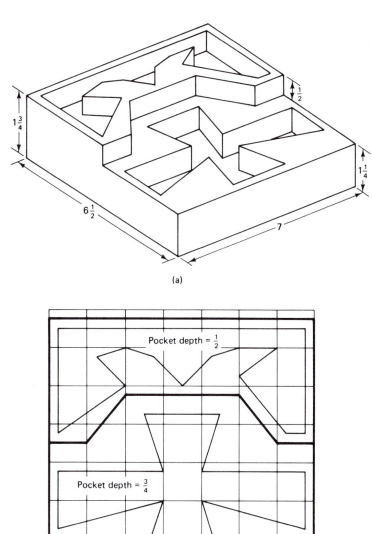

(a)

Pocket depth = $\frac{1}{2}$

Pocket depth = $\frac{3}{4}$

Dimensions in inches

(b)

Figure P11.8

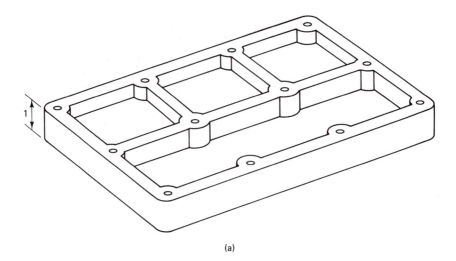

(a)

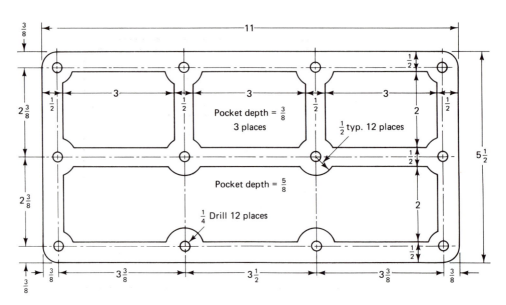

Dimensions in inches

(b)

Figure P11.9

11.9. Repeat Problem 11.8 for the part of Figure P11.9.

11.10. Repeat Problem 8.5 but use the POCKET command to remove material between the webs of the flange shown there.

11.11. Repeat Problem 11.10 for the part of Figure P8.2.

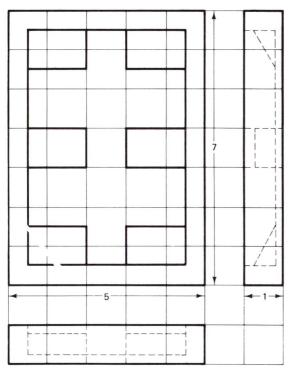

Dimensions in inches   **Figure P11.12**

11.12. The part of Figure P11.12 has canted planes for a portion of the recess bottom. Select a suitable tool size and write an APT part program with a selection of POCKET commands to produce the part. Estimate critical features to the nearest ¼ inch.

11.13. Repeat Problem 11.12 for the part of Figure P11.13.

11.14. The part of Figure P11.14 contains numerous recesses. Select a suitable tool size and write an APT part program to produce the part. Use POCKET commands for the recesses, contouring within the pockets with curved boundaries.

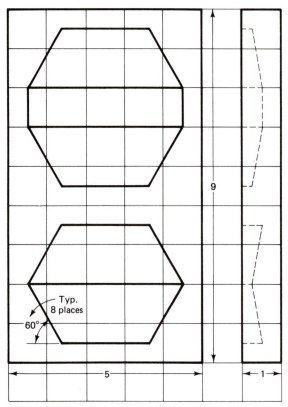

Dimensions in inches

Figure P11.13

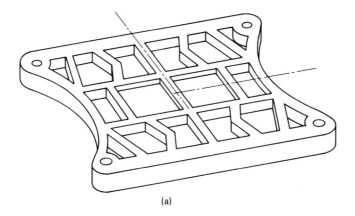

(a)

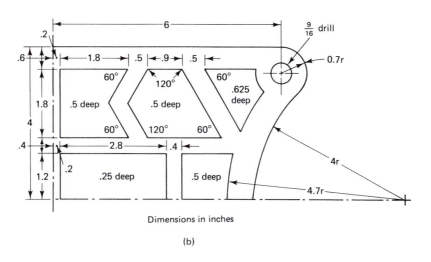

Dimensions in inches

(b)

Figure P11.14

CHAPTER **12**

Geometry of Three Dimensions

With the material of the preceding chapters we exercised well the APT features associated with plane geometry and, via transformations, we were able to perform rudimentary three-dimensional tasks. But to apply APT to three-dimensional work without restriction, we must have the means for describing geometric figures of arbitrary shape. Therefore, in preparation for the complex material of Chapters 13 and 14 we will now complete the presentation of the APT geometric description capability.

We begin by describing quadric surfaces—surfaces with an analytic representation—and then discuss the construction of ruled and polyconic surfaces. Until now we ignored point, line, and ellipse definitions dependent on three-dimensional geometry. The formal definitions of most vectors are covered before considering the remainder of these elementary figures. We will also discuss those matrix definitions dependent on three-dimensional geometry but reserve several vector definitions for discussion in Chapter 13 in connection with vector mathematics.

12.1 QUADRIC SURFACES

The graph of an equation in three variables is a **surface**. A surface is represented by an equation if the coordinates of every point on the surface satisfy the equation and if every point whose coordinates satisfy the equation lies on the surface. We have already discussed one kind of surface, the plane. In this section we discuss cylindrical surfaces, surfaces of revolution, and quadric surfaces defined as follows:

Def: The surface generated by a line moving along a given plane curve so that it is always parallel with a fixed line not in the plane of the given curve is a **cylindrical**

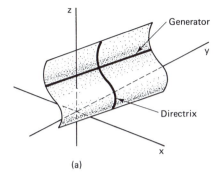

(a)

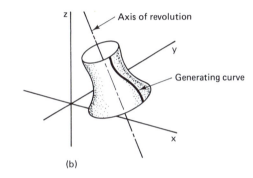

(b)

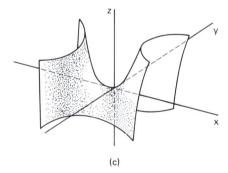

(c)

Figure 12.1   (a) Cylindrical surface; (b) surface of revolution; (c) quadric surface.

surface or **cylinder**. The moving line is called the **generator** of the cylinder, while the given plane curve is called the **directrix** of the cylinder. Any position of a generator is called a **ruling** of the cylinder. Figure 12.1(a) shows a cylindrical surface. Cylinders are the simplest of the quadric surfaces.

Def: The surface generated by revolving a plane curve about a fixed line in the plane of the curve is called a **surface of revolution**. The fixed line is called the **axis** of revolution, while the plane curve is called the **generating curve**. Figure 12.1(b) shows a surface of revolution.

Def: The surface corresponding to the graph of a second degree equation in three variables x, y, and z is called a **quadric surface**. Figure 12.1(c) shows such a surface.

12.1.1 General Quadrics

The general equation of the quadric surface defined above is of the form

$$Ax^2 + By^2 + Cz^2 + Fyz + Gzx + Hxy + Px + Qy + Rz + D = 0 \qquad (12.1)$$

where the coefficients are constants. The APT format is

QADRIC/A,B,C,F,G,H,P,Q,R,D

Coefficient values may be found for equation (12.1) that form imaginery surfaces (e.g., $x^2 + y^2 + z^2 + 1 = 0$). Such surfaces are not allowed in APT. Other degenerate forms are the point, line, and plane. Following are examples of these degenerate forms:

$$x^2 + y^2 + z^2 = 0 \qquad \text{One point, the origin}$$

$$y^2 = 0 \qquad \text{One plane, the } zx\text{-plane}$$

$$x^2 + z^2 = 0 \qquad \text{One line, the } y\text{-axis}$$

$$y^2 - z^2 = 0 \qquad \text{Two planes, } y - z = 0 \text{ and } y + z = 0$$

By translating and rotating the three-dimensional coordinate axes, it can be shown that equation (12.1) can be reduced to one of the following two simple forms:

$$A'x^2 + B'y^2 + C'z^2 + D' = 0 \qquad (12.2)$$

or

$$A'x^2 + B'y^2 + R'z = 0 \qquad (12.3)$$

These equations are often written as

$$\pm \frac{x^2}{a^2} \pm \frac{y^2}{b^2} \pm \frac{z^2}{c^2} = 1 \qquad (12.4)$$

or

$$\pm \frac{x^2}{a^2} \pm \frac{y^2}{b^2} = \frac{z}{c} \qquad (12.5)$$

with the sign chosen appropriate to the surface orientation. When given in the form of equations (12.2) through (12.5) the quadric surface exhibits symmetry with respect to various of the principal coordinate axes. The simple forms can always be defined from arbitrary quadric surface orientations via the REFSYS feature.

Quadric surfaces allowed in APT are shown in Figure 12.2. The examples conform with one sign assignment in equations (12.4) and (12.5) for each type of surface.

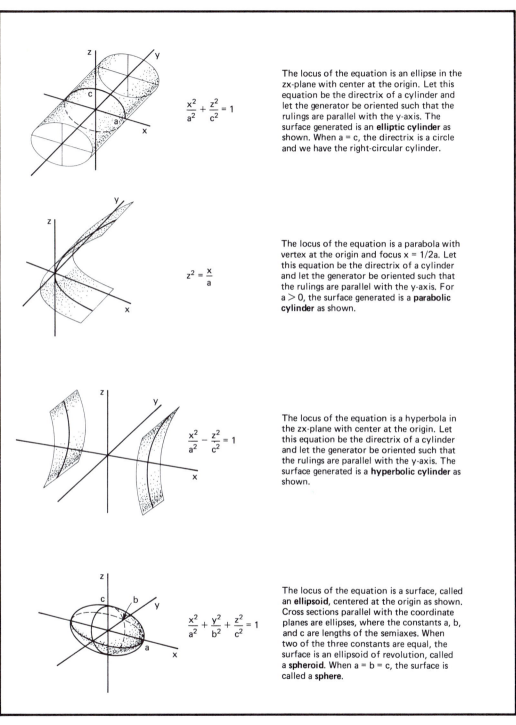

$$\frac{x^2}{a^2} + \frac{z^2}{c^2} = 1$$

The locus of the equation is an ellipse in the zx-plane with center at the origin. Let this equation be the directrix of a cylinder and let the generator be oriented such that the rulings are parallel with the y-axis. The surface generated is an **elliptic cylinder** as shown. When a = c, the directrix is a circle and we have the right-circular cylinder.

$$z^2 = \frac{x}{a}$$

The locus of the equation is a parabola with vertex at the origin and focus x = 1/2a. Let this equation be the directrix of a cylinder and let the generator be oriented such that the rulings are parallel with the y-axis. For a > 0, the surface generated is a **parabolic cylinder** as shown.

$$\frac{x^2}{a^2} - \frac{z^2}{c^2} = 1$$

The locus of the equation is a hyperbola in the zx-plane with center at the origin. Let this equation be the directrix of a cylinder and let the generator be oriented such that the rulings are parallel with the y-axis. The surface generated is a **hyperbolic cylinder** as shown.

$$\frac{x^2}{a^2} + \frac{y^2}{b^2} + \frac{z^2}{c^2} = 1$$

The locus of the equation is a surface, called an **ellipsoid**, centered at the origin as shown. Cross sections parallel with the coordinate planes are ellipses, where the constants a, b, and c are lengths of the semiaxes. When two of the three constants are equal, the surface is an ellipsoid of revolution, called a **spheroid**. When a = b = c, the surface is called a **sphere**.

Figure 12.2   Quadric surfaces allowed in APT.

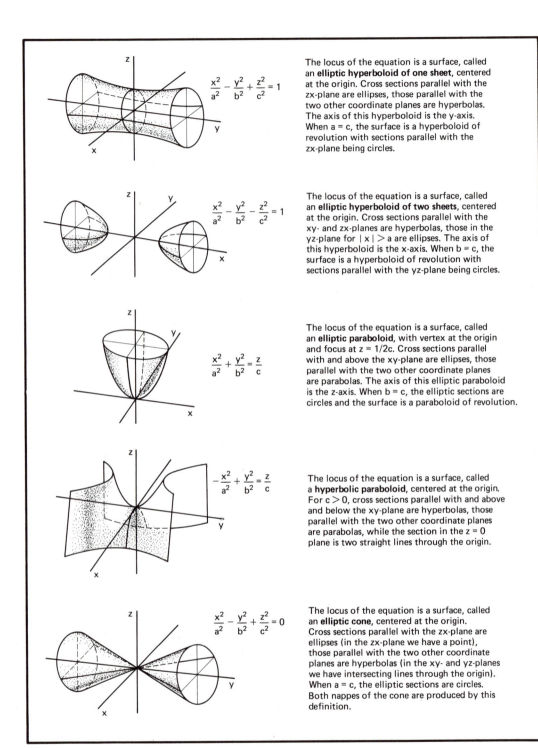

$$\frac{x^2}{a^2} - \frac{y^2}{b^2} + \frac{z^2}{c^2} = 1$$

The locus of the equation is a surface, called an **elliptic hyperboloid of one sheet**, centered at the origin. Cross sections parallel with the zx-plane are ellipses, those parallel with the two other coordinate planes are hyperbolas. The axis of this hyperboloid is the y-axis. When a = c, the surface is a hyperboloid of revolution with sections parallel with the zx-plane being circles.

$$\frac{x^2}{a^2} - \frac{y^2}{b^2} - \frac{z^2}{c^2} = 1$$

The locus of the equation is a surface, called an **elliptic hyperboloid of two sheets**, centered at the origin. Cross sections parallel with the xy- and zx-planes are hyperbolas, those in the yz-plane for $|x| > a$ are ellipses. The axis of this hyperboloid is the x-axis. When b = c, the surface is a hyperboloid of revolution with sections parallel with the yz-plane being circles.

$$\frac{x^2}{a^2} + \frac{y^2}{b^2} = \frac{z}{c}$$

The locus of the equation is a surface, called an **elliptic paraboloid**, with vertex at the origin and focus at z = 1/2c. Cross sections parallel with and above the xy-plane are ellipses, those parallel with the two other coordinate planes are parabolas. The axis of this elliptic paraboloid is the z-axis. When b = c, the elliptic sections are circles and the surface is a paraboloid of revolution.

$$-\frac{x^2}{a^2} + \frac{y^2}{b^2} = \frac{z}{c}$$

The locus of the equation is a surface, called a **hyperbolic paraboloid**, centered at the origin. For c > 0, cross sections parallel with and above and below the xy-plane are hyperbolas, those parallel with the two other coordinate planes are parabolas, while the section in the z = 0 plane is two straight lines through the origin.

$$\frac{x^2}{a^2} - \frac{y^2}{b^2} + \frac{z^2}{c^2} = 0$$

The locus of the equation is a surface, called an **elliptic cone**, centered at the origin. Cross sections parallel with the zx-plane are ellipses (in the zx-plane we have a point), those parallel with the two other coordinate planes are hyperbolas (in the xy- and yz-planes we have intersecting lines through the origin). When a = c, the elliptic sections are circles. Both nappes of the cone are produced by this definition.

Figure 12.2 (Continued)

The quadric surface is known from the drawing when defined. However, image surfaces for the hyperbolic cylinder, elliptic hyperboloid of two sheets, and elliptic cone must be visualized for tool positioning purposes to avoid problems with unexpected, but undesired, tool checking locations. These problems are identical to those discussed for general conics. The closed surface character of the quadric surfaces may lead to top-of-the-tool checking. This problem is addressed with tool motion considerations in Section 13.2.2.

The format of the canonical form of the quadric surface is

QADRIC/CANON,A,B,C,D,F/2,G/2,H/2,P/2,Q/2,R/2

The ordering of the coefficients differs from the general format given above with some of the coefficients divided by 2 for APT processor computational convenience.

When $C = F = G = R = 0$ in equation (12.1), we get the general conic of equation (6.1). Such equations should be defined as GCONICs rather than as QADRICs.

12.1.2 Cylinders

A cylinder was defined above as a general quadric surface. Because right-circular cylinders are used often, convenient definition formats for them are included in APT as shown in Table 12.1. The circle definitions of Table 3.3 are right-circular cylinders perpendicular to the xy-plane. With the REFSYS feature they may be defined at any orientation. The QADRIC format may also be used to define any cylinder.

12.1.3 Cones

The circular cone is a special case of the elliptic cone. Because circular cones are used often, convenient definition formats for them are included in APT as shown in Table 12.2. Unlike for the elliptic cone, only the nappe in the direction of the vector is defined by these formats. The QADRIC format may also be used to define any circular cone.

12.1.4 Spheres

The sphere is a special case of the ellipsoid. Because spheres are used often, convenient definition formats for them are included in APT as shown in Table 12.3. The QADRIC format may also be used to define any sphere.

12.2 RULED SURFACES

Certain geometric shapes cannot be adequately described by the surfaces discussed above. These may be transition surfaces between different geometric cross sections, surfaces with more asthetic appeal, or surfaces for clearance, structural integrity, or other demanding reasons. In APT, we often use a ruled surface to describe such shapes.

A **ruled surface** is the surface formed by **rulings** (straight lines) passing through

TABLE 12.1 CYLINDER DEFINITIONS

1	By a point on its axis, its axis orientation, and its radius (canonical form).

CYLNDR/CANON, x, y, z, a, b, c, *radius*

CYLNDR/ $\begin{Bmatrix} x,\,y,\,z \\ point \end{Bmatrix}$, $\begin{Bmatrix} a,\,b,\,c \\ vector \end{Bmatrix}$, *radius*

The right-circular cylinder defined has its axis passing through *point* (alternatively given by the coordinates x, y, z) oriented along *vector* (alternatively given by the components a, b, c) and with a radius value of *radius*. When CANON is used, the axis vector must be specified as a unit vector.

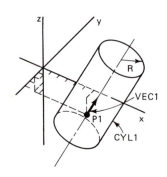

CYL1 = CYLNDR/6, −2, −2, 0, 2, 2, R
CYL1 = CYLNDR/P1, 0, 2, 2, R
CYL1 = CYLNDR/P1, VEC1, R
CYL1 = CYLNDR/6, −2, −2, VEC1, R
CYL1 = CYLNDR/P1, 0, .707, .707, R
CYL1 = CYLNDR/CANON, P1, 0, .707, .707, R

2	By two tangent planes and its radius.

CYLNDR/ $\begin{bmatrix} \text{XLARGE} \\ \text{XSMALL} \\ \text{YLARGE} \\ \text{YSMALL} \\ \text{ZLARGE} \\ \text{ZSMALL} \end{bmatrix}$, TANTO, *plane1*, $\begin{bmatrix} \text{XLARGE} \\ \text{XSMALL} \\ \text{YLARGE} \\ \text{YSMALL} \\ \text{ZLARGE} \\ \text{ZSMALL} \end{bmatrix}$, TANTO, *plane2*, RADIUS, *radius*

The right-circular cylinder defined is tangent to *plane1* and *plane2* and has a radius value of *radius*. Each modifier is associated with the immediately following tangent plane and identifies that side of the plane on which the cylinder is tangent. The modifiers are determined by visually slicing the cylinder with a coordinate plane (except for a plane containing the cylinder axis) and, for each tangent plane, selecting that modifier that best describes a dominant coordinate location of the axis pierce point in the slicing plane relative to the corresponding coordinate location at the tangency pierce point in the slicing plane.

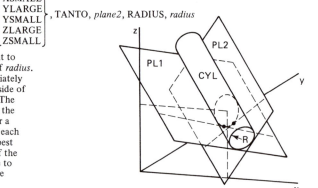

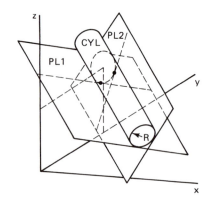

CYL = CYLNDR/ZLARGE, TANTO, PL1, ZLARGE, TANTO, $
      PL2, RADIUS, R
CYL = CYLNDR/YLARGE, TANTO, PL1, YSMALL, TANTO, $
      PL2, RADIUS, R
CYL = CYLNDR/XLARGE, TANTO, PL1, XSMALL, TANTO, $
      PL2, RADIUS, R

374

TABLE 12.2   CONE DEFINITIONS

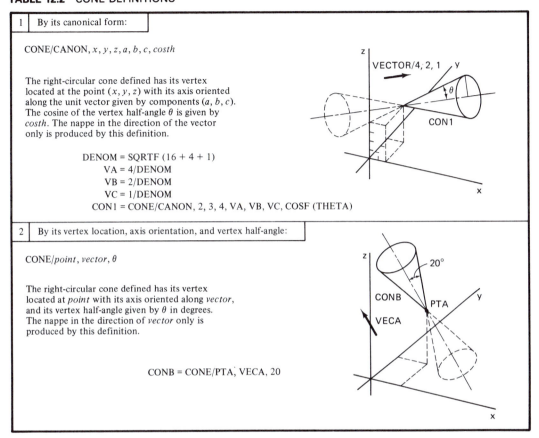

| 1 | By its canonical form: |

CONE/CANON, $x, y, z, a, b, c, costh$

The right-circular cone defined has its vertex located at the point (x, y, z) with its axis oriented along the unit vector given by components (a, b, c). The cosine of the vertex half-angle θ is given by *costh*. The nappe in the direction of the vector only is produced by this definition.

$$DENOM = SQRTF (16 + 4 + 1)$$
$$VA = 4/DENOM$$
$$VB = 2/DENOM$$
$$VC = 1/DENOM$$
$$CON1 = CONE/CANON, 2, 3, 4, VA, VB, VC, COSF (THETA)$$

| 2 | By its vertex location, axis orientation, and vertex half-angle: |

CONE/*point*, *vector*, θ

The right-circular cone defined has its vertex located at *point* with its axis oriented along *vector*, and its vertex half-angle given by θ in degrees. The nappe in the direction of *vector* only is produced by this definition.

$$CONB = CONE/PTA, VECA, 20$$

corresponding points on two planar curves. Each planar curve is formed by the intersection of a plane with a construction surface (the two construction surfaces need not be of the same type) and is defined within the ruled surface defining format (see Table 12.4). Allowed construction surfaces for this purpose are the line, circle, tabulated cylinder, ellipse, hyperbola, general conic, loft conic, plane, cylinder, cone, sphere, and quadric. One format in Table 12.4 is a degenerate case of the general form and requires a point to be specified as the second surface. Ruled surface examples are shown in Figure 12.3.

Geometric details for forming the planar curves are labeled in Figure 12.4. We now describe the procedure for forming one planar curve; the other is formed in an analogous manner. First, two points that are to become endpoints of the planar curve are located on the construction surface and a baseline (chord) is assumed to pass through them. The intersecting plane for the construction surface is then defined by the baseline and a third point (for planar curve 2) or by the baseline and a vector (for planar curve 1). The planar curve is now defined on the intersection of the plane with the

TABLE 12.3 SPHERE DEFINITIONS

| 1 | By its center point and radius (canonical form): |

$$\text{SPHERE/} \begin{cases} [\text{CONAN,}]\ x, y, z, r \\ point, r \\ \text{CENTER}, point, \text{RADIUS}, r \end{cases}$$

The sphere defined is centered at *point*, whose
coordinates are (x, y, z), with a radius of *r*.

SPHA = SPHERE/CANON, 3, 1, 3, 2
SPHA = SPHERE/3, 1, 3, 2
SPHA = SPHERE/PTA, 2
SPHA = SPHERE/CENTER, PTA, RADIUS, 2

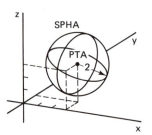

| 2 | By its center point and a point on its surface: |

SPHERE/CENTER, *point1*, *point2*

The sphere defined is centered at *point1* and
passes through *point2*.

SPHB = SPHERE/CENTER, PTB, PTS

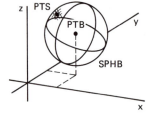

| 3 | By its center point and a tangent plane: |

SPHERE/CENTER, *point*, TANTO, *plane*

The sphere defined is centered at *point* and is
tangent to *plane* (*point* must not lie on *plane*).

SPHC = SPHERE/CENTER, PTC, TANTO, PLC

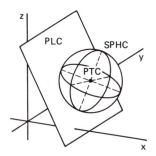

| 4 | As passing through four points: |

SPHERE/*point1*, *point2*, *point3*, *point4*

The sphere defined passes through *point1*,
point2, *point3*, and *point4*. The four points
must not be coplanar.

SPHD = SPHERE/PT1, PT2, PT3, PT4

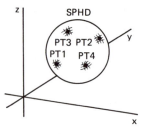

TABLE 12.4   RULED SURFACE DEFINITIONS

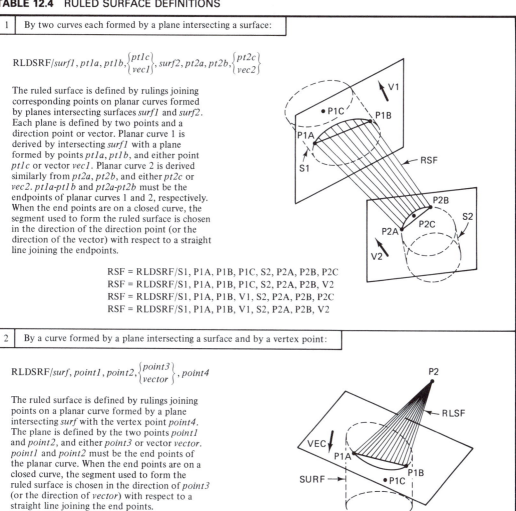

1 By two curves each formed by a plane intersecting a surface:

$$RLDSRF/surf1, pt1a, pt1b, \begin{Bmatrix} pt1c \\ vec1 \end{Bmatrix}, surf2, pt2a, pt2b, \begin{Bmatrix} pt2c \\ vec2 \end{Bmatrix}$$

The ruled surface is defined by rulings joining corresponding points on planar curves formed by planes intersecting surfaces *surf1* and *surf2*. Each plane is defined by two points and a direction point or vector. Planar curve 1 is derived by intersecting *surf1* with a plane formed by points *pt1a*, *pt1b*, and either point *pt1c* or vector *vec1*. Planar curve 2 is derived similarly from *pt2a*, *pt2b*, and either *pt2c* or *vec2*. *pt1a-pt1b* and *pt2a-pt2b* must be the endpoints of planar curves 1 and 2, respectively. When the end points are on a closed curve, the segment used to form the ruled surface is chosen in the direction of the direction point (or the direction of the vector) with respect to a straight line joining the endpoints.

RSF = RLDSRF/S1, P1A, P1B, P1C, S2, P2A, P2B, P2C
RSF = RLDSRF/S1, P1A, P1B, P1C, S2, P2A, P2B, V2
RSF = RLDSRF/S1, P1A, P1B, V1, S2, P2A, P2B, P2C
RSF = RLDSRF/S1, P1A, P1B, V1, S2, P2A, P2B, V2

2 By a curve formed by a plane intersecting a surface and by a vertex point:

$$RLDSRF/surf, point1, point2, \begin{Bmatrix} point3 \\ vector \end{Bmatrix}, point4$$

The ruled surface is defined by rulings joining points on a planar curve formed by a plane intersecting *surf* with the vertex point *point4*. The plane is defined by the two points *point1* and *point2*, and either *point3* or vector *vector*. *point1* and *point2* must be the end points of the planar curve. When the end points are on a closed curve, the segment used to form the ruled surface is chosen in the direction of *point3* (or the direction of *vector*) with respect to a straight line joining the end points.

RLSF = RLDSRF/SURF, P1A, P1B, P1C, P2
RLSF = RLDSRF/SURF, P1A, P1B, VEC, P2

construction surface between the endpoints. The plane is not necessarily parallel with a coordinate plane.

The direction point and vector are also used for computing corresponding points while generating rulings. The procedure is illustrated in Figure 12.5. Rulings are formed by the "percent of chords" method, whereby corresponding points on the planar curves are located on a percentage basis relative to the baselines. Each baseline is declared to

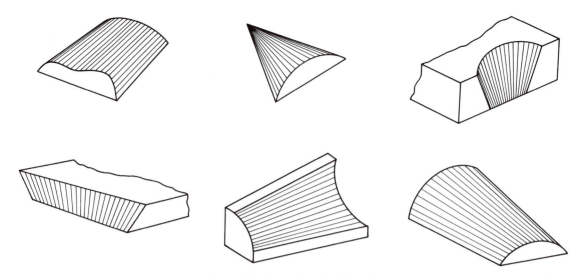

Figure 12.3   Examples of ruled surface geometry.

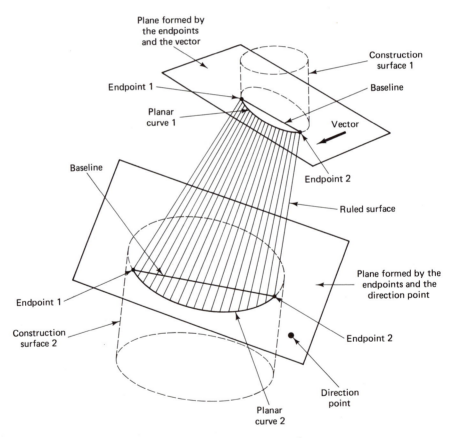

Figure 12.4   Geometry of ruled surface planar curves.

378

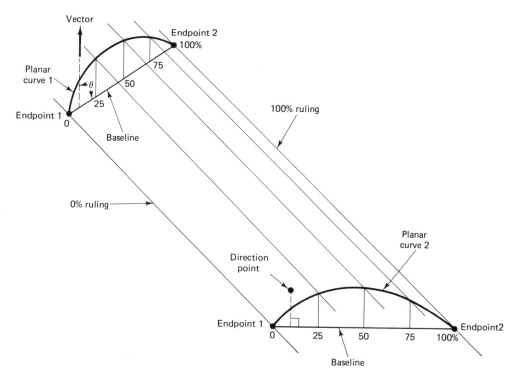

Figure 12.5   Generating rulings.

be 100 percent in length as marked in the figure, where other points on the baseline are marked appropriately. The 0% and 100% points are associated with the first and second points, respectively, in the formats of Table 12.4. Lines connecting them are rulings. Their relative order in the format is important to avoid a distorted orientation of the ruled surface. The direction point or vector is used to generate intermediate rulings by projecting *from* the baseline *to* the planar curve. If a direction point is given in the format, the projections are perpendicular to the baseline and on the same side of the baseline as the direction point. If a vector is given, projections are parallel with and in the same direction as the vector.

When the endpoints lie on a closed planar curve the direction point, or vector, is used to choose the desired curve from the two possibilities in the manner described above. An example is shown in Figure 12.6, where point PA and vector VA select curve A (above the baseline) while point PB and vector VB select curve B (below the baseline.

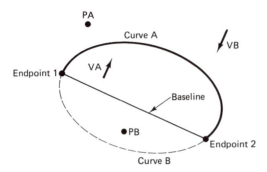

Figure 12.6   Planar curve choices for a ruled surface derived from a closed surface.

The following restrictions should be observed to avoid difficulty in defining and using the ruled surface.

1. When a direction point is specified, the three points defining the intersecting plane must not be collinear nor must two or more of the points be coincident.
2. When a vector is used to define the intersecting plane, it must not be parallel with the baseline nor should it be at a small angle, say $< 10°$, to the baseline [see projection vector 1 in Figure 12.7(a)].
3. A planar curve tangent line must not be perpendicular to the baseline [see Figure 12.7(a)].
4. The planar curve must not intersect the baseline between endpoints since the ruled surface is defined only on the projection direction side of the baseline [see planar curve 1 in Figure 12.7(b)].
5. It must not be possible to multiply intersect the planar curve in the specified projection direction from any point on the baseline [see planar curve 2 in Figure 12.7(b)].

The restrictions above suggest that multiple ruled surfaces may be required to represent a surface not otherwise singly defined.

The description and artistic rendition of a ruled surface suggests that rulings are generated and stored, then used in computing cut vectors. Contrarily, they are generated only as needed to position the cutter relative to the drive, part, and check surfaces.

The ruled surface extends longitudinally beyond the planar surfaces. In theory, the rulings extend to infinity but extrapolating only a short distance beyond the planar curves is advised to avoid computational difficulties in the APT processor. Consequently, what may otherwise have been a difficult-to-define planar curve now becomes a compromise construction or an exact cross section but displaced relative to adjacent surfaces of the workpiece. The ruled surface, however, does not extend laterally beyond the endpoints.

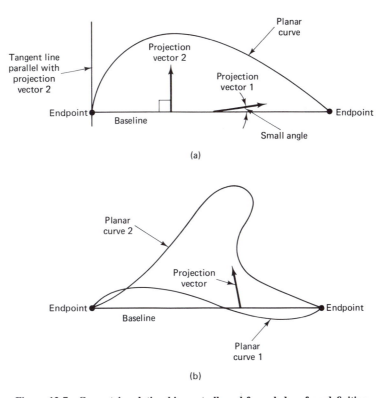

Figure 12.7   Geometric relationships not allowed for ruled surface definition.

Example 12.1

Figure 12.8 shows a ruled surface formed between a circular segment and a straight line. Given the dimensions in the figure, we choose to define the ruled surface as follows:

MAT1	= MATRIX/ZXROT,90	100
REFSYS/MAT1		110
C1	= CIRCLE/0,5,0,2	120
L1	= LINE/YAXIS, − 1	130
P1	= POINT/YSMALL,INTOF,L1,C1	140
P2	= POINT/YLARGE,INTOF,L1,C1	150
P3	= POINT/ − 2,5,0	160
REFSYS/NOMORE		170
P4	= POINT/6,0,1	180
P5	= POINT/(6 + 5*COSF(30)),(5*SINF(30)),3	190
L2	= LINE/P4,P5	200
RSF	= RLDSRF/C1,P1,P2,P3,L2,P4,P5,P3	210

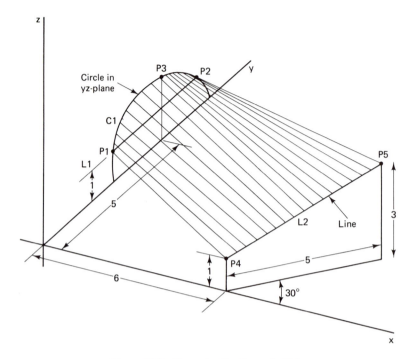

Figure 12.8   Geometry for Example 12.1.

Circular segment C1 and the other parameters of the ruled surface associated with it are defined in the xy-plane, then rotated about the y-axis. Line L2 is defined in a straight-forward manner via computation. Because the second surface of RSF is a line, its direction point, or vector, must not be collinear with it nor in the same direction, respectively. In this case P3 is used as a direction point for both surfaces.

12.3 POLYCONIC SURFACES

Many shapes are used in the design of physical objects. We have been able to closely model, if not exactly describe, important classes of shapes with the quadric and ruled surfaces. A shape commonly used will now be modeled with the polyconic surface. This shape has a cross section adequately modeled as a conic and has a polynomial profile when sectioned along its longitudinal axis. The polyconic surface shares some limitations with the ruled surface.

The APT polyconic surface is a complicated assembly derived from a set of poly-nomials in parametric form, a transformation matrix, and an origin and normalization specification. We first describe the geometry of the polyconic surface as interpreted by the APT system, then present its defining format.

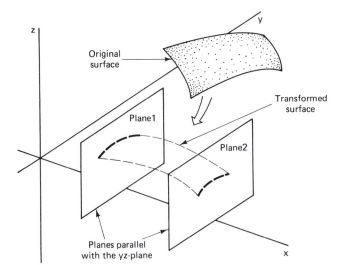

Figure 12.9   Candidate polyconic surface and its transformed location.

Assume that the "original surface" of Figure 12.9 is a candidate for description as a polyconic surface. Its arbitrary orientation conflicts with the requirements for defining it as a polyconic surface so we relocate it to become the "transformed surface." Its longitudinal axis is now parallel with the x-axis and it is delimited in this direction by planes parallel with the yz-plane (marked plane1 and plane2). Were we to view the surface in these planes, or in any other plane between these planes and parallel with the yz-plane, we would see a planar curve that could be described as a general conic (actually, a loft conic). This curve, for one such plane, is shown in Figure 12.10.

The cross section of the transformed surface in Figure 12.10 shows the curve delimited by endpoints p_1 and p_2. Tangents to this curve at the endpoints intersect at the point marked p_3. A parallelogram with a diagonal from p_3 can now be completed as shown. The coordinates of these points and the ratio $r = k_1/k_2$ are related to the poly-

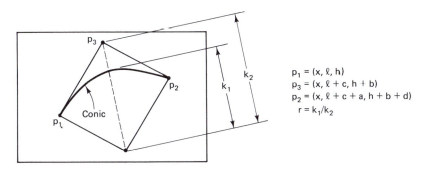

$$p_1 = (x, \ell, h)$$
$$p_3 = (x, \ell + c, h + b)$$
$$p_2 = (x, \ell + c + a, h + b + d)$$
$$r = k_1/k_2$$

Figure 12.10   Plane1 viewed in the direction of the positive x-axis.

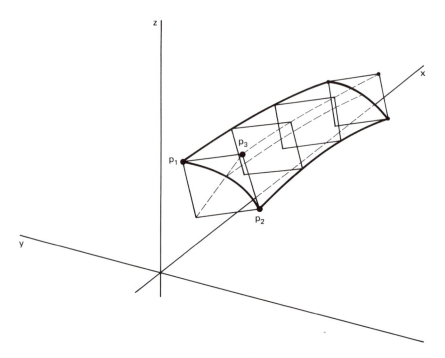

Figure 12.11   Visualization of the parallelogram construction of the conics.

nomials to be described. The points p_1 and p_2, their tangents, and the point that can be derived from r define a conic as described for the loft conic (Section 6.2). Figure 12.11 shows the transformed surface with the conics and parallelograms for several intersecting planes.

The profile of the transformed surface in its longitudinal direction, the x-axis, must be described by a set of polynomials, the surface control equations. Seven polynomials are sufficient for this purpose. They really represent the construction points of the parallelograms. Two of these polynomials, p_l and p_h, are projections onto the xy-plane and zx-plane of the curve formed by connecting all points p_1. Similarly, for points p_2 and p_3, four more polynomials are derived, while the seventh polynomial is the ratio curve in the rx-plane. These polynomials appear in Figure 12.12 with subscripts correlating them to the coordinates of the points of the parallelogram shown in Figure 12.10. They are also shown as functionally dependent on the parameter t, rather than on coordinate x. The points p_1, p_2, and p_3 in Figure 12.12 appear there only to permit associating the polynomials with the geometry of Figure 12.10. Each polynomial is limited to the following form:

$$p(t) = p_0 + p_q t^{1/2} + p_1 t + p_2 t^2 + p_3 t^3 + p_4 t^4$$
$$+ p_5 t^5 + p_6 t^6 + p_7 t^7 \qquad 0 \leq t \leq 1 \qquad (12.6)$$

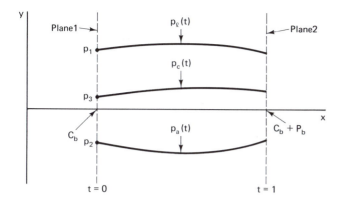

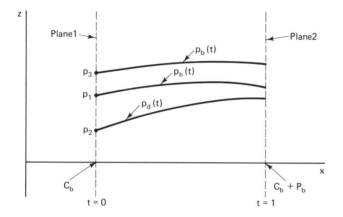

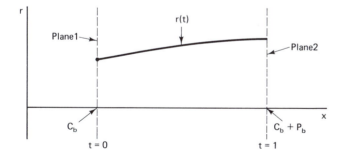

Figure 12.12 Parametric polynomials formed from the construction points of the parallelograms.

The coefficients must be constants. The parameter range for t is a normalization of the physical range $C_b \leq x \leq C_b + P_b$. Polynomials p_a, p_b, p_c and p_d are derived as follows. First, project the curve formed by connecting all points p_3 onto the ty- and zt-planes to form polynomials p_{lc} and p_{hb}, respectively. Similarly, for p_2 form polynomials p_{lca} and p_{hbd}. Then perform the subtractions,

$$\left.\begin{array}{l} p_a(t) = p_{lca}(t) - p_{lc}(t) \\ p_b(t) = p_{hb}(t) - p_h(t) \\ p_c(t) = p_{lc}(t) - p_l(t) \\ p_d(t) = p_{hbd}(t) - p_{hb}(t) \end{array}\right\} \tag{12.7}$$

The ratio curve $r(t)$ is used by APT to compute that point on the parallelogram diagonal through which the conic passes, tangent in the manner shown in Figure 12.10.

The APT format for the polyconic surface is

```
POLCON/CANON,jroots,m11,m12,m13,m14,m21,m22,m23,m24,      $
        m31,m32,m33,m34,thk,Cb,Pb,                        $
        A0,A1,A2,A3,A4,A5,A6,A7,                          $
        B0,B1,B2,B3,B4,B5,B6,B7,                          $
        C0,C1,C2,C3,C4,C5,C6,C7,                          $
        D0,D1,D2,D3,D4,D5,D6,D7,                          $
        L0,L1,L2,L3,L4,L5,L6,L7,                          $
        H0,H1,H2,H3,H4,H5,H6,H7,                          $
        R0,R1,R2,R3,R4,R5,R6,R7,                          $
        Aq,Bq,Cq,Dq,Lq,Hq,Rq
```

Parameter *jroots*, when set equal to 2, indicates that nonzero square root coefficients are present in the format; when set equal to 1, all coefficients are zero. Parameters *m11* through *m34*, the transformation matrix elements, are discussed below. Parameter *thk* is the thickness between the defined surface and the working surface [see "COPY using the SAME modifier" in Section 7.3.1], *Cb* is the x-coordinate value of the origin for the polynomials, and *Pb* is the physical range of x that polynomial parameter t normalizes (see Figure 12.12). Parameters *A0–A7*, *B0–B7*, . . . , *Aq–Rq* are the coefficients of the seven polynomials, each of the form shown in equation (12.6).

In terms of the parameters of the defining format above, the transformation matrix is as follows:

$$\begin{bmatrix} m11 & m12 & m13 & m14 \\ m21 & m22 & m23 & m24 \\ m31 & m32 & m33 & m34 \end{bmatrix}$$

This matrix conforms with that of equation (7.14) of Section 7.1.2 and with Matrix-1 definition in Table 7.1. The first three columns are the linear transformation, the fourth column is for translation. With reference to Figure 12.9, this matrix can be interpreted as transforming the "original surface" *to* the "transformed surface" location, at which it is then described as defined above. This is the inverse action of the matrix when used

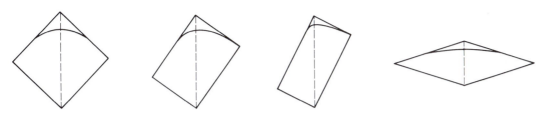

Figure 12.13   Various parallelogram shapes with associated conics.

with the REFSYS feature. The matrix must be constructed so as to position the surface at the desired $x = Cb$ location.

The following restrictions and precautions should be observed to avoid difficulty in defining and using the polyconic surface.

1. The polyconic surface exists only between its lateral defining polynomials and in the region from $x = Cb$ to $x = CB + Pb$. No attempt should be made to drive or check on this surface outside these bounds.
2. The shape of a conic derived from the polynomials is influenced by the aspect ratio of the parallelogram as implied in Figure 12.13. This conic reaches its greatest ''height'' as it crosses the diagonal.
3. The value of the ratio polynomial $r(t)$ for a given conic must cause the conic to cross the diagonal at a ''height'' greater than points p_1 and p_2. Otherwise, the conic cannot be defined.
4. A polyconic surface may be defined with the shape shown in Figure 12.14. However, inversion of tangents at the lateral polynomials requires that $r(t) = 1$ for some value of t and the tangent lines be coincident (p_1, p_2, and p_3 must be collinear). This condition will be mathematically difficult to achieve with the polynomials and should be avoided for the reason given in item 3. Separate adjoining polyconics should be defined for this surface.

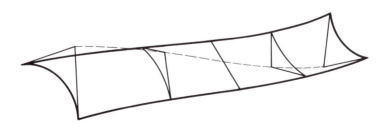

Figure 12.14   Shape difficult to describe as a polyconic surface.

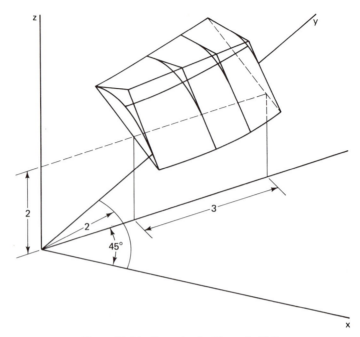

Figure 12.15   Geometry for Example 12.2.

Example 12.2

Figure 12.15 shows a surface whose geometry will be described as a polyconic. Let's assume that its projections onto the yt-plane and zt-plane after rotation yield the polynomials of Figure 12.16, where the range of t conforms with the requirements of the polyconic definition. The surface was translated 1 unit in the $-z$ direction and rotated by $-45°$ about the z-axis.

We now perform the polynomial subtractions of equation (12.7) and bring forth polynomials $p_l(t)$, $p_h(t)$, and $r(t)$ to get

$$p_a(t) = -1.25 - 0.7t + 0.55t^2$$
$$p_b(t) = 1 + t - 1.25t^2$$
$$p_c(t) = -1.25 - 0.2t$$
$$p_d(t) = -1 - 1.3t + 1.55t^2$$
$$p_l(t) = 1.5 + 0.5t + 0.25t^2$$
$$p_h(t) = 1 - 0.5t + 0.5t^2$$
$$p_r(t) = 0.8 + 0.1t$$

Coefficients for the $\sqrt{t}$ terms are all zero. We form the polyconic definition as follows:

RESERV/VAL,12	100
MAT1 = MATRIX/XYROT,-45	110
MAT2 = MATRIX/TRANSL,0,0,-1	120

```
MAT3 = MATRIX/MAT2,MAT1                                        130
       OBTAIN,MATRIX/MAT3,VAL(1,THRU,12)                      140
POL  = POLCON/CANON,1,VAL(1,THRU,12), 0,2,5,          $       150
              −1.25,−.7,.55,0,0,0,0,0,                $       160
              1,1,−1.25,0,0,0,0,0,0,                  $       170
              −1.25,−.2,0,0,0,0,0,0,0,                $       180
              −1,−1.3,1.55,0,0,0,0,0,0,               $       190
              1.5,.5,.25,0,0,0,0,0,0,                 $       200
              1,−.5,.5,0,0,0,0,0,0,                   $       210
              .8,.1,0,0,0,0,0,0,0,                    $       220
              0,0,0,0,0,0,0,0                                 230
```

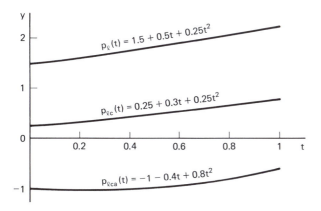

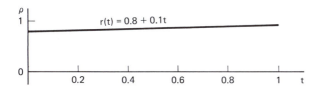

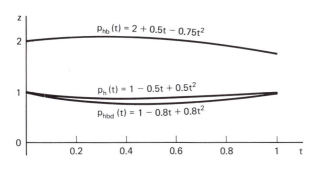

Figure 12.16   Polynomials obtained by projecting Figure 12.15 onto the axes shown.

12.4 VECTOR DEFINITIONS

Vectors were first introduced in Section 2.3.3 in connection with the INDIRV statement as part of a startup procedure. We used this statement numerous times since then and will find it useful for three-dimensional work also. A more formal introduction to vectors appeared in Section 7.1.1 in connection with coordinate transformations. Its concept is basic to the preparation of transformation matrices. Finally, we found that we could aid the APT system in determining the TO/PAST surfaces with vectors in the SRFVCT statement of Section 8.2.6.

In the chapters that follow we will find vectors important for three-dimensional work, where they will be used to define geometry, orient the tool for multiaxis work, establish directions, and so on. A selection of vector defining formats for these purposes is given in Table 12.5. We are still involved in defining geometry and will need vectors for this purpose in Section 12.5. Vector formats for computational applications are given in Section 13.1.

12.5 MORE MATRIX DEFINITIONS

We used matrices extensively since their introduction in Chapter 7, where a selection of defining formats is given in Table 7.1. The rest of the APT matrix definitions, given in Table 12.6, require advanced mathematical and geometrical concepts for their understanding and construction. These formats are useful for three-dimensional work.

The direction of the unit normals of the planes defining Matrix-8 (Table 12.6) determine the positive axes directions for the local coordinate system being defined by the matrix. We discussed in Section 8.4 the storage representation of the plane. From that discussion we know that the unit normal vector may be negative of that needed for the plane definition. This is compensated for by storing the negative of d_i, the distance of the plane from the origin. We suggested that care be taken before using plane parameters extracted with the OBTAIN statement. Such care must also be used when defining Matrix-8 to ensure that the desired transformation be represented in the local coordinate system established by the plane unit normal vectors. The sense of the unit normal for various plane definitions is discussed in Section 13.1.5.

The geometric concept underlying the transformation defined by Matrix-9 is suggested by the nomenclature of the format. The examples accompanying the definition clearly show that one matrix defined as the inverse of another implements a transformation that retraces the effect of the direct matrix. We can demonstrate this mathematically with a special case as follows.

Because division by a matrix is not defined, we premultiply both sides of the matrix equation $\mathbf{AB} = \mathbf{C}$, where $\mathbf{A}$ is an $n \times n$ matrix and $\mathbf{B}$ and $\mathbf{C}$ are each $n \times 1$ column vectors, by $\mathbf{A}'$, an $n \times n$ matrix that is the inverse of matrix $\mathbf{A}$, to get

$$\mathbf{A'AB} = \mathbf{A'C}$$

TABLE 12.5   VECTOR DEFINITIONS

1	By x, y, z components (canonical form):

VECTOR/[CANON,] dx, dy, dz

The vector defined has values dx, dy, and dz that
are components along the x, y, and z axes,
respectively. Equivalently, this is a vector directed
<u>from</u> the origin (0, 0, 0) <u>to</u> the point (dx, dy, dz).
Its length is determined from the given components.

   VEC1A = VECTOR/9, −4, −6
   VEC1B = VECTOR/−9, 4, 6
   VEC2A = VECTOR/5, 7, 7
   VEC2B = VECTOR/−5, −7, −7

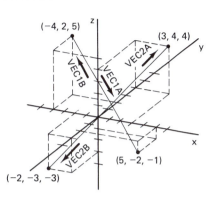

2	By two endpoints:

$$\text{VECTOR/} \begin{Bmatrix} x1, y1, z1, x2, y2, z2 \\ point1, point2 \end{Bmatrix}$$

The vector defined is directed <u>from</u> the first point,
given by coordinates $(x1, y1, z1)$ or symbolically as
$point1$, <u>to</u> the second point, given by coordinates
$(x2, y2, z2)$ or symbolically as $point2$. Its length is
the distance between the points.

   VEC1A = VECTOR/−3, −2, 2, 3, 4, 4
   VEC1A = VECTOR/P2, P3
   VEC2A = VECTOR/−6, 3, 3, 5, −4, −4
   VEC2A = VECTOR/P1, P4

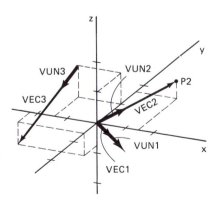

3	By normalizing a vector, a point, or components:

$$\text{VECTOR/UNIT,} \begin{Bmatrix} vector \\ point \\ dx, dy, dz \end{Bmatrix}$$

The vector defined is of unit length derived from
$vector$, from a vector defined from the origin to
$point$, or from a vector whose components are
dx, dy, dz.

   VUN1 = VECTOR/UNIT, VEC1
   VUN2 = VECTOR/UNIT, P2
   VUN3 = VECTOR/UNIT, −.8, −1.5, −1.7

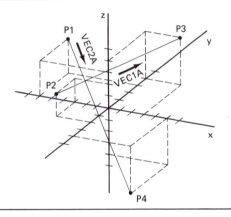

TABLE 12.5 (CONTINUED)

4	In the *xy*-plane at an angle to a line:

$$\text{VECTOR/ATANGL}, \theta, \textit{line}, \begin{Bmatrix} \text{POSX} \\ \text{XLARGE} \\ \text{POSY} \\ \text{YLARGE} \\ \text{NEGX} \\ \text{XSMALL} \\ \text{NEGY} \\ \text{YSMALL} \end{Bmatrix}$$

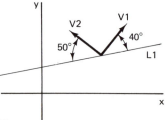

The defined vector is of unit length in the xy-plane at an angle θ to *line*. Angle θ is measured <u>from</u> *line* toward the vector. It is positive when measured in a counter-clockwise direction, negative in a clockwise direction. The modifier is chosen in conjunction with θ to select the desired vector from the two possibilities. POSX and XLARGE have the same meaning, namely, that the defined vector has a positive x-component. Similarly for the other pairs of modifiers.

$$V1 = \text{VECTOR/ATANGL}, 40, L1, \text{POSX}$$
$$V1 = \text{VECTOR/ATANGL}, -140, L1, \text{POSY}$$
$$V1 = \text{VECTOR/ATANGL}, 220, L1, \text{XLARGE}$$
$$V1 = \text{VECTOR/ATANGL}, -320, L1, \text{YLARGE}$$
$$V2 = \text{VECTOR/ATANGL}, -50, L1, \text{NEGX}$$
$$V2 = \text{VECTOR/ATANGL}, 130, L1, \text{POSY}$$
$$V2 = \text{VECTOR/ATANGL}, -230, L1, \text{YLARGE}$$
$$V2 = \text{VECTOR/ATANGL}, 310, L1, \text{XSMALL}$$

5	As perpendicular to a plane:

$$\text{VECTOR/PERPTO}, \textit{plane}, \begin{Bmatrix} \text{POSX} \\ \text{POSY} \\ \text{POSZ} \\ \text{NEGX} \\ \text{NEGY} \\ \text{NEGZ} \end{Bmatrix}$$

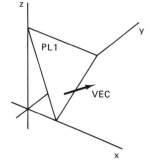

The defined vector is of unit length and perpendicular to *plane*. From among two possible directions, its orientation is determined by the choice of modifier. Each modifier specifies the polarity of a vector component along a coordinate axis. For example, POSX denotes a positive x-component value. The modifier is usually chosen on the basis of the dominant direction for the vector.

$$VEC = \text{VECTOR/PERPTO}, PL1, \text{POSX}$$
$$VEC = \text{VECTOR/PERPTO}, PL1, \text{POSY}$$
$$VEC = \text{VECTOR/PERPTO}, PL1, \text{POSZ}$$

TABLE 12.5 (CONTINUED)

---

6 | By its length and an angle in a coordinate plane:

$$\text{VECTOR/LENGTH}, l, \text{ATANGL}, \theta, \begin{Bmatrix} \text{XYPLAN} \\ \text{YZPLAN} \\ \text{ZXPLAN} \end{Bmatrix}$$

The vector defined is of length l at an angle θ to
a coordinate axis in the coordinate plane given
by the modifier. Angle θ is measured from the
first axis given in the modifier name (x-axis for
XYPLAN, y-axis for YZPLAN, and z-axis for
ZXPLAN). It is positive when measured in a
counterclockwise direction <u>from</u> that axis <u>to</u> the
vector, negative in a clockwise direction. To
establish proper vector direction, (a) mentally
translate the vector so its tail is at the origin,
then (b) view the plane from the positive axis not
included in the modifier name and measure θ
<u>from</u> the positive axis given in the modifier.

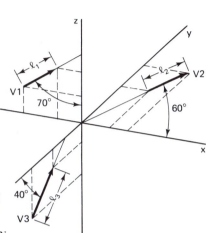

V1 = VECTOR/LENGTH, L1, ATANGL, −70, ZXPLAN
V1 = VECTOR/LENGTH, L1, ATANGL, 290, ZXPLAN
V2 = VECTOR/LENGTH, L2, ATANGL, 60, XYPLAN
V2 = VECTOR/LENGTH, L2, ATANGL, −300, XYPLAN
V3 = VECTOR/LENGTH, L3, ATANGL, 40, YZPLAN
V3 = VECTOR/LENGTH, L3, ATANGL, −320, YZPLAN

---

7 | As parallel with the intersection of two planes:

$$\text{VECTOR/PARLEL}, \text{INTOF}, plane1, plane2, \begin{Bmatrix} \text{POSX} \\ \text{POSY} \\ \text{POSZ} \\ \text{NEGX} \\ \text{NEGY} \\ \text{NEGZ} \end{Bmatrix}$$

The defined vector is of unit length and parallel
with the line of intersection of *plane1* with *plane2*.
Its directional orientation is determined by the
choice of modifier. Each modifier specifies the
polarity of a vector component along a coordinate
axis. For example, POSX denotes a positive
x-component value. The modifier should be chosen
on the basis of the dominant direction for the
vector.

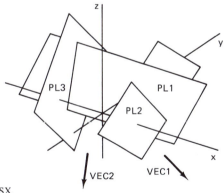

VEC1 = VECTOR/PARLEL, INTOF, PL1, PL2, POSX
VEC2 = VECTOR/PARLEL, INTOF, PL1, PL3, NEGY

TABLE 12.6   MATRIX DEFINITIONS

8	By three mutually perpendicular planes:

MATRIX/*plane1*, *plane2*, *plane3*

The matrix defined represents a transformation from a local system into the base system. The mutually perpendicular coordinate planes of the local system are *plane1*, the yz-plane, *plane2*, the zx-plane, and *plane3*, the xy-plane. For planes whose canonical forms are

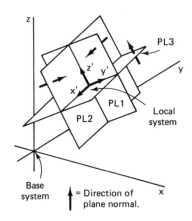

$$\text{plane1}: \; a_1, b_1, c_1, d_1$$
$$\text{plane2}: \; a_2, b_2, c_2, d_2$$
$$\text{plane3}: \; a_3, b_3, c_3, d_3$$

the matrix generated will be

$$\begin{bmatrix} a_1 & a_2 & a_3 & (a_1 d_1 + a_2 d_2 + a_3 d_3) \\ b_1 & b_2 & b_3 & (b_1 d_1 + b_2 d_2 + b_3 d_3) \\ c_1 & c_2 & c_3 & (c_1 d_1 + c_2 d_2 + c_3 d_3) \end{bmatrix}$$

The positive axes of the local system are determined by the (a_i, b_i, c_i), the unit normals of each plane. The direction of the unit normal must be determined in conjunction with the sign of d_i.

↑ = Direction of plane normal.

$$\text{MAT1} = \text{MATRIX/PL1, PL2, PL3}$$

9	As the inverse of a given matrix:

MATRIX/INVERS, *matrix*

The matrix defined performs the inverse transformation of that performed by *matrix*. If *matrix* is defined as

$$\begin{bmatrix} a_1 & b_1 & c_1 & d_1 \\ a_2 & b_2 & c_2 & d_2 \\ a_3 & b_3 & c_3 & d_3 \end{bmatrix}$$

The inverse matrix generated is

$$\begin{bmatrix} a_1 & a_2 & a_3 & -(a_1 d_1 + a_2 d_2 + a_3 d_3) \\ b_1 & b_2 & b_3 & -(b_1 d_1 + b_2 d_2 + b_3 d_3) \\ c_1 & c_2 & c_3 & -(c_1 d_1 + c_2 d_2 + c_3 d_3) \end{bmatrix}$$

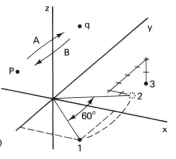

Example: Given that matrix A transforms point p into point q, then B = MATRIX/INVERS, A transforms point q into point p.

Example: Given, M1 = MATRIX/TRANSL, −2, 4, −2
                      M2 = MATRIX/XYROT, 60
                      M3 = MATRIX/M1, M2   (path 1–2–3)

           then, M4 = MATRIX/INVERS, M3   (path 3–2–1)

           functions as

                      M5 = MATRIX/TRANSL, 2, −4, 2
                      M6 = MATRIX/XYROT, −60
                      M7 = MATRIX/M6, M5   (path 3–2–1)

TABLE 12.6 (CONTINUED)

| 10 | By a point and two vectors: |

MATRIX/*point*, *vector1*, *vector2*

The matrix defined represents a transformation
from a local system into the base system. Let
vector1 and *vector2* originate at *point*, the origin
of the local system. Then the positive x-axis is
in the direction of *vector1* while *vector2*, in
conjunction with *vector1*, determines the xy-plane
of the local system (both vectors lie in this plane).
The y-axis of the local system lies on that side of
vector1 for which *vector2* makes an angle less
than 180° when measured from *vector1*. The
positive z-axis is then constructed perpendicular
to the xy-plane to form a right-hand coordinate
system.

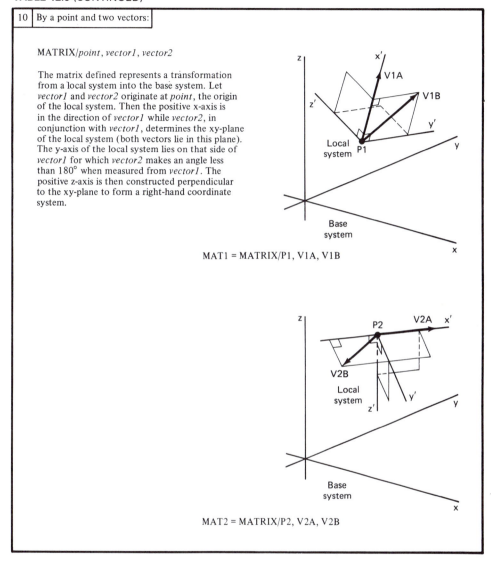

MAT1 = MATRIX/P1, V1A, V1B

MAT2 = MATRIX/P2, V2A, V2B

Since $\mathbf{A'A} = \mathbf{U}$, the unit matrix, the result is

$$\mathbf{B} = \mathbf{A'C}$$

For $n = 3$, matrix $\mathbf{A}$ transforms point $\mathbf{B}$ into point $\mathbf{C}$. Conversely, matrix $\mathbf{A'}$ transforms point $\mathbf{C}$ into point $\mathbf{B}$. More specifically, for rotation in the xy-plane the transformation matrix is

$$\mathbf{A} = \begin{bmatrix} \cos\theta & -\sin\theta & 0 \\ \sin\theta & \cos\theta & 0 \\ 0 & 0 & 1 \end{bmatrix}$$

Its inverse is

$$\mathbf{A'} = \begin{bmatrix} \cos\theta & \sin\theta & 0 \\ -\sin\theta & \cos\theta & 0 \\ 0 & 0 & 1 \end{bmatrix}$$

and

$$\mathbf{A'A} = \begin{bmatrix} \cos\theta & \sin\theta & 0 \\ -\sin\theta & \cos\theta & 0 \\ 0 & 0 & 1 \end{bmatrix} \begin{bmatrix} \cos\theta & -\sin\theta & 0 \\ \sin\theta & \cos\theta & 0 \\ 0 & 0 & 1 \end{bmatrix} = \begin{bmatrix} 1 & 0 & 0 \\ 0 & 1 & 0 \\ 0 & 0 & 1 \end{bmatrix}$$

The Matrix-9 definition gives us a convenient way to compute the inverse of a matrix for whatever purpose it is required. Vectors are used in the definition of Matrix-10. However, we must still be able to visualize the orientation of the z'-axis as shown in the figures of that definition. This is conveniently done by applying the right-hand rule as described in Section 1.5.

12.6 MISCELLANEA

Now that we have a supply of vector definitions and have all the APT matrices described, we can present more definitions of entities discussed earlier. In particular, we will define points, lines, and ellipses that use three-dimensional entities in their formats. We also complete the discussion of tabulated cylinders when defined in other than the xy-plane.

12.6.1 Points

The point definitions are completed in Table 12.7, where vectors and planes are used in a three-dimensional orientation to define the points. Numbering of the definitions is continued from Table 3.1. We have presented 16 point definitions, not necessarily the number available for a given APT processor.

12.6.2 Lines

The last line definition is given in Table 12.8. It is defined via planes and its number, Line-17, is continued from Table 3.2. There may be more or less than 17 definitions for a given APT processor.

TABLE 12.7 POINT DEFINITIONS

14	By the intersection of a plane and a vector:

POINT/INTOF, *plane*, *vector*, *point*

The point is defined where *vector* pierces *plane*.
Vector is positioned in space so as to first pass
through *point*.

PA = POINT/INTOF, PLN, VEC1, P1
PB = POINT/INTOF, PLN, VEC2, P2

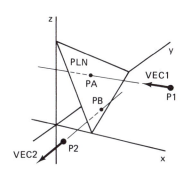

15	By a vectorial displacement from a point:

POINT/*point*, DELTA, *scalar*, TIMES [, UNIT], *vector*

The point defined is located according to the
orientation of *vector* with respect to *point*, a
reference location, at a distance related to the
magnitude of *vector* via the numerical value
scalar. This distance may be the product of
scalar by the magnitude of *vector*, or by the
value of *scalar* alone (when UNIT is included
to signify that *vector* is to be applied as a unit
vector for this calculation).

P1 = POINT/PREF, DELTA, 2, TIMES, VEC
P2 = POINT/PREF, DELTA, −5, TIMES, UNIT, VEC

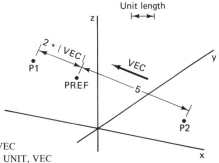

16	As the intersection of three planes:

POINT/INTOF, *plane1*, *plane2*, *plane3*

The point defined is at the intersection of
plane1, *plane2*, and *plane3*.

PP = POINT/INTOF, PL1, PL2, PL3

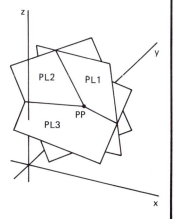

TABLE 12.8 LINE DEFINITIONS

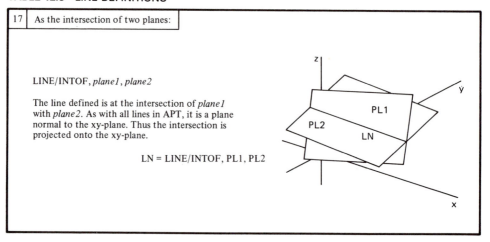

| 17 | As the intersection of two planes: |

LINE/INTOF, *plane1*, *plane2*

The line defined is at the intersection of *plane1* with *plane2*. As with all lines in APT, it is a plane normal to the xy-plane. Thus the intersection is projected onto the xy-plane.

LN = LINE/INTOF, PL1, PL2

12.6.3 Ellipses

The two ellipse definitions of Table 12.9 supplement the definition given in Section 6.1.3. These remaining definitions reference three-dimensional entities. It is emphasized that the ellipses are defined in the *xy*-plane. Thus the PRINT/3 listing of their arguments should be interpreted according to the GCONIC printout description.

12.6.4 Tabulated Cylinders

The tabulated cylinder definition of Section 6.3 was limited to the *xy*-plane only. Its transformation in the *xy*-plane via the TRFORM,*matrix* couplet was illustrated in Example 7.2. Not shown there, however, were the TABCYL points in another principal coordinate plane, nor were they transformed in an arbitrary manner. In this paragraph we complete the TABCYL discussion in the context of such orientations.

The TABCYL format below is repeated from Section 6.3.1.

$$name = \text{TABCYL}/\begin{Bmatrix} \text{NOX} \\ \text{NOY} \\ \text{NOZ} \\ \text{RTHETA} \\ \text{THETAR} \end{Bmatrix}, \text{SPLINE,[TRFORM,}matrix,]data$$

This discussion concerns the NOX and NOY coordinate-pair modifiers. Another modifier, XYZ, allows defining with arbitrary point locations. This modifier is not discussed in this book. NOX specifies that the coordinate-pair values of *data* lie in the *yz*-plane (y,z pairs, in that order) while NOY specifies that they are in the *zx*-plane (z,x pairs, in that order). These values are subject to transformation by *matrix*. Options for

TABLE 12.9 ELLIPSE DEFINITIONS

| 2 | By the intersection of a plane and a cylinder: |

ELLIPS/INTOF, *plane*, *cylinder*

The ellipse is defined by the projection onto the
xy-plane of the intersection of *plane* with *cylinder*.
The cylinder axis vector must not be perpendicular
to the plane normal vector. When the cylinder axis
vector is perpendicular to the xy-plane, the ellipse
degenerates into a circle.

ELP2 = ELLIPS/INTOF, PLN2, CYL

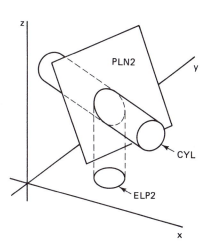

| 3 | By the intersection of a plane and a cone: |

ELLIPS/INTOF, *plane*, *cone*

The ellipse is defined by the projection onto the
xy-plane of the intersection of *plane* with *cone*.
The angle of the plane with respect to the cone
axis must be greater than the half angle of the
vertex of the cone. When the cone axis vector is
perpendicular to the xy-plane and the plane is
parallel with the xy-plane, the ellipse degenerates
into a circle.

ELP3 = ELLIPS/INTOF, PLN3, CON

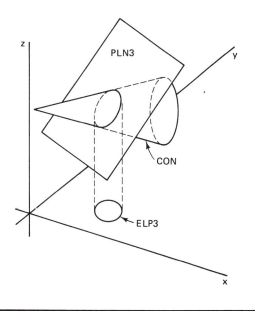

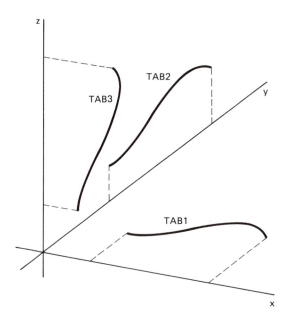

Figure 12.17   **TABCYLs defined in the three principal coordinate planes.**

data entry follow the description of Section 6.3.1 but are modified to allow *y-z* and *z-x* values. Rulings to generate TABCYLs with the NOX or NOY modifiers are parallel with the *x*- or *y*-axis, respectively.

The TABCYLs of Figure 12.17 are defined with the same data as follows:

     TAB1 = TABCYL/NOZ,SPLINE,1.5,1.14, 2.75,1.65, 3.75,2.25, $ \\
                 5,3, 6.5,3.6, 7.75,3.7, 8.75,3.3 \\
     TAB2 = TABCYL/NOX,SPLINE,1.5,1.14, 2.75,1.65, 3.75,2.25, $ \\
                 5,3, 6.5,3.6, 7.75,3.7, 8.75,3.3 \\
     TAB3 = TABCYL/NOY,SPLINE,1.5,1.14, 2.75,1.65, 3.75,2.25, $ \\
                 5,3, 6.5,3.6, 7.75,3.7, 8.75,3.3

We know from Section 7.2.1 that TABCYLs are not transformed by the REFSYS feature. To use them within the REFSYS bounding statements, it is first necessary to transform the points in the TABCYL definition. This is done with the TRFORM,*matrix* couplet. The action is as follows. First, the points are transformed according to *matrix*. Then, depending on the modifier specified (NOZ, NOX, or NOY), these points are projected onto the respective plane for use in defining the TABCYL. The TABCYL is defined *only* in the plane corresponding to the chosen modifier even though arbitrary transformation with the TRFORM,*matrix* couplet is possible.

PROBLEMS

12.1. Write APT geometric statements for the object of Figure P12.1. Use the PRINT/3 statement
to verify selected geometric definitions and also use motion statements with a cutter diameter
of zero to trace out key geometric features. Auxiliary geometric definitions may be required
for the latter purpose.

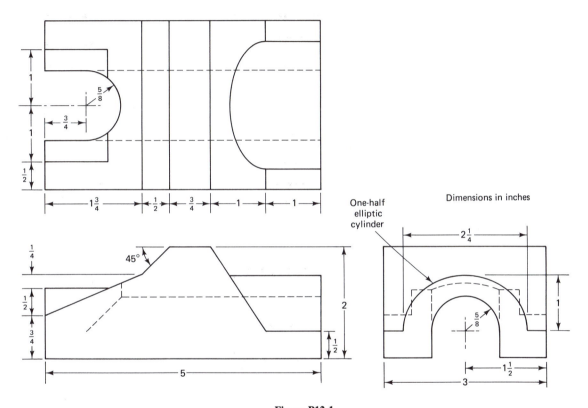

Figure P12.1

12.2. Repeat Problem 12.1 for Figure P12.2.

12.3. Repeat Problem 12.1 for Figure P12.3. Use the grid to determine dimensions not given.

12.4. Repeat Problem 12.1 for Figure P12.4.

12.5. Repeat Problem 12.1 for Figure P12.5.

12.6. Repeat Problem 12.1 for Figure P12.6.

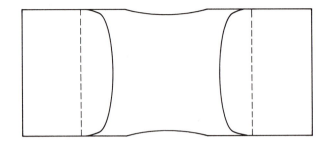

Dimensions in inches

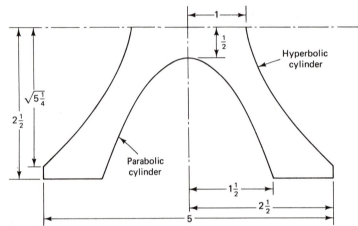

Figure P12.2

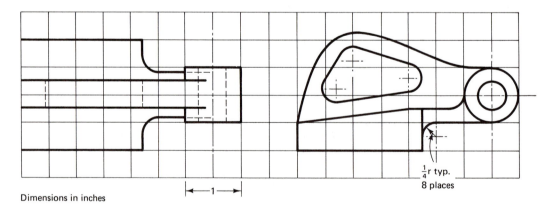

Dimensions in inches

$\frac{1}{4}$r typ.
8 places

Figure P12.3

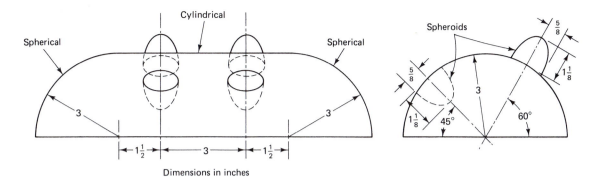

Dimensions in inches

Figure P12.4

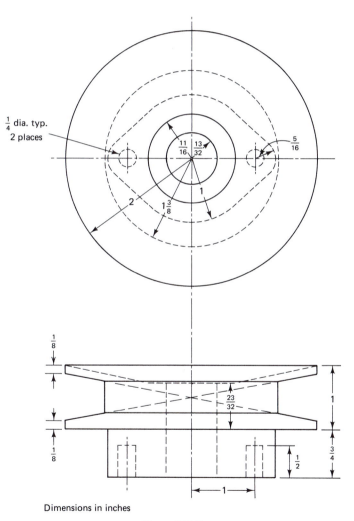

Dimensions in inches

Figure P12.5

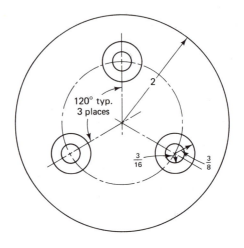

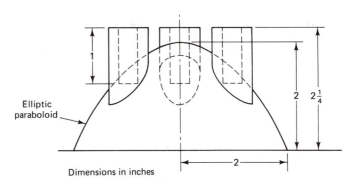

Figure P12.6

12.7. Repeat Problem 12.1 for Figure P12.7.

12.8. Repeat Problem 12.1 for Figure P12.8.

12.9. Repeat Problem 12.1 for Figure P12.9.

12.10. Repeat Problem 12.1 for Figure P12.10. Use the ruled surface definition for that part of the object where it is indicated.

12.11. Repeat Problem 12.1 for Figure P12.11. The large cylindrical section has two raised patches, one smaller one on top of another somewhat larger one. The raised patches are to be faired into the cylinder as indicated in the front and side views with ruled surfaces (not sketched in). The upper peripheries of the two patches govern protruding dimensions. Additional fairing is permitted as shown in the front and side views.

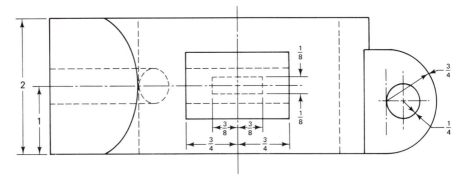

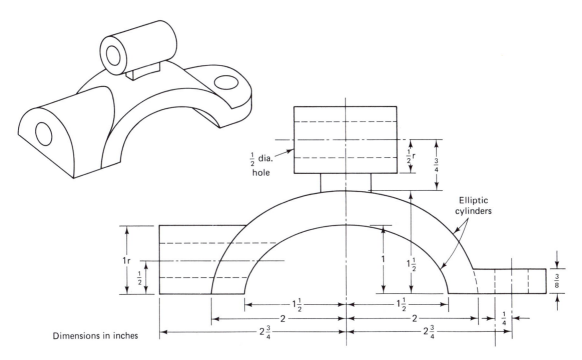

Dimensions in inches

Figure P12.7

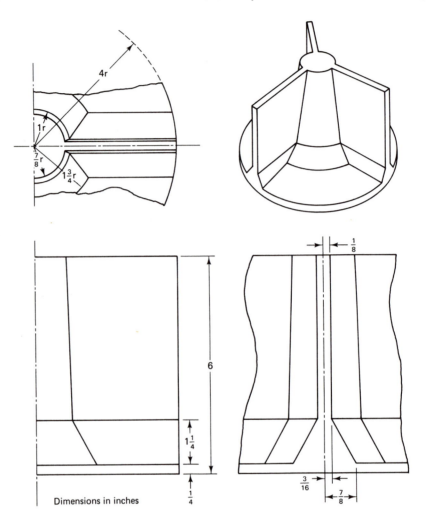

Dimensions in inches

Figure P12.8

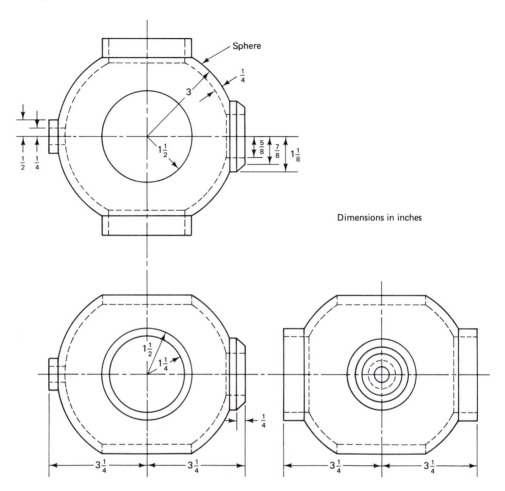

Dimensions in inches

Figure P12.9

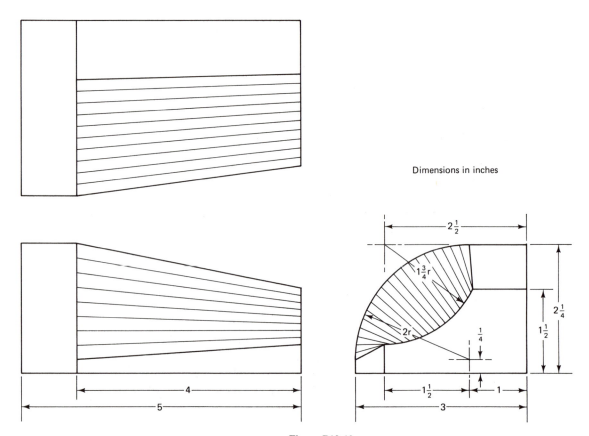

Dimensions in inches

Figure P12.10

12.12. Repeat Problem 12.1 for Figure P12.12. However, describe the scooped-out section as a polyconic surface. The shape of the section near the center of the object is shown in the top view. Use the grid to determine dimensions not given.

12.13. Repeat Problem 12.1 for Figure P12.13. One blade of a propeller is shown. Describe it as polyconic surfaces. Let R, the radius of the blade, be 4 in. and ψ, the rake angle, be 15°. Use the grid to determine dimensions not given.

12.14. Repeat Problem 12.13 for the blades of Figure P12.14. Except for the shape of the blades, other geometric particulars are as shown in Figure P12.13.

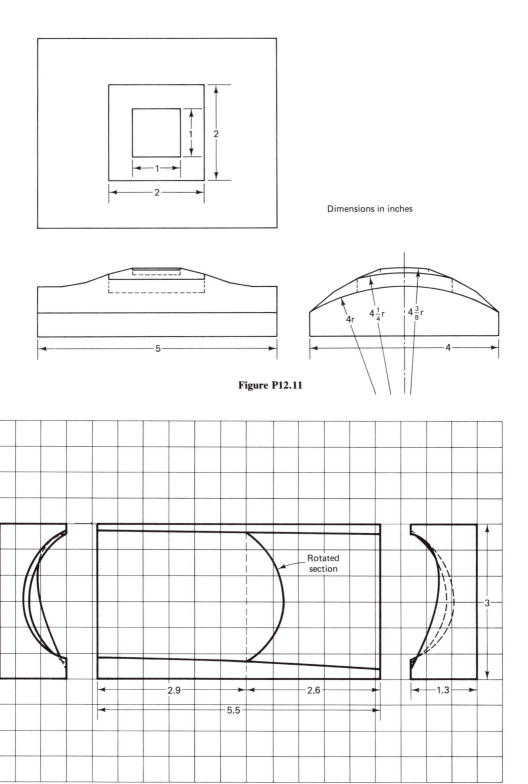

Dimensions in inches

Figure P12.11

Rotated section

Dimensions in inches

Figure P12.12

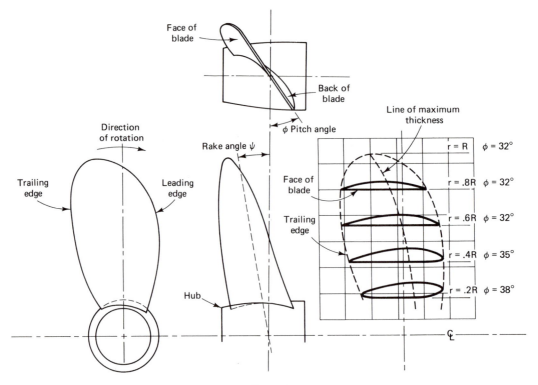

Figure P12.13

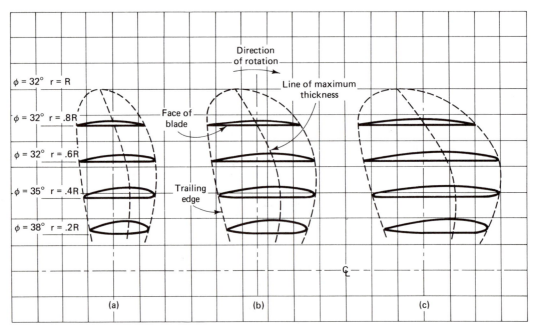

Figure P12.14

CHAPTER 13

Three-Axis Contouring

The APT geometric definitions of Chapter 12 completed the essential material for describing parts for contouring under three-axis control. Three-axis control implies a fixed tool axis parallel with the z-axis. The material of Chapter 14 will allow the tool axis to be arbitrarily oriented within the coordinate system under what is there referred to as multiaxis control, of which three-axis control is a special case.

In this chapter we present the essentials of vector mathematics applicable to three-axis and multiaxis applications. We also present the expanded cutter statement, from which more complex cutter envelopes are created. Because we admit movement along the three principal axes, it is appropriate that we allow cutter shapes more suited to the job at hand. Finally, we do not claim that APT will perform miracles by always computing without difficulty cut vectors for arbitrarily complex tool-to-surface relationships but instead we alert the part programmer to issues contributing to computational difficulty. All of this material will apply to the multiaxis control issues discussed in Chapter 14.

13.1 VECTOR MATHEMATICS

We defined vectors for use in geometric definitions and for directing the tool. Their use in this manner remains important for three-dimensional and multiaxis work. However, we must be prepared to compute with vectors as the need arises. Therefore, we now present those APT features that involve vector mathematics. The remaining vector definitions given in Table 13.1 are suitable for this purpose.

TABLE 13.1 VECTOR DEFINITIONS

8	By the addition or subtraction of vectors or points:

$$\text{VECTOR/}\begin{Bmatrix}vect1\\point1\end{Bmatrix}, \begin{Bmatrix}\text{PLUS}\\\text{MINUS}\end{Bmatrix}, \begin{Bmatrix}vect2\\point2\end{Bmatrix}$$

The vector defined is formed by adding (modifier PLUS) or subtracting (modifier MINUS) some combination of vectors (*vect1* and/or *vect2*) and/or points (*point1* and/or *point2*). The selection of *point1* or *point2* assumes a vector from the origin to the point.

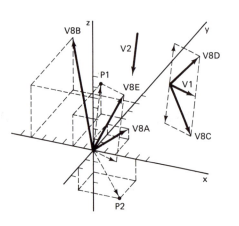

$$V8A = \text{VECTOR/P1, PLUS, P2}$$
$$= (-2i + 4j + 3k) + (4i - 2j - 2k)$$
$$= 2i + 2j + k$$
$$V8B = \text{VECTOR/P1, MINUS, P2}$$
$$= (-2i + 4j + 3k) - (4i - 2j - 2k)$$
$$= -6i + 6j + 5k$$
$$V8C = \text{VECTOR/V1, PLUS, V2}$$
$$V8D = \text{VECTOR/V1, MINUS, V2}$$
$$V8E = \text{VECTOR/V1, PLUS, P1}$$

9	By a scalar times a vector or point:

$$\text{VECTOR/}scalar, \text{TIMES,}\begin{Bmatrix}vector\\point\end{Bmatrix}$$

The vector defined is formed by multiplying each component of *vector* (a vector) by the value *scalar* or by multiplying each coordinate value of *point* (a point assumed to be a vector from the origin to the point) by *scalar*.

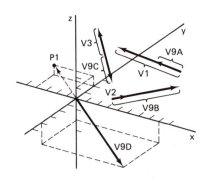

$$V9A = \text{VECTOR/.4, TIMES, V1}$$
$$V9B = \text{VECTOR/5, TIMES, V2}$$
$$V9C = \text{VECTOR/}{-1}, \text{TIMES, V3}$$
$$V9D = \text{VECTOR/}{-2}, \text{TIMES, P1}$$
$$= (-2)(-4i + 2j + k) = 8i - 4j - 2k$$

10	By the cross product of vectors or points:

$$\text{VECTOR/}\begin{Bmatrix}vect1\\point1\end{Bmatrix}, \text{CROSS,}\begin{Bmatrix}vect2\\point2\end{Bmatrix}$$

The vector defined is formed by taking the cross product in the form A × B, where A is represented by a vector or a point (*vect1* or *point1*, respectively) and where B is similarly represented by *vect2* or *point2*. The selection of *point1* or *point2* assumes a vector from the origin to the point. The magnitude of the resultant vector is $|A||B|\sin\theta$, where θ is the included angle. Its direction is determined according to the right-hand rule applied to vector products.

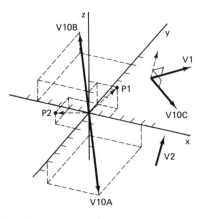

$$V10A = \text{VECTOR/P1, CROSS, P2}$$
$$= (2i + j + 2k) \times (-2i - 2j + k)$$
$$= -5i - 6j - 2k$$
$$V10B = \text{VECTOR/P2, CROSS, P1}$$
$$= -5i + 6j + 2k$$
$$V10C = \text{VECTOR/V1, CROSS, V2}$$

13.1.1 Addition and Subtraction

Unit vectors along the coordinate axes were defined in Section 7.1.2 as follows:

$$\hat{\mathbf{e}}_1 = \{1, 0, 0\}$$

$$\hat{\mathbf{e}}_2 = \{0, 1, 0\}$$

$$\hat{\mathbf{e}}_3 = \{0, 0, 1\}$$

For convenience, we now adopt the engineering notation and define them as follows:

$$\mathbf{i} = [1, 0, 0]$$

$$\mathbf{j} = [0, 1, 0]$$

$$\mathbf{k} = [0, 0, 1]$$

The vectors **a** and **b** can then be written as

$$\mathbf{a} = [a_1, a_2, a_3] = a_1\mathbf{i} + a_2\mathbf{j} + a_3\mathbf{k}$$

$$\mathbf{b} = [b_1, b_2, b_3] = b_1\mathbf{i} + b_2\mathbf{j} + b_3\mathbf{k}$$

The operation of vector addition is given by

$$\mathbf{a} + \mathbf{b} = (a_1\mathbf{i} + a_2\mathbf{j} + a_3\mathbf{k}) + (b_1\mathbf{i} + b_2\mathbf{j} + b_3\mathbf{k})$$

$$= (a_1 + b_1)\mathbf{i} + (a_2 + b_2)\mathbf{j} + (a_3 + b_3)\mathbf{k}$$

$$= [a_1 + b_1, a_2 + b_2, a_3 + b_3]$$

The sum of two vectors, therefore, is another vector whose components are formed by adding the corresponding components of the contributing vectors. A geometric interpretation is given in Figure 13.1. Vector **b** is moved parallel with itself so that its initial point coincides with the final point of **a**. The sum of vectors **a** and **b** is another vector from the initial point of **a** to the final point of **b**. It is the diagonal of a parallelogram.

Vector addition is commutative, that is, $\mathbf{a} + \mathbf{b} = \mathbf{b} + \mathbf{a}$. The operation of vector subtraction is given by

$$\mathbf{a} - \mathbf{b} = (a_1\mathbf{i} + a_2\mathbf{j} + a_3\mathbf{k}) - (b_1\mathbf{i} + b_2\mathbf{j} + b_3\mathbf{k})$$

$$= (a_1\mathbf{i} + a_2\mathbf{j} + a_3\mathbf{k}) + (-b_1\mathbf{i} - b_2\mathbf{j} - b_3\mathbf{k})$$

$$= (a_1 - b_1)\mathbf{i} + (a_2 - b_2)\mathbf{j} + (a_3 - b_3)\mathbf{k}$$

$$= [a_1 - b_1, a_2 - b_2, a_3 - b_3]$$

The difference between two vectors, therefore, is another vector whose components are formed by subtracting the components of the second vector from the corresponding components of the first vector. A geometric interpretation is given in Figure 13.2. Vector **b** is moved parallel with itself so that its initial point coincides with the final point of **a**, then made negative (reversed in direction). The difference between vectors **a** and **b** is

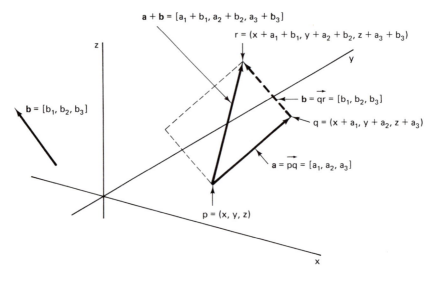

Figure 13.1   Vector addition.

another vector from the initial point of **a** to the final point of −**b**. It is the diagonal of a parallelogram. The APT format for vector addition and subtraction is shown as Vector-8 in Table 13.1.

We denote a zero vector (one having zero magnitude) by **0**. It represents a single point for which direction is wholly ambiguous. It is simplest to think of it as having all

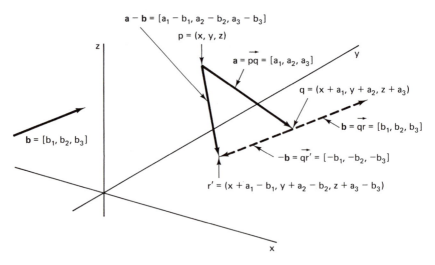

Figure 13.2   Vector subtraction.

directions. Therefore, the vector **0** is both parallel with and perpendicular to every vector **a**.

13.1.2 Scalar Times a Vector

Sometimes we must multiply a vector **a** by a scalar c. The product of c and **a**, denoted by c**a**, is a vector given by

$$c\mathbf{a} = c(a_1\mathbf{i} + a_2\mathbf{j} + a_3\mathbf{k})$$

$$= ca_1\mathbf{i} + ca_2\mathbf{j} + ca_3\mathbf{k}$$

$$= [ca_1, ca_2, ca_3]$$

For $c > 0$, then c**a** is in the same direction as **a**; for $c < 0$, then c**a** is in the direction opposite to **a**. In effect, we have scaled the length of vector **a** by a factor of c. This we show as follows: For $c\mathbf{a} = [ca_1, ca_2, ca_3]$ we have

$$|c\mathbf{a}| = \sqrt{(ca_1)^2 + (ca_2)^2 + (ca_3)^2}$$

$$= c\sqrt{a_1^2 + a_2^2 + a_3^2}$$

$$= c\,|\mathbf{a}|$$

The APT function for computing the length (magnitude) of a vector is

LNTHF(*vector*)

where *vector* is a single vector or a vector expression.

The APT format for multiplying a vector by a scalar is shown as Vector-9 in Table 13.1. With this format we may define one vector as opposite in direction to another by setting the value of scalar c to -1. In APT it is not possible to define such a vector in an arithmetic assignment statement in the form VECT1 $= -$VECT2 because geometric entities cannot be treated as scalar variables.

13.1.3 Dot Product

Although we will find it necessary to multiply vectors, we must distinguish between two kinds of vector multiplication operations. One operation, the dot product, produces a scalar value. The other, the cross product, produces another vector. In this section we discuss the dot product.

Given two vectors **a** and **b** as follows:

$$\mathbf{a} = [a_1, a_2, a_3] \qquad \mathbf{b} = [b_1, b_2, b_3]$$

we form the dot product, also known as the scalar product or inner product, as follows:

$$\mathbf{a} \cdot \mathbf{b} = (a_1\mathbf{i} + a_2\mathbf{j} + a_3\mathbf{k}) \cdot (b_1\mathbf{i} + b_2\mathbf{j} + b_3\mathbf{k})$$

$$= a_1b_1 + a_2b_2 + a_3b_3$$

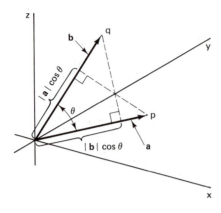

Figure 13.3 Construction for interpreting the dot product.

The dot product operator is denoted by the "•" multiplication symbol. The result of applying it is a scalar value.

A geometric interpretation of the dot product is obtained from Figure 13.3. In the plane of **a** and **b** we apply the law of cosines and get

$$|\vec{pq}|^2 = |\mathbf{a}|^2 + |\mathbf{b}|^2 - 2|\mathbf{a}|\,|\mathbf{b}|\cos\theta$$

from which

$$\cos\theta = \frac{|\mathbf{a}|^2 + |\mathbf{b}|^2 - |\vec{pq}|^2}{2|\mathbf{a}|\,|\mathbf{b}|}$$

$$= \frac{(a_1{}^2 + a_2{}^2 + a_3{}^2) + (b_1{}^2 + b_2{}^2 + b_3{}^2) - [(a_1 - b_1)^2 + (a_2 - b_2)^2 + (a_3 - b_3)^2]}{2|\mathbf{a}|\,|\mathbf{b}|}$$

$$= \frac{2a_1b_1 + 2a_2b_2 + 2a_3b_3}{2|\mathbf{a}|\,|\mathbf{b}|}$$

$$= \frac{a_1b_1 + a_2b_2 + a_3b_3}{|\mathbf{a}|\,|\mathbf{b}|}$$

Hence

$$\cos\theta = \frac{\mathbf{a}\cdot\mathbf{b}}{|\mathbf{a}|\,|\mathbf{b}|}$$

from which we obtain

$$\mathbf{a}\cdot\mathbf{b} = |\mathbf{a}|\,|\mathbf{b}|\cos\theta \qquad 0° \leq \theta \leq 180°$$

Thus the dot product of two vectors is the product of their magnitudes and the cosine of the angle between the vectors. If $\mathbf{a} = \mathbf{u}$, the unit vector, then

$$\mathbf{u} \cdot \mathbf{b} = |\mathbf{u}| \, |\mathbf{b}| \cos \theta = |\mathbf{b}| \cos \theta$$

Thus $|\mathbf{b}| \cos \theta$ is the scalar projection of $\mathbf{b}$ onto $\mathbf{a}$ and is the component of $\mathbf{b}$ in the direction of $\mathbf{a}$. We arrive at an analogous result if we let $\mathbf{b} = \mathbf{u}$. In general, we can say that the dot product is the product of the magnitude of one vector by the scalar projection of the second vector onto the first.

In APT the DOTF function is used to compute the dot product as follows:

dot-product $= vect1 \cdot vect2 = \text{DOTF}(vect1, vect2)$

where *vect1* and *vect2* are the two vectors for which the dot product is desired.

If we apply the dot product to the unit vectors, we obtain

$$\mathbf{i} \cdot \mathbf{i} = \mathbf{j} \cdot \mathbf{j} = \mathbf{k} \cdot \mathbf{k} = 1$$

and

$$\mathbf{i} \cdot \mathbf{j} = \mathbf{j} \cdot \mathbf{i} = \mathbf{i} \cdot \mathbf{k} = \mathbf{k} \cdot \mathbf{i} = \mathbf{j} \cdot \mathbf{k} = \mathbf{k} \cdot \mathbf{j} = 0$$

The dot product obeys the commutative law,

$$\mathbf{a} \cdot \mathbf{b} = \mathbf{b} \cdot \mathbf{a}$$

and when applied to a single vector, say $\mathbf{a}$, yields the magnitude squared,

$$\mathbf{a} \cdot \mathbf{a} = |\mathbf{a}|^2$$

The dot product $\mathbf{a} \cdot \mathbf{b} = ab \cos \theta$ will be zero when a or b or $\cos \theta$ is zero. If $\cos \theta = 0$, $\mathbf{a}$ and $\mathbf{b}$ are perpendicular. Since we said the $\mathbf{0}$ vector is perpendicular to every vector, we can summarize all cases in the rule

$$\mathbf{a} \cdot \mathbf{b} = 0 \qquad \text{if and only if } \mathbf{a} \perp \mathbf{b}$$

This property is the basis for most geometric applications of the dot product.

13.1.4 Cross Product

The cross product, also known as the vector product, is a vector multiplication operation that produces a vector result. We illustrate its computation as follows.

Given two vectors $\mathbf{a}$ and $\mathbf{b}$ as follows:

$$\mathbf{a} = [a_1, a_2, a_3] \qquad \mathbf{b} = [b_1, b_2, b_3]$$

we form the cross product according to the following definition:

$$\mathbf{a} \times \mathbf{b} = [a_2 b_3 - a_3 b_2, \; a_3 b_1 - a_1 b_3, \; a_1 b_2 - a_2 b_1]$$

For a numerical example, let

$$\mathbf{a} = [-1, 3, 2] \qquad \mathbf{b} = [4, -2, 1]$$

Then

$$\mathbf{a} \times \mathbf{b} = [-1, 3, 2] \times [4, -2, 1]$$

$$= [(3)(1) - (2)(-2), (2)(4) - (-1)(1), (-1)(-2) - (3)(4)]$$

$$= [3 + 4, 8 + 1, 2 - 12]$$

$$= [7, 9, -10]$$

$$= 7\mathbf{i} + 9\mathbf{j} - 10\mathbf{k}$$

An alternative computation uses concepts from determinants. First, prepare a third-order determinant as shown below, then expand it as the sum of second-order minors of the first row.

$$\mathbf{a} \times \mathbf{b} = \begin{vmatrix} \mathbf{i} & \mathbf{j} & \mathbf{k} \\ a_1 & a_2 & a_3 \\ b_1 & b_2 & b_3 \end{vmatrix} = \begin{vmatrix} a_2 & a_3 \\ b_2 & b_3 \end{vmatrix} \mathbf{i} - \begin{vmatrix} a_1 & a_3 \\ b_1 & b_3 \end{vmatrix} \mathbf{j} + \begin{vmatrix} a_1 & a_2 \\ b_1 & b_2 \end{vmatrix} \mathbf{k}$$

This third-order determinant is unique in that the elements of its first row are vectors. For the numerical values given above,

$$\mathbf{a} \times \mathbf{b} = \begin{vmatrix} 3 & 2 \\ -2 & 1 \end{vmatrix} \mathbf{i} - \begin{vmatrix} -1 & 2 \\ 4 & 1 \end{vmatrix} \mathbf{j} + \begin{vmatrix} -1 & 3 \\ 4 & -2 \end{vmatrix} \mathbf{k}$$

$$= \{(3)(1) - (-2)(2)\}\mathbf{i} - \{(-1)(1) - (4)(2)\}\mathbf{j} + \{(-1)(-2) - (4)(3)\}\mathbf{k}$$

$$= (3 + 4)\mathbf{i} - (-1 - 8)\mathbf{j} + (2 - 12)\mathbf{k}$$

$$= 7\mathbf{i} + 9\mathbf{j} - 10\mathbf{k}$$

The APT format for computing the cross product is given as Vector-10 in Table 13.1.

To interpret the cross product geometrically, we first establish the following identity:

$$|\mathbf{a} \times \mathbf{b}|^2 = |\mathbf{a}|^2|\mathbf{b}|^2 - (\mathbf{a} \cdot \mathbf{b})^2$$

For the left-hand side we have

$$|\mathbf{a} \times \mathbf{b}|^2 = (a_2 b_3 - a_3 b_2)^2 + (a_3 b_1 - a_1 b_3)^2 + (a_1 b_2 - a_2 b_1)^2$$

$$= a_2^2 b_3^2 - 2a_2 a_3 b_2 b_3 + a_3^2 b_2^2 + a_3^2 b_1^2 - 2a_1 a_3 b_1 b_3$$

$$+ a_1^2 b_3^2 + a_1^2 b_2^2 - 2a_1 a_2 b_1 b_2 + a_2^2 b_1^2$$

while for the right-hand side we have

$$|\mathbf{a}|^2|\mathbf{b}|^2 - (\mathbf{a} \cdot \mathbf{b})^2 = (a_1^2 + a_2^2 + a_3^2)(b_1^2 + b_2^2 + b_3^2)$$

$$- (a_1 b_1 + a_2 b_2 + a_3 b_3)^2$$

$$= a_1^2 b_2^2 + a_1^2 b_3^2 + a_2^2 b_1^2 + a_2^2 b_3^2 + a_3^2 b_1^2 + a_3^2 b_2^2$$

$$- 2a_1 a_2 b_1 b_2 - 2a_1 a_3 b_1 b_3 - 2a_2 a_3 b_2 b_3$$

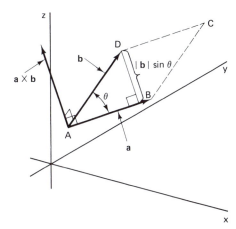

Figure 13.4   Construction for interpreting the cross product.

Since the two expressions are equal, the identity is established. We further manipulate the right-hand side as follows:

$$|\mathbf{a} \times \mathbf{b}|^2 = |\mathbf{a}|^2|\mathbf{b}|^2 - (\mathbf{a} \cdot \mathbf{b})^2$$

$$= |\mathbf{a}|^2|\mathbf{b}|^2 - |\mathbf{a}|^2|\mathbf{b}|^2 \cos^2\theta$$

$$= |\mathbf{a}|^2|\mathbf{b}|^2(1 - \cos^2 \theta)$$

$$= |\mathbf{a}|^2|\mathbf{b}|^2 \sin^2\theta$$

Because $0° \leq \theta \leq 180°$, $\sin \theta > 0$. Therefore,

$$|\mathbf{a} \times \mathbf{b}| = |\mathbf{a}||\mathbf{b}|\sin \theta$$

From Figure 13.4 we compute the area of parallelogram *ABCD* as base x altitude $= |\mathbf{a}|\ |\mathbf{b}|\ \sin \theta$. Thus the magnitude of the cross product is the area of the parallelogram formed by the constituent vectors.

To show that the vector $\mathbf{a} \times \mathbf{b}$ is perpendicular to $\mathbf{a}$ and perpendicular to $\mathbf{b}$, and thus normal to the plane of $\mathbf{a}$ and $\mathbf{b}$, we compute the following.

$$(\mathbf{a} \times \mathbf{b}) \cdot \mathbf{a} = a_1a_2b_3 - a_1a_3b_2 + a_2a_3b_1 - a_1a_2b_3 + a_1a_3b_2 - a_2a_3b_1 = 0$$

$$(\mathbf{a} \times \mathbf{b}) \cdot \mathbf{b} = a_2b_1b_3 - a_3b_1b_2 + a_3b_1b_2 - a_1b_2b_3 + a_1b_2b_3 - a_2b_1b_3 = 0$$

From the results of Paragraph 13.1.3 we conclude that the perpendicularity condition is established.

The product $\mathbf{a} \times \mathbf{b}$ is defined such that the constituent vectors and the resultant vector are oriented according to the direction established by applying the right-hand rule, provided that the rotation is taken from $\mathbf{a}$ to $\mathbf{b}$ through the angle θ as shown in Figure 13.5.

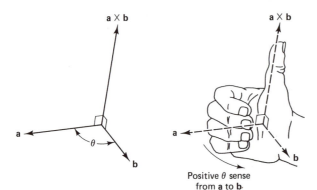

Positive θ sense
from **a** to **b**.

Figure 13.5 Applying the right-hand rule to the cross product.

The following relationships are noted for future reference.

For any vector **a** in three-dimensional space,

$$\mathbf{a} \times \mathbf{a} = \mathbf{0}, \text{ the zero vector}$$

Applying the cross product to the unit coordinate axis vectors gives

$$\mathbf{i} \times \mathbf{i} = \mathbf{j} \times \mathbf{j} = \mathbf{k} \times \mathbf{k} = \mathbf{0}$$

$$\mathbf{i} \times \mathbf{j} = \mathbf{k} \qquad \mathbf{j} \times \mathbf{k} = \mathbf{i} \qquad \mathbf{k} \times \mathbf{i} = \mathbf{j}$$

$$\mathbf{j} \times \mathbf{i} = -\mathbf{k} \qquad \mathbf{k} \times \mathbf{j} = -\mathbf{i} \qquad \mathbf{i} \times \mathbf{k} = -\mathbf{j}$$

In particular we have $\mathbf{i} \times \mathbf{j} \neq \mathbf{j} \times \mathbf{i}$, but since $\mathbf{i} \times \mathbf{j} = \mathbf{k}$ and $\mathbf{j} \times \mathbf{i} = -\mathbf{k}$, then $\mathbf{i} \times \mathbf{j} = -(\mathbf{j} \times \mathbf{i})$. In general, $\mathbf{a} \times \mathbf{b} = -(\mathbf{b} \times \mathbf{a})$ so the cross product is not commutative.

To illustrate the foregoing relationships numerically, for the vectors of the preceding example, we have

$$\mathbf{b} \times \mathbf{a} = \begin{vmatrix} b_2 & b_3 \\ a_2 & a_3 \end{vmatrix} \mathbf{i} - \begin{vmatrix} b_1 & b_3 \\ a_1 & a_3 \end{vmatrix} \mathbf{j} + \begin{vmatrix} b_1 & b_2 \\ a_1 & a_2 \end{vmatrix} \mathbf{k}$$

$$= \begin{vmatrix} -2 & 1 \\ 3 & 2 \end{vmatrix} \mathbf{i} - \begin{vmatrix} 4 & 1 \\ -1 & 2 \end{vmatrix} \mathbf{j} + \begin{vmatrix} 4 & -2 \\ -1 & 3 \end{vmatrix} \mathbf{k}$$

$$= (-4 - 3)\mathbf{i} - (8 + 1)\mathbf{j} + (12 - 2)\mathbf{k}$$

$$= -7\mathbf{i} - 9\mathbf{j} + 10\mathbf{k}$$

We note that $\mathbf{a} \times \mathbf{b} = \mathbf{0}$ only when a or b or $\sin \theta$ is 0. If $\sin \theta = 0$, **a** and **b** are parallel. Since we said that the **0** vector is parallel with every vector, we can summarize all cases in the rule

$$\mathbf{a} \times \mathbf{b} = \mathbf{0} \qquad \text{if and only if } \mathbf{a} \text{ and } \mathbf{b} \text{ are parallel}$$

This property is the basis for most geometric applications of the cross product.

13.1.5 Applications to Geometry

The vector principles developed above are illustrated here with geometric applications. Because planes are used in several of the examples, it is essential that the orientation of the unit normal as derived from the plane definition be known. Accordingly, we define in Table 13.2 the sense of the unit normal for each plane defined in Table 8.1. Not all APT processors may define the orientation as given in the table for the IBM APT-AC processor. Therefore, the sense of the unit normal for a given APT processor should be verified before using values extracted from the canonical form. This precaution was emphasized in Section 8.4. It also applies to the canonical form for the line, which is stored as a plane perpendicular to the *xy*-plane.

TABLE 13.2   UNIT NORMAL DIRECTIONS FOR THE PLANE DEFINITIONS

	Definition	Sense of unit normal
Plane-1	By the coefficients of the plane equation (normal form) PLANE/*a*,*b*,*c*,*d*	As given by *a*, *b*, *c*
Plane-2	By three noncollinear points PLANE/*point1*,*point2*,*point3*	In direction of cross product (VECTOR/*point1*,*point2*) × (VECTOR/*point1*,*point3*)
Plane-3	Through a point and parallel with another plane PLANE/*point*,PARLEL,*plane*	Same as *plane*
Plane-4	Parallel with but offset from another plane PLANE/PARLEL,*plane,* $\left\{ \begin{matrix} \text{XLARGE} \\ \cdot \\ \cdot \\ \cdot \end{matrix} \right\}$,*d*	From *plane* to plane being defined
Plane-5	Through a point and perpendicular to a vector PLANE/*point*,PERPTO,*vector*	Same as *vector*
Plane-6	Through two points and perpendicular to another plane PLANE/PERPTO,*plane*,*point1*,*point2*	In direction of cross product (VECTOR/*point1*,*point2*) × (normal of *plane*)
Plane-7	Through a point and perpendicular to two intersecting planes PLANE/*point*,PERPTO,*plane1*,*plane2*	In direction of cross product (Normal of *plane1*) × (normal of *plane2*)
Plane-8	Through a point and tangent to a cylinder PLANE/*point,* $\left\{ \begin{matrix} \text{XLARGE} \\ \cdot \\ \cdot \\ \cdot \end{matrix} \right\}$,TANTO,*cylinder*	Radially outward from *cylinder* axis through point of tangency

Example 13.1

At times we must determine whether two lines are perpendicular or parallel. By representing the lines as nonzero vectors, we can compute their dot product ($\mathbf{a} \cdot \mathbf{b} = 0$ when perpendicular) or their cross product ($\mathbf{a} \times \mathbf{b} = \mathbf{0}$ when parallel), then test the result for zero. We use the DOTF function and test for a zero scalar value and apply definition Vector-10 and test the resulting vector for zero length with the LNTHF function.

Example 13.2

We can use vector principles to determine whether two planes are perpendicular or parallel. By extracting the unit normals from their canonical forms, say $\mathbf{n}_1$ and $\mathbf{n}_2$, we determine perpendicularity by testing for $\mathbf{n}_1 \cdot \mathbf{n}_2 = 0$ in a manner analogous to that of Example 13.1. The planes are parallel if their unit normals are parallel. We determine this by extracting the unit normal components from the canonical forms, then testing them for equality of corresponding components. Or we can form vectors, take their dot product, and test the result for $+1$.

Example 13.3

The unit vector perpendicular to two nonparallel lines is computed as follows: $\mathbf{u} = (\mathbf{a} \times \mathbf{b})/\left|\mathbf{a} \times \mathbf{b}\right|$. Vectors $\mathbf{a}$ and $\mathbf{b}$ are formed from the lines, the LNTHF function is used to obtain the magnitude of the cross product. The lines referred to are not APT lines because for them the answer is trivial; it is $\pm\mathbf{k}$ (APT lines are defined in the xy-plane).

From Figure 13.6 we form $\mathbf{a} = -4\mathbf{i} + 2\mathbf{j} + 3\mathbf{k}$ and $\mathbf{b} = -6\mathbf{i} + \mathbf{j} + \mathbf{k}$. Then

$$\mathbf{a} \times \mathbf{b} = \begin{vmatrix} \mathbf{i} & \mathbf{j} & \mathbf{k} \\ -4 & 2 & 3 \\ -6 & 1 & 1 \end{vmatrix} = -\mathbf{i} - 14\mathbf{j} + 8\mathbf{k}$$

and

$$\left|\mathbf{a} \times \mathbf{b}\right| = \sqrt{1^2 + 14^2 + 8^2} = 16.155$$

Therefore,

$$\mathbf{u} = \frac{-\mathbf{i} - 14\mathbf{j} + 8\mathbf{k}}{16.155} = -0.062\mathbf{i} - 0.876\mathbf{j} + 0.495\mathbf{k}$$

Example 13.4

We determine the coordinates of projection of the endpoint of a line onto a vector. For this example, from Figure 13.6, let the endpoint of a line be the point p. The coordinates to be determined are at point q on an extension of the vector $\mathbf{u}$ determined in Example 13.3. We have

$$\text{length of } oq = \mathbf{p} \cdot \mathbf{u} = \begin{bmatrix} -5, & 12, & 3 \end{bmatrix} \begin{bmatrix} -0.062 \\ -0.867 \\ 0.495 \end{bmatrix}$$

$$= 0.31 - 10.404 + 1.485 = -8.609$$

The minus sign indicates that it is positioned on the negative side of the origin with respect to vector $\mathbf{u}$. Then

$$q = (oq)\mathbf{u} = (-8.609)(-0.062\mathbf{i} - 0.867\mathbf{j} + 0.495\mathbf{k}) = 0.534\mathbf{i} + 7.464\mathbf{j} - 4.261\mathbf{k}$$

Therefore,

$$\mathbf{q} = [0.534, 7.464, -4.261]$$

Should this computation have been made in APT, we would have used Vector-9 to compute **q**.

Example 13.5

The angle between two lines is readily computed with relationships developed for the dot product and cross product. We compute angle θ between the lines passing through the origin and points a and b of Figure 13.6 as follows:

$$\cos \theta = \frac{\mathbf{a} \cdot \mathbf{b}}{|\mathbf{a}| \, |\mathbf{b}|} \quad \text{and} \quad \sin \theta = \frac{|\mathbf{a} \times \mathbf{b}|}{|\mathbf{a}||\mathbf{b}|}$$

Therefore,

$$\theta = \arctan \frac{\sin \theta}{\cos \theta} = \arctan \frac{|\mathbf{a} \times \mathbf{b}|}{\mathbf{a} \cdot \mathbf{b}}$$

With $|\mathbf{a} \times \mathbf{b}| = 16.155$ from Example 13.3 and

$$\mathbf{a} \cdot \mathbf{b} = [-4, 2, 3] \begin{bmatrix} -6 \\ 1 \\ 1 \end{bmatrix} = 29$$

Then

$$\theta = \arctan \frac{16.155}{29} = 29.12°$$

Example 13.6

The true shape of an arbitrarily oriented plane is found in descriptive geometry by drawing auxiliary views. Vector mathematics allows us to designate two lines in the plane as vectors and to compute their cross product to obtain the vector along our desired line of sight. For

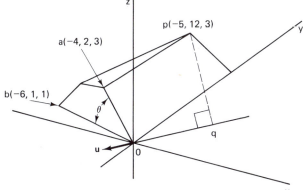

Figure 13.6   Geometry for Examples 13.3 through 13.5.

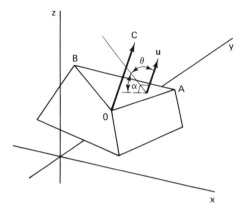

Figure 13.7   Geometry for Examples 13.6 and 13.7.

example, in Figure 13.7 we designate edges *OA* and *OB* as vectors. Then our line of sight is opposite to the vector $\mathbf{OC} = \mathbf{OA} \times \mathbf{OB}$.

Example 13.7

The angle between a line and plane in APT is computed according to the procedure of Example 13.5. In this case we form a vector to represent the line as shown in Figure 13.7. The second line is represented by the unit normal to the plane, after adjusting for its sense of direction if necessary, so that the angle θ in the figure may be computed. Angle α, the desired angle, equals $90° - \theta$.

13.2 TOOL SHAPE DESCRIPTION

Tools described by the cutter format of Section 2.2 are adequate for the APT work introduced thus far. This is because the part surfaces are planes, perhaps canted. We can install a cutter whose shape differs from that described by the cutter statement to achieve a purpose in an unorthodox manner. For example, an undersized tool can be used for roughing purposes, then, with the same cut vectors, the correct size tool can be used for the finishing cut. We choose not to fool the system in this manner.

For three-dimensional contouring the surfaces may assume any spatial orientation. The complex surface relationships involved require the whole cutter envelope to be known to make the computations for the cutter location. The expanded form of the CUTTER statement allows the whole envelope to be described. The cutter format of Section 2.2 is a special case of the expanded form. Below we describe the expanded form and discuss the significance of the tool height as it relates to three-dimensional contouring.

13.2.1 The Expanded CUTTER Statement

The expanded form of the CUTTER statement is as follows:

CUTTER/d,r,e,f,α,β,h

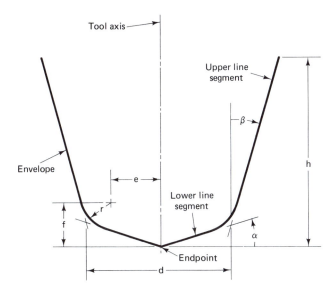

Figure 13.8 Parameters of the expanded CUTTER statement and the cutter envelope created.

Of the seven parameters, the first two only are used in the CUTTER statement of Section 2.2. All seven are used to create the internal cutter envelope used for calculations to position the cutter even though not all are given in the simpler format. There are default values as described below. The geometric interpretation of all parameters is given in Figure 13.8, where the bold outline represents a cross section of the cutter envelope created (actually, it is a surface of revolution). The parameters have the following meaning.

d The tool diameter, which is twice the radial distance from the tool axis to the intersection of the lower and upper line segments.

r The radius of the corner circle; it can be zero or larger than $d/2$.

e The radial distance from the tool axis to the center of the corner circle; it can be positive (corner circle and its center on the same side of the tool axis) or negative (corner circle and its center on opposite sides of the tool axis).

f The distance from the tool endpoint to the center of the corner circle measured parallel with the tool axis.

α The angle from a radial line through the tool endpoint to the lower line segment; it is in degrees, positive, and in the range $0° \leq \alpha < 90°$.

β The angle between the upper segment and the tool axis; it is in degrees, positive (when sloping outward from the intersection point of the lower and upper line segments) or negative (when sloping inward from the intersection point), and in the range $-90° < \beta < 90°$.

h The cutter height measured from the tool endpoint along the tool axis.

Tool shapes shown in Figure 13.9 are representative of those that can be described by some combination of values for the parameters. Default values are assigned to omitted

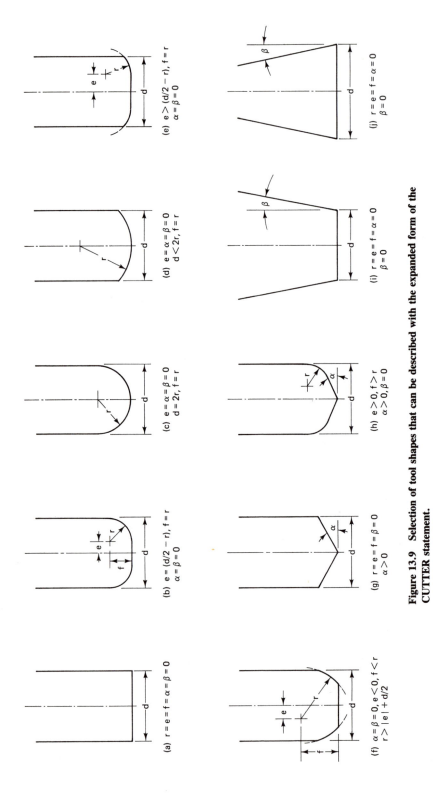

Figure 13.9 Selection of tool shapes that can be described with the expanded form of the CUTTER statement.

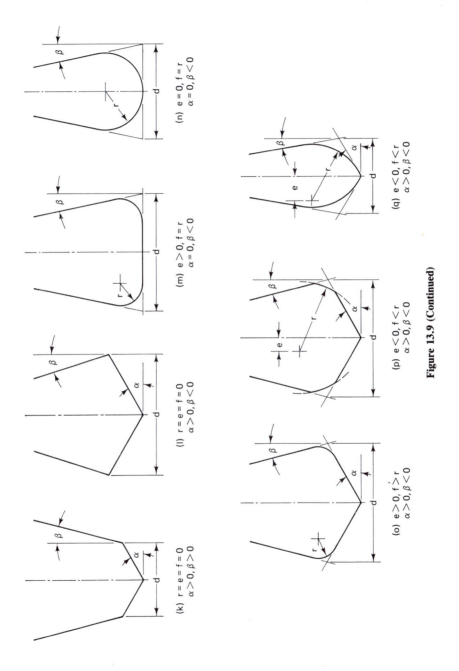

(k) $r = e = f = 0$
$\alpha > 0, \beta > 0$

(l) $r = e = f = 0$
$\alpha > 0, \beta < 0$

(m) $e > 0, f = r$
$\alpha = 0, \beta < 0$

(n) $e = 0, f = r$
$\alpha = 0, \beta < 0$

(o) $e > 0, f > r$
$\alpha > 0, \beta < 0$

(p) $e < 0, f < r$
$\alpha > 0, \beta < 0$

(q) $e < 0, f < r$
$\alpha > 0, \beta < 0$

Figure 13.9 (Continued)

427

parameters when the simpler format of Section 2.2 is used. Such values are assigned on the following basis.

1. $\alpha = \beta = 0$
2. $f = r$; when r is omitted, $f = r = 0$
3. $e = d/2 - r$
4. $h = 5$ (for $r \leqslant 5$ or when omitted) or $h = r$ (for $r > 5$)

Parameter values for the expanded format must be consistent among themselves so that allowed geometries only are described. Shapes not allowed violate one or more of the following restrictions on parameter values.

1. Tool surface normals cannot intersect each other. Values for e, f, and r must be consistent to avoid violating this restriction [see Figure 13.10, shapes (a) and (b)].
2. The corner circle must be tangent to or intersect each line segment. Values for e, f, and r must be consistent to avoid violating this restriction [see Figure 13.10, shapes (c), (d), and (e)]. To ensure corner circle tangent conditions when α and/or β are nonzero, the d and f values should be computed by the program with the following expressions:

$$d = 2(e + r \cos(45 + (\beta - \alpha)/2)/\cos(45 - (\beta + \alpha)/2))$$
$$f = r \sin(45 + (\beta - \alpha)/2)/\cos(45 - (\beta + \alpha)/2) + (d \tan \alpha)/2$$

When $r = 0$, zero e and f values are not inconsistent because APT will compute the corner at the intersection of the upper and lower line segments [see Figure 13.9, shapes (g), (i), (j), (k), and (l)].

3. The value for h must be greater than the calculated tool height, $(d \tan \alpha)/2$ [see Figure 13.10, shape (f), for a tool with d, h, and α inconsistent].
4. Angles α and β must be chosen such that $\alpha + |\beta| < 90°$ [see Figure 13.10, shape (g)].
5. Only parameters e and β may have negative values; all others must have zero or positive values [see Figure 13.10, shape (h)].

13.2.2 Cutter Height Significance

The cutter height was not a factor in the two-dimensional work of previous chapters but its importance was implied in the discussion of tool positioning during startup (see Section 8.2.2, especially Figure 8.3). Because we are not working with bounded surfaces in APT, part geometries may cause us considerable grief in three-dimensional contouring. This may result from image surfaces originating from mathematical representations. Therefore, the cutter height cannot be casually specified.

The cutter height, parameter h, is usually irrelevant because the contact point (the point between the tool and the part) is generally on the bottom or side of the tool (see

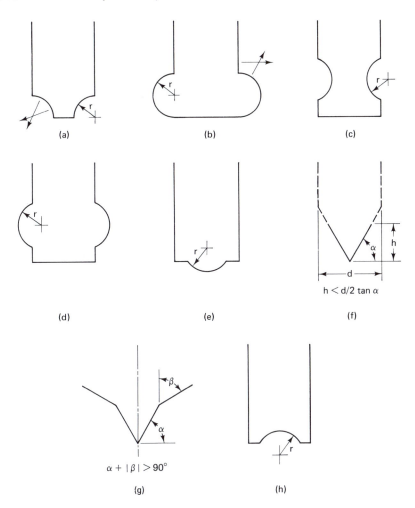

Figure 13.10   Cutter shapes that cannot be described in APT.

Figure 13.11). Furthermore, this height is not necessarily the same as the physical length of the tool (usually specified by a postprocessor statement). Thus it is common to accept its default value of 5 units or to assign some innocuous value consistent with other CUTTER statement parameters. The internal cutter envelope defined in Figure 13.8 does not permit contact points on the top of the tool as shown in Figure 13.11(d).

Undesired tool check surfaces, and hence undesired tool contact points, are often caused by geometric constraints while working within closed figures or by geometric image surfaces. The first case is illustrated in Figure 13.12(a), where the upper inside surface of the cylinder causes an undesired contact point. The second case is illustrated

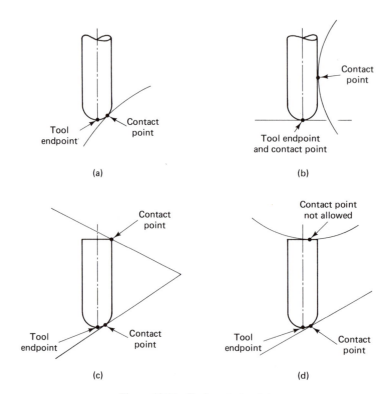

Figure 13.11 Tool contact points.

in Figure 13.12(b), where the undesired contact point is caused by the upper (image) sheet of the hyperbolic cylinder. Other quadric surfaces can cause similar problems.

One remedy for the problems of Figure 13.12 is to reduce the cutter height to h' as shown in Figure 13.12(a). Problems during startup may be avoided by selecting an initial position within the geometric object. An alternative remedy may be to perform a transformation of cut vectors similar to that of Example 8.3.

Depending on cutter shapes, problems such as shown in Figure 13.13 may be handled with the cutter option statement. Its format is as follows:

$$\text{CUTTER/OPTION},\begin{Bmatrix}1\\2\end{Bmatrix},\begin{Bmatrix}ra,hi\\\text{OFF}\end{Bmatrix}$$

where a zero thickness ring of radius ra is defined a distance hi above the tool endpoint as shown in Figure 13.14(a). Actually, parameters ra and hi define a point on the tool that becomes a ring as the tool revolves. The ring functions as a surface of contact between the tool and the control surface, which may be the part surface (parameter = 1) or drive surface (parameter = 2). There is no top or bottom to the ring.

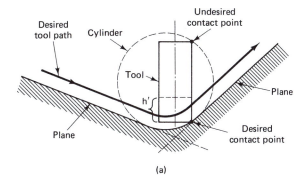

(a)

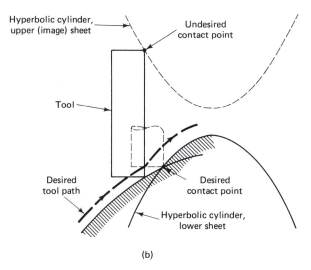

(b)

Figure 13.12   Examples of undesired tool contact points.

The ring cutter is used to compute offset positions from the control surface when the CUTTER/OPTION statement is specified. It is modal, remaining in effect until countermanded by another CUTTER/OPTION statement or until canceled by a similar statement with the OFF keyword, after which time the cutter statement in effect prior to use of the ring is restored.

The ring is usually defined with a radius identical to that of the real cutter. Such is the case for Figure 13.13 where a ring is shown to eliminate the difficulty. The ring used there is described as shown in Figure 13.14(b). A different application of the ring cutter appears in Example 14.5. A fillet-end tool may be replaced with the ring of Figure 13.14(d), while in Figure 13.14(c) we have a degenerate case whereby the ring is reduced to the tool endpoint.

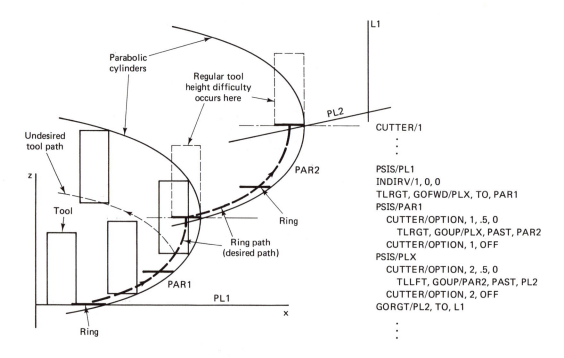

```
CUTTER/1
        .
        .
PSIS/PL1
INDIRV/1, 0, 0
TLRGT, GOFWD/PLX, TO, PAR1
PSIS/PAR1
   CUTTER/OPTION, 1, .5, 0
      TLRGT, GOUP/PLX, PAST, PAR2
   CUTTER/OPTION, 1, OFF
PSIS/PLX
   CUTTER/OPTION, 2, .5, 0
      TLLFT, GOUP/PAR2, PAST, PL2
   CUTTER/OPTION, 2, OFF
GORGT/PL2, TO, L1
        .
        .
```

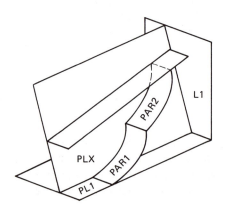

Figure 13.13   Problem geometry handled with the ring cutter applied to part and drive surfaces.

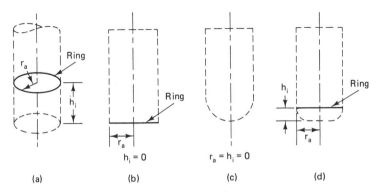

Figure 13.14   Ring locations defined by the CUTTER/OPTION statement.

13.3 TOOL-TO-SURFACE RELATIONSHIPS

While contouring, we have been required to maintain the proper tool-to-surface relationships. Failure to have done so led to fatal diagnostics or to an incorrectly shaped part. In either case, additional delay, hence cost, was incurred. Shapes requiring three-dimensional contouring merely compound such problems. Convoluted part surfaces, up or down motion commands, reassignments among part, drive, and check surfaces, and other factors cause problems that require constant attention to tool-to-surface relationships. We now review and introduce some concepts, some refinements, and some dynamics of tool positioning, all for the purpose of improving our understanding of tool-to-surface relationships.

13.3.1 Review and Refinements

Tolerance specifications discussed in Section 2.4 did not address the possibility of multiple check surfaces. When such tolerance extensions are available, difficulty in tool positioning may be avoided during the program checkout phase. Because it is important to avoid diagnostics to complete calculations of APT Sections 2 and 3, a relaxation of the tolerance extensions may be in order. Apart from this requirement, rarely would it be necessary to specify different tolerances for each of the surfaces.

   Subsequent tightening of the tolerances may not solve all our problems. For example, an extreme curvature change over a small portion of the curve may cause a cut vector to be output that is within tolerance of the curve at its endpoints, yet out of tolerance at some other part of the cut vector. This condition is shown in Figure 13.15. An APT processor may have a special computational routine that eliminates the error condition. The GOUGCK (gouge check) statement for the IBM APT-AC system is one such routine.

It is a modal statement of the form

$$\text{GOUGCK/}\begin{Bmatrix} \text{ON} \\ \text{OFF} \end{Bmatrix}$$

A special calculation phase is entered from the point where this feature is turned ON to the place where it is turned OFF. This feature should be used sparingly since it is time consuming.

It is wise to review the effect of tool positioning during startup, especially with respect to the PAST modifier as discussed in Section 8.2.2. There we showed unwanted tool positioning effects. Also, for three-dimensional contouring, the tool-to-part-surface modifier of the general motion command deserves special attention. This modifier was discussed in Section 8.2.5. There we showed for the first time the check surface modifier PSTAN. This modifier is used when the check surface, rather than the drive surface, is tangent to the part surface. Its use is demonstrated in examples of Sections 13.4 and 14.4.

13.3.2 Some Problem Configurations

It is easy to believe that tool-positioning problems can arise from complex geometrical relationships in three-dimensional space. Computational difficulties often are related to the vector alignment of the tool and its tangent surfaces as well as to the precision of computation. This topic is of concern here because APT is implemented on computers of varying characteristics, most notable of which is the word length. Before singling out some special problems, we review an expected important outcome during a tool-positioning computation.

Given that deviations from precise tangency are permitted within the limits of the tolerance specifications, the tool must be simultaneously tangent to both the part and drive surfaces at the ends of cut vectors. To achieve this condition, the computational

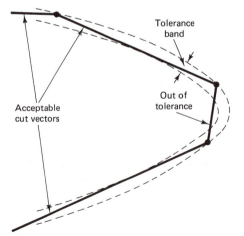

Figure 13.15 Geometry requiring special computation to eliminate the out of tolerance cut vector.

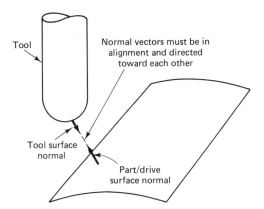

Figure 13.16 Surface normals for tool-to-part/drive surface positioning.

task is to locate points on the tool profile (surface) and on the workpiece surface such that the surface normals are directed toward each other and are in alignment. Adjustments in tool positioning are made until opposing normals meet these conditions and the surfaces are within tolerance specifications. Failure to reach these positioning requirements leads to cut vector computation failure. An exaggerated view of this interface for one of the tangent surfaces is shown in Figure 13.16.

The following examples illustrate the potential for tool positioning problems.

Nonintersecting Normals

Not all surfaces are unbounded for surface normal computation. For example, tabulated cylinders, ruled surfaces, and polyconic surfaces have bounded geometry relative to part orientation as discussed in Sections 6.3, 12.2, and 12.3, respectively. Thus tool and surface normals will not intersect where the surfaces do not exist. The remedy is to observe the geometry bounds and position the tool accordingly, possibly aided with the SRFVCT feature.

Nearly Parallel Surfaces

Space curves are formed by the intersection of two surfaces. A special case is the line formed by intersecting planes, thus allowing the planes to function as part-drive, part-check, or drive-check surfaces. Nearly parallel planes may cause problems from lack of precision in computation and from limitations of the algorithms for positioning the tool as described above. Factors that function to limit the closeness to parallelism computed for such surfaces include the shape of the cutter profile, the location of the tool relative to the surfaces, the distance of the surfaces from the origin of the coordinate system, and the choice of tool position modifiers.

An appreciation for this problem can be obtained from Figure 13.17(a), where intersecting points for two sets of nearly parallel lines differ widely in location when the coefficient of one line is varied by just 5 parts in 10,000 (note the small angle of

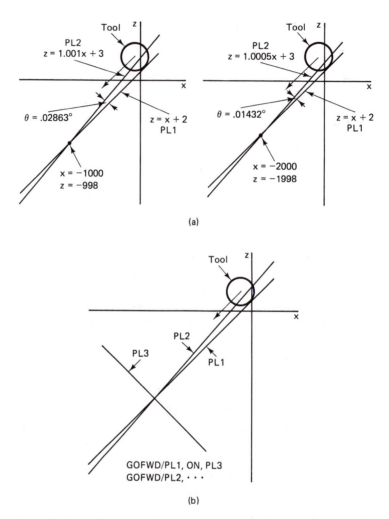

Figure 13.17 (a) Nearly parallel planes; (b) auxiliary check surface to avoid problems with the command GOFWD/PL1,PAST,PL2 when used in part (a).

intersection). These lines could just as well be the end view of intersecting planes. Both sets of lines might cause computational failure when used as drive and check surfaces, although the IBM APT-AC system had no difficulty with these planes when used as drive-check and part-check surfaces. Planes intersecting more than one degree from parallelism should cause no difficulty. When problems do occur, one remedy is to introduce an auxiliary check surface as shown in Figure 13.17(b).

A condition analogous to that described above can be created from geometry similar to that of Figure 13.18(a), where a plane nearly tangent to a cylinder intersects the cylinder. Let the tool be driven along the intersection, with the cylinder as the part surface

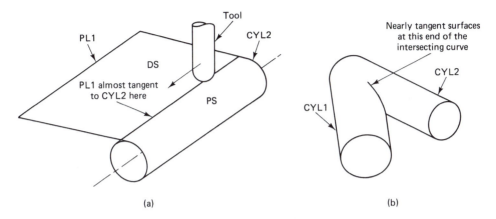

Figure 13.18   **Nearly parallel surfaces formed from a plane and a cylinder (a)
and from two intersecting cylinders (b).**

and the plane as the drive surface. A plane tangent to the cylinder is created from which
a surface normal is derived for the part surface. Since the drive surface is already a plane,
we now have the equivalent of two nearly parallel planes for which computational failure
may ensue, especially when the cutter contacts the surfaces near the intersection curve.
A similar problem may develop where the intersecting cylinders of Figure 13.18(b) merge
such that nearly parallel planes form from tangent planes. A remedy for both cases,
although perhaps computationally difficult to implement, is to define a drive surface such
that with a TLON position modifier the tool follows the desired cutter path. For Figure
13.18(a) this is a plane perpendicular to the given plane, for Figure 13.18(b) this is a
plane passing through the intersection curve.

Relative Size Predicaments

The algorithm for computing the position of the tool is also sensitive to the size of
the tool relative to the surfaces to which it is being positioned. Difficulties may arise
when the tool is much larger than the surfaces presented to it. Some examples are shown
in Figure 13.19.

Parts (a) and (b) of Figure 13.19 should be contrasted. In both cases the surface is
enclosed within the tool envelope. In part (a), where the tool is offset from the drive
surface, the cut vectors cannot be computed for the second statement of the code while
in part (b), where the tool is on the drive surface, proper cut vectors are computed.

The very small surfaces presented by the tip of the cone and the small cylinder in
parts (c) and (d) of Figure 13.19 may also cause positioning difficulty for the geometry
given.

Some tool positioning problems may be avoided by (1) reducing the cutter size to
more nearly match the geometry, (2) replacing the actual surface with a substitute surface
of greater apparent size but oriented such that cut vectors equivalent to those desired are

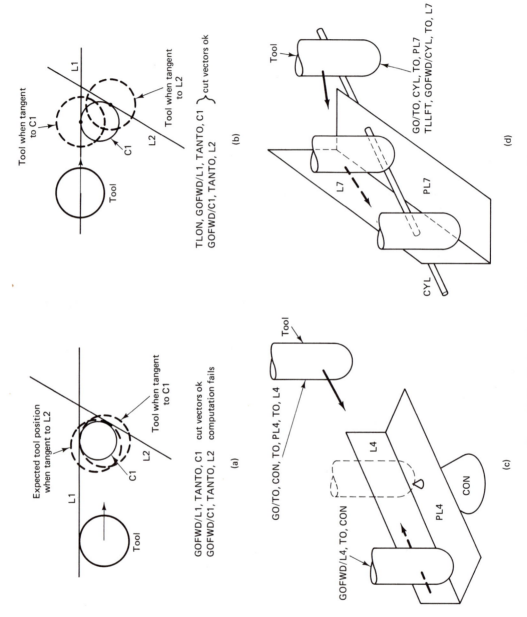

Expected tool position
when tangent to L2

Tool when tangent
to C1

L1

C1

L2

Tool

GOFWD/L1, TANTO, C1   cut vectors ok
GOFWD/C1, TANTO, L2   computation fails

(a)

Tool when tangent
to C1

Tool when tangent
to L2

L1

C1

L2

Tool

TLON, GOFWD/L1, TANTO, C1 ⎱ cut vectors ok
GOFWD/C1, TANTO, L2   ⎰

(b)

Tool

GO/TO, CON, TO, PL4, TO, L4

L4

PL4

CON

GOFWD/L4, TO, CON

(c)

Tool

GO/TO, CYL, TO, PL7
TLLFT, GOFWD/CYL, TO, L7

L7

PL7

CYL

(d)

Figure 13.19 Tool positioning subject to relative size difficulties.

438

computed, (3) bringing the tool closer to the surface before issuing the startup command (perhaps applying the DNTCUT-CUT feature), or (4) carefully computing INDIRP or INDIRV parameters for more accurately directing the tool to the desired surfaces during startup.

13.4 EXAMPLES

The first example illustrates contouring on a nonplanar part surface and the use of the PSTAN check modifier in a multiple check motion statement, the second example illustrates motion commands on a ruled surface, and the third example uses vectors to define geometric figures.

Example 13.8

Three different part surfaces are used in this example. Transition from one part surface to another is via the PSTAN check modifier in a multiple check surface motion statement. The part is shown in Figure 13.20(a). We are concerned only with machining the surface bounded by C1, L1, L3, and L2, shown in the top view of Figure 13.20(b). The general plan for controlling the tool appears in Figure 13.20(c).

A macro will be used while contouring between C1 and L3. We will dynamically define the line LL on which to direct the tool. A ball-end tool will be used with a height equal to its radius because we will contour within cylinder CY1. We are precomputing the x-coordinate of line LL for loop testing purposes to ensure that the TRANTO labels will be written on the CL-file during Section 2 processing. The macro is as follows:

```
     $$*************************************************************
     DRIVS = MACRO/TLPOS = TLRGT,TLDIR = GORGT,SURF1,SURF2,CK        100
             X = X + DX                                               110
             LL = LINE/L2,CANON,,,,X                                  120
             PSIS/SF(J)                                               130
             TLPOS,TLDIR/SURF1,ON,LL,LM1,PSTAN,SF(J + 1),LM2          140
     LM1)    TRANTO/LM3                                               150
     LM2)    PSIS/SF(J + 1)                                           160
             GOFWD/SURF1,ON,LL                                        170
     LM3)    TLON,TLDIR/LL,CK,SURF2                                   180
             TERMAC                                                   190
     $$*************************************************************
```

Subscripted variables will be used for convenience in the loop to specify the part surface and to determine loop exit conditions. Geometric definitions, computations, and tool movement are as follows.

```
        D = .5                                                        200
        R = D/2                                                       210
        CUTTER/D,R,0,0,0,0,R                                          220
        CANON/ON                                                      230
        RESERV/SF,4,XX,3                                              240
SF(1)   = PLANE/0,0,1,1                                               250
```

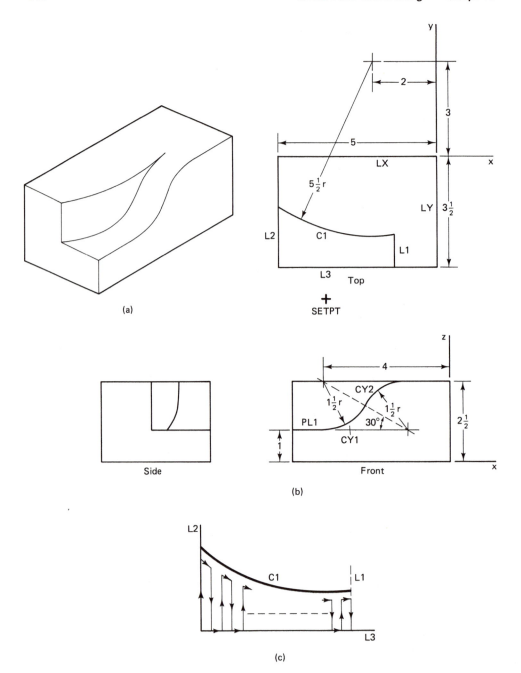

(a)

(b)

(c)

Figure 13.20   Part geometry for Example 13.8.

SF(4)	= PLANE/0,0,1,2.5	260
C1	= CIRCLE/−2,3,5,5	270
SF(2)	= CYLNDR/−4,0,2.5,0,1,0,1.5	280
CCX2	= −4 + 3*COSF(30)	290
SF(3)	= CYLNDR/CCX2,0,1,0,1,0,1.5	300
L2	= LINE/YAXIS,−5	310
L3	= LINE/XAXIS,−3.5	320
XX(1)	= −4	330
XX(2)	= −4 + (1.5 − R)*COSF(30)	340
XX(3)	= CCX2	350
SETPT	= POINT/−3.5,−4.5,3	360
	DXD = .09	370
	N = ABSF(INTGF((−5 − CCX2)/DXD)) + 1	380
	DX = ABSF(−5 − CCX2)/N	390
	X = − 5	400
	XND = CCX2	410
	FROM/SETPT	420
	GO/ON,L2,TO,SF(1)	430
	TLON,GORGT/L2,TO,C1	440
	LOOPST	450
	J = 1	460
LP1)	CALL/DRIVS,TLPOS = TLRGT,TLDIR = GORGT,SURF1 = C1,SURF2 = L3,CK = ON	470
	IF (X − XND) LP2, LP7, LP7	480
LP2)	IF (X − XX(J)) LP4, LP3, LP3	490
LP3)	J = J + 1	500
LP4)	CALL/DRIVS,TLPOS = TLON,TLDIR = GOLFT,SURF1 = L3,SURF2 = C1,CK = TO	510
	IF (X − XND) LP5, LP7, LP7	520
LP5)	IF (X − XX(J)) LP1, LP6, LP6	530
LP6)	J = J + 1	540
	JUMPTO/LP1	550
LP7)	LOOPND	560
	GOTO/SETPT	570

Tool stepover is given by variable DXD in line 370. This is a starting value that is modified to the value in DX so that an integral number N equally stepped passes are made.

Example 13.9

This example illustrates motion commands in the direction of, but not exactly along, the rulings of a ruled surface which is the part surface. The ruled surface is shown in Figure 13.21. It is defined in a manner similar to that of Example 12.1.

The plan is to drive the tool on a line in the general direction of the rulings. The line is dynamically defined with CANON in the MACRO. The first pass takes the tool on L3A toward line L2, thereafter in a back-and-forth manner with the last pass on L4A toward L2. A ball-end tool ½ in. in diameter is assumed. Computation of the scallop is ignored; it was discussed in Example 8.2.

The following part program includes the geometric definitions (lines 100 through 470) and the MACRO and motion statements (lines 500 through 790). A substantial part of each

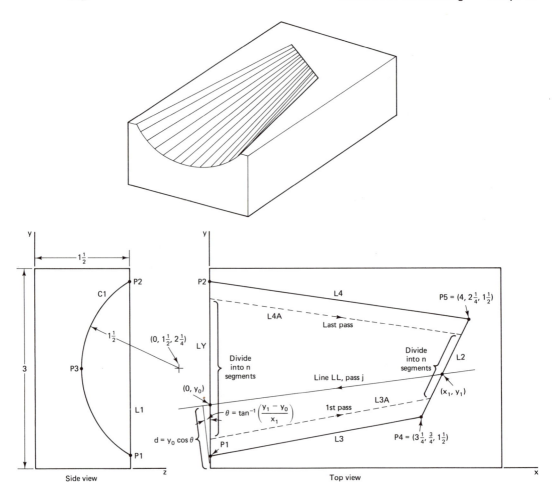

Figure 13.21 Part for Example 13.9.

end of the ruled surface is divided into *n* equal segments for use in defining line LL, along which the tool is commanded.

MAT1 = MATRIX/ZXROT, − 90	100
REFSYS/MAT1	110
C1 = CIRCLE/2.25,1.5,0,1.5	120
L1 = LINE/YAXIS,1.5	130
P1 = POINT/YSMALL,INTOF,L1,C1	140
P2 = POINT/YLARGE,INTOF,L1,C1	150
P3 = POINT/.75,1.5,0	160
REFSYS/NOMORE	170
P4 = POINT/3.25,.75,1.5	180
P5 = POINT/4,2.25,1.5	190

```
L2        = LINE/P4,P5                                            200
RSRF      = RLDSRF/C1,P1,P2,P3,L2,P4,P5,P3                        210
L3        = LINE/P1,P4                                            220
L4        = LINE/P2,P5                                            230
L3A       = LINE/PARLEL,L3,YLARGE,.25                             240
L4A       = LINE/PARLEL,L4,YSMALL,.25                             250
LY        = LINE/YAXIS                                            260
P1A       = POINT/INTOF,LY,L3A                                    270
P4A       = POINT/INTOF,L2,L3A                                    280
P2A       = POINT/INTOF,LY,L4A                                    290
P5A       = POINT/INTOF,L2,L4A                                    300
          OBTAIN,POINT/P1A,,P1AY                                  310
          OBTAIN,POINT/P2A,,P2AY                                  320
          OBTAIN,POINT/P4A,P4AX,P4AY                              330
          OBTAIN,POINT/P5A,P5AX,P5AY                              340
$$   SEGMENT COMPUTATIONS FOR L2 END OF RSRF.                     350
          N = 24        $$ MUST BE EVEN NUMBER                    360
          IHYP = SQRTF((P5AX−P4AX)**2 + (P5AY−P4AY)**2)/N         370
          GAMA = ATANF(1.5/.75)                                   380
          D2X = IHYP*COSF(GAMA)                                   390
          D2Y = IHYP*SINF(GAMA)                                   400
          X1 = P4AX                                               410
          Y1 = P4AY                                               420
$$   SEGMENT COMPUTATIONS FOR C1 END OF RSRF.                     430
          D1Y = (P2AY − P1AY)/N                                   440
          X0 = 0                                                  450
          Y0 = P1AY                                               460
          LL = LINE/0,P1AY,P4AX,P4AY                              470
$$***************************************************             500
DRVLIN = MACRO                                                    510
          Y0 = Y0 + D1Y                                           520
          X1 = X1 + D2X                                           530
          Y1 = Y1 + D2Y                                           540
          THETA = ATANF((Y1−Y0)/X1)                               550
          A = COSF(90 + THETA)                                    560
          B = COSF(THETA)                                         570
          D = Y0*B                                                580
          LL = LINE/LL,CANON,A,B,0,D                              590
TERMAC                                                            600
$$***************************************************             610
SETPT    = POINT/1,1.5,2.25                                       620
          FROM/SETPT                                              630
          INDIRP/P1A                                              640
          GO/ON,LY,TO,RSRF,ON,LL                                  650
          INDIRP/P4A                                              660
          TLON,GOFWD/LL,ON,L2                                     670
          J = 0                                                   680
          LOOPST                                                  690
```

BEG)	CALL/DRVLIN	700
	GOLFT/L2,ON,LL	710
	GOLFT/LL,ON,LY	720
	CALL/DRVLIN	730
	GORGT/LY,ON,LL	740
	GORGT/LL,ON,L2	750
	J = J + 2	760
	IF (N − J) ENDD, ENDD, BEG	770
ENDD)	LOOPND	780
	GOTO/SETPT	790

Example 13.10

This example shows how vectors are used to define geometric figures. The part, shown in Figure 13.22, consists of two unequal diameter spheres, connected with a cone tangent to

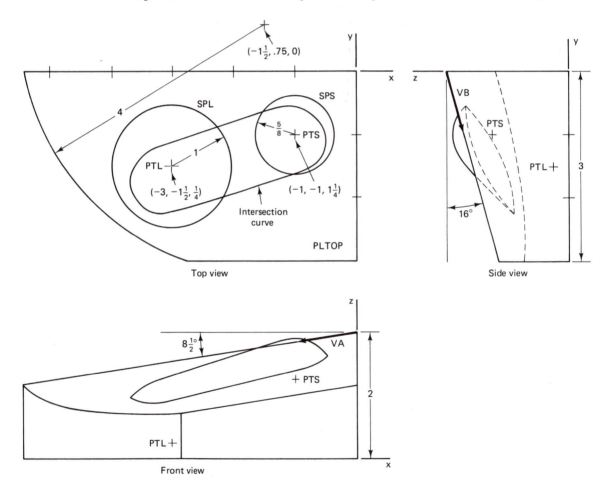

Figure 13.22   Part for Example 13.10.

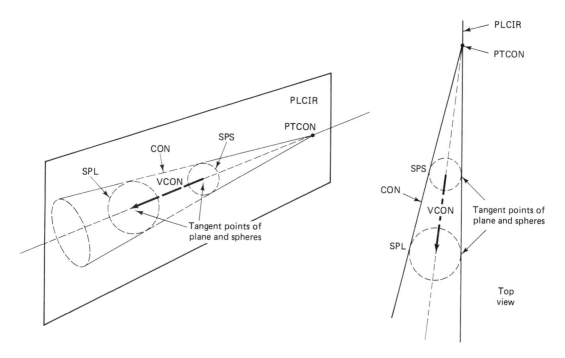

Figure 13.23   Geometry for defining the cone.

each sphere, placed such that the sphere centers lie below a canted plane. The part program is shown below.

The plane is defined with definition Plane-5 of Table 8.1 (through a point and perpendicular to a vector). To get the vector for this plane (PLTOP), it is convenient first to define two vectors in the plane (VA, VB), then define the needed vector (VPN) by their cross product.

To define the cone CON (definition Cone-2, Table 12.2, by its vertex location, axis orientation, and vertex half-angle) we first define the sphere projections on the *xy*-plane as circles (CSM,CLG), with line LCIR tangent to them. We then extract line LCIR parameters with an OBTAIN statement and use them to define plane PLCIR tangent to the spheres (because line and plane canonic definitions are identical). Vector VCON through the sphere centers pierces plane PLCIR at the cone vertex. The vertex half-angle THETA is then computed and CON defined. Figure 13.23 shows the geometry for defining the cone.

A zero-diameter tool traces the intersection curve of the spheres and cone assembly with the canted plane. Line LXM is used for intermediate checking. Contouring this part with a nonzero diameter tool under three-axis control requires computation for checking the tool at the tangent points of the cone with the spheres because the plane in which the tangent points lie for each sphere is not parallel with the tool axis.

```
$$    ***   CANTED PLANE DEFINITION  ****                         100
VA    = VECTOR/LENGTH,1,ATANGL, −(90+8.5),ZXPLAN                  110
VB    = VECTOR/LENGTH,1,ATANGL,(180+16),YZPLAN                    120
```

VPN = VECTOR/VA,CROSS,VB	130
PTOP = POINT/0,0,2	140
PLTOP = PLANE/PTOP,PERPTO,VPN	150
$$ *** SPHERE DEFINITIONS ****	160
PTS = POINT/ $-1, -1, 1.25$	170
PTL = POINT/ $-3, -1.5, .25$	180
SPS = SPHERE/CENTER,PTS,RADIUS,.625	190
SPL = SPHERE/CENTER,PTL,RADIUS,1	200
$$ *** CONE TANGENT TO SPHERES ****	210
CSM = CIRCLE/CENTER,PTS,RADIUS,.625	220
CLG = CIRCLE/CENTER,PTL,RADIUS,1	230
LCIR = LINE/RIGHT,TANTO,CLG,RIGHT,TANTO,CSM	240
OBTAIN,LINE/LCIR,A,B,C,D	250
PLCIR = PLANE/A,B,C,D	260
VCON = VECTOR/PTS,PTL	270
PTCON = POINT/INTOF,PLCIR,VCON,PTS	280
THETA = ATANF((1 − .625)/DISTF(PTS,PTL))	290
CON = CONE/PTCON,VCON,THETA	300
$$ *** *** ***	310
LXM = LINE/XAXIS, -1.5	320
SETPT = POINT/ $-5, -3, 3$	330
FROM/SETPT	340
GO/ON,LXM,ON,PLTOP	350
TLON,GORGT/LXM,ON,SPL	360
GORGT/SPL,ON,CON	370
GOFWD/CON,ON,SPS	380
GOFWD/SPS,ON,LXM	390
GOFWD/SPS,ON,CON	400
GOFWD/CON,ON,SPL	410
GOFWD/SPL,ON,LXM	420
GOTO/SETPT	430

PROBLEMS

13.1. Write an APT part program to machine the object of Figure P12.1. Ignore the cylindrical cutout accessible from the bottom of the object. Select a suitable shape and size of cutter.

13.2. Write an APT part program to machine the object of Figure P12.2. Ignore the parabolic cylindrical cutout accessible from the bottom of the object. Select a suitable shape and size of cutter.

13.3. Figure P13.3 shows a three-dimensional representation of the equation defined there. The object is obtained by revolving the curve shown in the positive right half-plane of the cross-sectional view. Write an APT part program to machine the object for A = 2 in. and B = 2.5 in. Select a suitable shape and size of cutter.

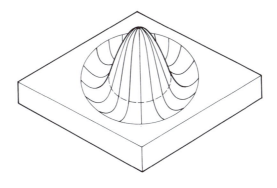

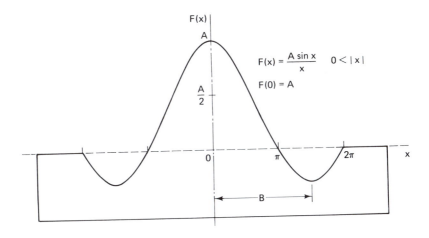

Figure P13.3

13.4. Write an APT part program to machine the object of Figure P12.6. Select a suitable shape and size of cutter.

13.5. Repeat Problem 13.4 for the object of Figure P12.10.

13.6. Repeat Problem 13.4 for the object of Figure P12.12.

CHAPTER 14

Multiaxis Part Programming

We now face the greatest challenge to applying the APT language, namely, directing the tool for proper orientation in space for producing parts of arbitrary complexity in three-dimensional space. In this chapter we describe the rich set of constructs for controlling the tool axis. We assume that proficiency in applying the APT language has been acquired and that a keen sense of visualizing complex part relationships has been developed.

14.1 NOMENCLATURE

Variable tool axis orientation is associated with **multiaxis** part programming. However, not all parts require continuous variation of the axis during tool movement. The axis may be fixed at an angle to one or more of the principal axes and thereafter unchanged during part production. Some possible tool orientations with respect to the part are shown in Figure 14.1.

The tool axis may be tangent to the controlling surface [Figure 14.1(a)], it may be normal to the surface [Figure 14.1(b) and (d)], or it may be normal to the surface but controlled along its axis [Figure 14.1(c)] (obvious alignment of the parts with respect to the coordinate axes assumed).

We now are concerned with multiaxis control of the tool. Control in this context implies simultaneous control of the tool axis variation as well as control of the machine along the principal axes as described in Chapter 13 for three-dimensional work. Cut vectors computed by APT in a multiaxis mode reflect relative motion between the tool and the part as previously described. Directed movements of the tool and/or part are resolved by the postprocessor for a specified machine/controller combination.

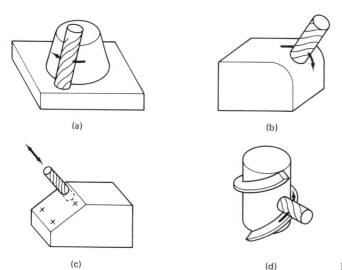

(a)

(b)

(c)

(d)   **Figure 14.1   Possible tool orientations.**

Axis is a term we use to describe each coordinate being controlled. For three-axis work, we control the tool along the three coordinate axes. For some machines, it is possible to rotate the tool about one or more of the coordinate axes. Hence six-axis control might be possible (three along the coordinate axes and three more because of the rotations; see Figure 14.2). For three-axis control there is no rotation about the coordinate axes (tool rotation about its own axis is excluded). Figure 14.3(a) shows a part requiring three-axis control.

Four-axis control is three-axis control with rotation about one of the coordinate axes. Parts requiring no more than four-axis control are shown in Figures 14.1(b) and 14.3(b). Five-axis control is three-axis control with rotation about two of the coordinate axes. The parts of Figure 14.1(a) and (c) require five-axis control.

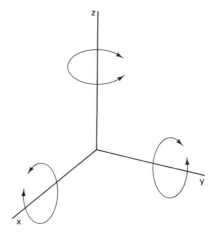

Figure 14.2   The coordinate axes and additional axes for tool control (the rotations).

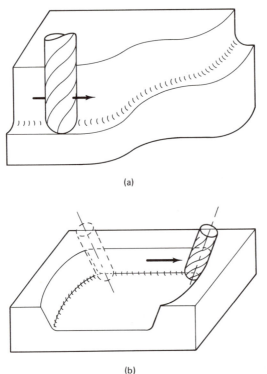

(a)

(b)

Figure 14.3　(a) Three-axis control;
(b) four-axis control.

14.2 TOOL AXIS SPECIFICATION

Thus far, the tool axis orientation for three-axis contouring was treated no differently than for two-axis contouring as discussed earlier in the book—it was maintained parallel with the z-axis. The part geometries for four-axis and five-axis contouring require that the arbitrary tool axis orientation capability of the machine be exploited. Therefore, we now address the capability in APT for orienting the tool to satisfy these contouring requirements. At times we must fix the tool axis in a predetermined orientation, or we must allow it to vary in a predetermined manner relative to the drive and/or part surfaces, or we must cause it to vary in some yet precisely unknown way but to be determined by conditions occurring while contouring. The issues regarding tool axis orientation in APT are discussed below.

14.2.1 Preliminaries

Machine tools with multiaxis control capability require tool end coordinates and tool axis vector components to control the rotational axes. Therefore, we must cause to be placed on the CL-file the tool axis vector components as well as the tool end coordinates for use by the postprocessor. The tool axis is defined as a unit vector originating at the tool endpoint and directed toward the top of the tool. Until otherwise specified, this vector defaults to the z-axis unit vector (with components 0, 0, 1).

The multiaxis statement causes tool axis vector information to be output. It is implemented by either of the following forms:

$$\text{MULTAX}$$

$$\text{MULTAX/}\left\{\begin{matrix} \text{ON} \\ \text{OFF} \end{matrix}\right\}$$

These statements control output of the vectors, not their values. The MULTAX form causes vectors to be output from the location in the program where it appears until the end of the program. The tool axis vectors and the tool end coordinates are output to the CL-file and appear on the CL-printout. This is a modal command which, when turned on, cannot be turned off in the part program.

The second form of the statement selectively turns this feature on and off for those applications that require multiaxis control for portions of the part program only. The vectors are output for motion commands bounded by the MULTAX/ON and MULTAX/OFF statements. Should both forms of the multiaxis statement appear in the same part program, their combined effect is implementation dependent (the MULTAX statement takes precedence for the remainder of the program in the IBM APT-AC system).

The postprocessor must recognize the multiaxis statement and compute commands for control of the machine tool. Therefore, the part programmer must know the exact interpretation of the two statement forms by the postprocessor before incorporating them in the program.

Tool axis vectors in the TRACUT and COPY regions are modified by the associated transformation matrices.

14.2.2 In Motion Commands: Fixed Orientation

Tool axis orientation may be specified in the motion commands described below. This is a fixed orientation in an absolute sense that stays in effect until changed. The multiaxis statement must be in effect for the tool axis vectors to be output. The motion commands described below are modifications of those described in Section 9.2.

1. $\text{FROM/}\left\{\begin{matrix} point \\ x,y,z \end{matrix}\right\}\left[\,,\left\{\begin{matrix} vector \\ i,j,k \end{matrix}\right\}\,\right][\,,f\,]$

Tool axis orientation is specified by the first optional field, where it may be given symbolically by *vector*, or by the vector components i,j,k. The second optional field is the feed rate (see Section 10.3). Normalized tool axis vector components are output with this statement.

2. $\text{GOTO/}\left\{\begin{matrix} point \\ x,y,z \end{matrix}\right\}\left[\,,\left\{\begin{matrix} vector \\ i,j,k \end{matrix}\right\}\,\right][\,,f\,]$

The parameters of this statement have the same meaning as those given for the FROM statement.

3. $\text{GODLTA}/\begin{Bmatrix} vector1 \\ dx_1,dy_1,dz_1 \end{Bmatrix}, \begin{Bmatrix} vector2 \\ dx_2,dy_2,dz_2 \end{Bmatrix}[,f]$

Tool axis orientation is specified by the second field, where it may be given symbolically by *vector2* or by the vector components dx_2,dy_2,dz_2. The optional field is the feed rate. This form of the GODLTA statement does not permit the scalar *dtlax* (see Section 9.2) to be used when specifying the tool axis orientation. Normalized tool axis vector components are output for the new tool location.

Example 14.1

Figure 14.4 shows the geometry of a part into which we will drill eight holes. Symmetry of the part suggests use of the COPY feature. A part program without postprocessor statements that performs this task is as follows:

```
        $$***********************************           300
        DRVC = MACRO/P,V,D                              400
                GOTO/P,V                                500
                GODLTA/ − D                             600
                GODLTA/D                                700
                TERMAC                                  800
        $$***********************************           900
        P1    = POINT/0, − 2.125,.5                     1100
        L1    = LINE/XAXIS, − 2.125                     1200
        L2    = LINE/YAXIS,2.125                        1300
        L3    = LINE/1, − 2,2, − 1                      1400
        L4    = LINE/PARLEL,L3,XLARGE,.125              1500
        L5    = LINE/0,0,1.5, − 1.5                     1600
                ZSURF/.5                                1700
        P2    = POINT/INTOF,L1,L4                       1800
        P3    = POINT/INTOF,L5,L4                       1900
        P4    = POINT/INTOF,L2,L4                       2000
        V1    = VECTOR/0, − 1,0                         2100
        V2    = VECTOR/1, − 2,0                         2200
        V3    = VECTOR/1.5, − 1.5,0                     2300
        V4    = VECTOR/2, − 1,0                         2400
        SETPT = POINT/ − 1, − 3,2                       2500
                MULTAX                                  2600
                FROM/SETPT                              2700
                INDEX/5                                 2800
                  CALL/DRVC,P = P1,V = V1,D = .625      2900
                  GOTO/P2,V2                            3000
                  CALL/DRVC,P = P3,V = V3,D = .625      3100
                  GOTO/P4,V4                            3200
                COPY/5,XYROT,90,3                       3300
                GOTO/SETPT                              3400
```

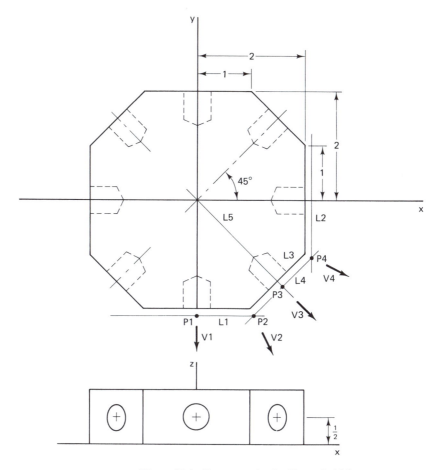

Figure 14.4   Part geometry for Example 14.1.

A portion of the CL-printout appears in Figure 14.5. It shows the tool axis vector being output with the tool endpoint coordinates. The effect of the COPY transformation on the tool axis vector can be seen. The GODLTA/*dtlax* form of this statement is used to cause motion for the drilling operation to be parallel with the tool axis, regardless of its orientation.

14.2.3 TLAXIS: Fixed or Variable Orientation

The TLAXIS (tool axis) statement allows the part programmer to place the tool axis in a fixed or variable orientation mode. The fixed mode is an absolute orientation in space. The variable mode fixes the axis in a relative sense while allowing variation in orientation

Figure 14.5 Part of the CL-printout for Example 14.1.

ISN	Statement	Ref / Comment	Tool endpoint coordinates			Tool axis vector components			LABEL	REC M	CARD
0001	PARTNO/	EXAMPLE 14.2-1.								00002	00000100
0026	MULTAX/	ON								00004	00002600
0027	FROM/	SETPT	-1.00000000	5.00000000						00006	00002700
0028	INDEX /		-3.00000000	2.00000000		0.0	0.0	1.00000000		00008	00002800
0005	GOTO/	P1	-2.12500000	0.50000000		0.0	-1.00000000	0.0		00010	00000500
0006	GOTO/	TLAXIS	-1.50000000	0.50000000		0.0	-1.00000000	0.0		00012	00000600
0007	GOTO/	TLAXIS	-2.12500000	0.50000000		0.0	-1.00000000	0.0		00014	00000700
0030	GOTO/	P2	-2.12500000	0.50000000		0.44721360	-0.89442719	0.0		00016	00003000
0005	GOTO/	P3	-1.58838835	0.50000000		0.70710678	-0.70710678	0.0		00018	00000500
0006	GOTO/	TLAXIS	-1.14644661	0.50000000		0.70710678	-0.70710678	0.0		00020	00000600
0007	GOTO/	TLAXIS	-1.58838835	0.50000000		0.70710678	-0.70710678	0.0		00022	00000700
0032	GOTO/	P4	-1.05177670	0.50000000		0.89442719	-0.44721360	0.0		00024	00003200
0033	COPY /	XYROT *** THIS STARTS COPY/ 5 PASS 1 ***	2.12500000	5.00000000		90.00000000	3.00000000			00025	00003300
0033	GOTO/	P1	0.0	0.50000000		1.00000000	0.0	0.0		00027	00003300
0006	GOTO/	TLAXIS	0.0	0.50000000		1.00000000	0.0	0.0		00029	00000600
0007	GOTO/	TLAXIS	0.0	0.50000000		1.00000000	0.0	0.0		00031	00000700
0030	GOTO/	P2	2.12500000	0.50000000		0.89442719	0.44721360	0.0		00033	00003000
0005	GOTO/	P3	1.58838835	0.50000000		0.70710678	0.70710678	0.0		00035	00000500
0006	GOTO/	TLAXIS	1.14644661	0.50000000		0.70710678	0.70710678	0.0		00037	00000600
0007	GOTO/	TLAXIS	1.58838835	0.50000000		0.70710678	0.70710678	0.0		00039	00000700
0032	GOTO/	P4	1.05177670	0.50000000		0.44721360	0.89442719	0.0		00041	00003200
0033	COPY /	XYROT *** THIS COMPLETES COPY/ 5 PASS 1 ***	2.12500000	5.00000000		90.00000000	3.00000000			00042	00003300
0033	GOTO/	P1	2.12500000	0.50000000		0.0	1.00000000	0.0		00044	00003300
0006	GOTO/	TLAXIS	1.50000000	0.50000000		0.0	1.00000000	0.0		00046	00000600
0007	GOTO/	TLAXIS	2.12500000	0.50000000		0.0	1.00000000	0.0		00048	00000700
0030	GOTO/	P2	-1.05177670	0.50000000		-0.44721360	0.89442719	0.0		00050	00003000

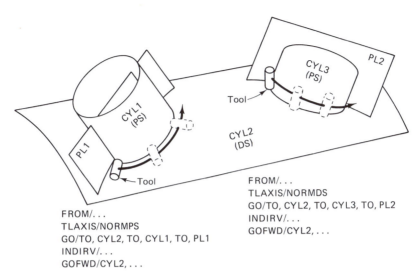

Figure 14.6   Variable tool axis orientation normal to a surface.

contingent on other variables controlling the tool motion. The TLAXIS statement allows the part programmer to control the tool axis in the following ways.

1. Fixed in absolute orientation in space (fixed mode):

$$\text{TLAXIS/}\begin{Bmatrix} vector \\ a,b,c \end{Bmatrix}$$

This statement fixes the tool axis along *vector*, or alternatively along a vector whose components are [a,b,c] but not necessarily direction cosines. The tool orientation is effective with the next motion command. The effect of this statement is the same as that produced by the statements of Section 14.2.2.

2. Fixed normal to the part surface or drive surface (variable mode):

$$\text{TLAXIS/}\begin{Bmatrix} \text{NORMPS} \\ \text{NORMDS} \end{Bmatrix}$$

This statement causes the tool axis to be made normal to the part surface, NORMPS, or to the drive surface, NORMDS. During the cut sequence, the tool axis is automatically changed as needed to maintain the specified relationship (see Figure 14.6).

3. Fixed in its current orientation (switches from variable to fixed mode):

TLAXIS/1

This statement causes the tool axis to be fixed as oriented when this statement is processed. This statement switches the tool axis from a variable axis mode to a fixed axis mode. It occurs at point A in Figure 14.7.

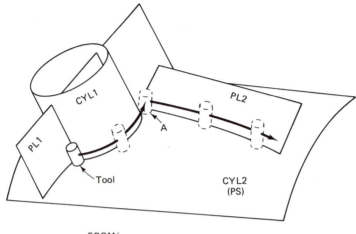

```
FROM/. . .
TLAXIS/NORMPS
GO/TO, CYL1, TO, CYL2, TO, PL1
INDIRV/. . .
GOFWD/CYL1, TO, PL2
TLAXIS/1
GORGT/PL2, . . .
```

Figure 14.7   Switching from a variable tool axis orientation to a fixed orientation at point A.

4. Parallel with the rulings of a ruled surface (variable mode):

$$\text{TLAXIS/PARLEL,} \left\{ \begin{matrix} 1 \\ 2 \end{matrix} \right\}$$

This statement causes the tool axis to be oriented parallel with the rulings of a ruled surface that serves as the part surface (parameter = 1) or as the drive surface (parameter = 2) for the following motion command. A new tool axis statement must be issued when continuing to contour after leaving the ruled surface. Both forms of this statement are applied to the tool axis orientation of Figure 14.8 while contouring along the ruled surface. The choice was to fix the tool axis before continuing with a contouring command.

5. Maintained at an angle from the normal to the part surface or drive surface in a plane normal to the direction of motion (variable mode):

$$\text{TLAXIS/ATANGL,} \left\{ \begin{matrix} 1 \\ 2 \end{matrix} \right\}, \alpha \qquad -90° \leq \alpha \leq 90°, \quad \alpha \neq 0°$$

This statement causes the tool axis to be oriented at an angle α from the normal to the *controlling surface*, which is the part surface (parameter = 1) or the drive surface (parameter = 2). The *noncontrolling surface* is the other surface constraining the tool

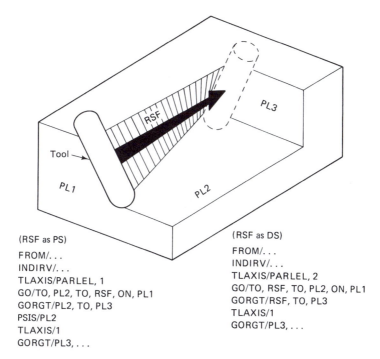

Figure 14.8   Alignment of the tool axis parallel with the rulings of a ruled surface when it is used as a part surface and as a drive surface.

position while contouring. Angle α (in degrees) is made positive when it is measured *from* the controlling surface normal *toward* the noncontrolling surface normal; it is made negative when it is measured *away from* the noncontrolling surface normal. The two possible orientations of the tool axis, corresponding to the two signs of α, are shown in Figure 14.9, where the part surface is chosen to be the controlling surface.

Angle α defines the half-angle of a cone whose axis is the normal to the controlling surface and whose vertex is the tool endpoint near the intersection of the two constraining surfaces as shown in Figure 14.9. The tool axis lies in the nappe of the cone. However, this form of the TLAXIS statement further restricts the tool so that it also lies in the plane normal to the direction of tool motion. The sign of α is used to differentiate between the two possible orientations of the tool under these constraints.

6. Maintained at an angle from the normal to the part surface or drive surface in a fixed plane (variable mode):

$$\text{TLAXIS/ATANGL,}\begin{Bmatrix}1\\2\end{Bmatrix}\text{,}\alpha\text{,}vector \qquad -90° \le \alpha \le 90°, \quad \alpha \ne 0°$$

This statement is interpreted almost identically to the statement described in paragraph 5. The exception is the influence of *vector*, a vector used to define a plane in which

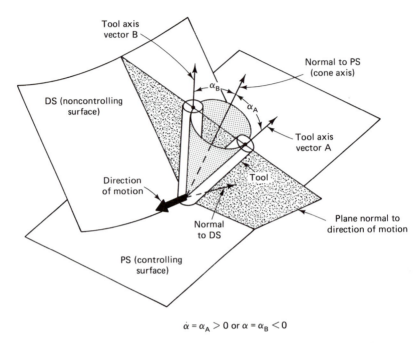

$$\dot{\alpha} = \alpha_A > 0 \text{ or } \alpha = \alpha_B < 0$$

Figure 14.9   Two possible tool axis orientations for the statement TLAXIS/ ATANGL,1,α (part surface as controlling surface).

the tool must lie. The plane so defined is normal to *vector* and is fixed in this orientation regardless of tool direction of motion. The geometry is illustrated in Figure 14.10. The numeric parameter and angle α are interpreted as for the statement of paragraph 5.

The cutter axis must lie in the plane defined by *vector*. As contouring proceeds, angle α may be modified by the APT processor to maintain this relationship. This form of tool axis statement is used for four-axis contouring—the tool rotates about one axis in the plane determined by *vector*, yet it also moves relative to the coordinate axes. An example is shown in Figure 14.11, where a tool axis reorientation occurs at the point where the tool axis plane is tangent to the drive surface.

7. Maintained at an angle from the normal to the part surface or drive surface and at an angle to the direction of motion (variable mode):

$$\text{TLAXIS/ATANGL,} \begin{Bmatrix} 1 \\ 2 \end{Bmatrix} , \alpha, \text{CUTANG}, \beta \qquad \begin{aligned} &-90° \leq \alpha \leq 90°, \quad \alpha \neq 0° \\ &-90° \leq \beta \leq 90° \end{aligned}$$

This statement causes tool axis orientation similar to that of paragraph 5. Rather than maintaining the tool axis in a plane normal to the direction of the motion, this statement allows the tool to lean into or away from the direction of motion. This leaning action is specified by angle β. The numeric parameter and angle α are interpreted as for the statement of paragraph 5. The corresponding geometry is shown in Figure 14.12.

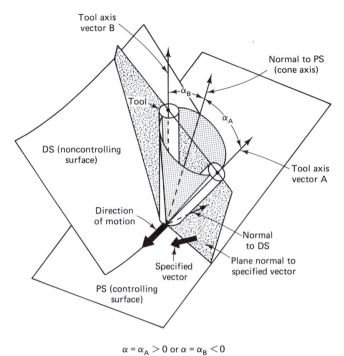

Figure 14.10 Two possible tool axis orientations for the statement TLAXIS/ATANGL,1,α,*vector* (part surface as controlling surface).

$\alpha = \alpha_A > 0$ or $\alpha = \alpha_B < 0$

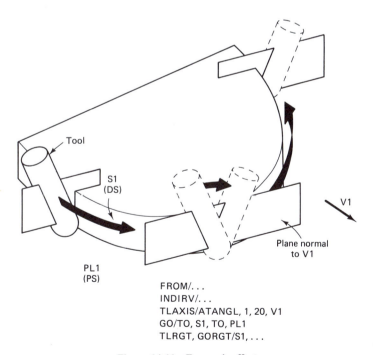

```
FROM/...
INDIRV/...
TLAXIS/ATANGL, 1, 20, V1
GO/TO, S1, TO, PL1
TLRGT, GORGT/S1, ...
```

Figure 14.11 Four-axis effect.

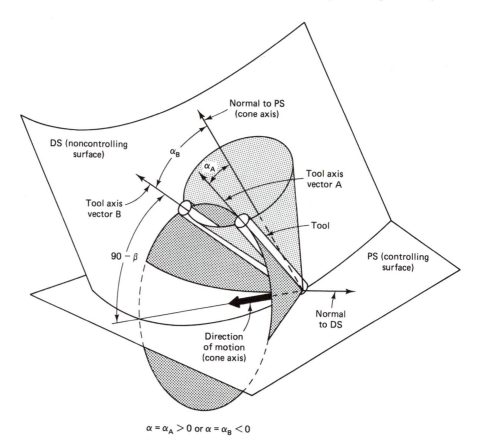

$$\alpha = \alpha_A > 0 \text{ or } \alpha = \alpha_B < 0$$

Figure 14.12   Two possible tool axis orientations for the statement TLAXIS/ ATANGL,1,α,CUTANG,β (part surface as controlling surface, β as a lead angle).

The plane of paragraph 5 is replaced by a cone with half-angle $90 - \beta$ (β in degrees) and axis coincident with the direction of motion. Because β can be made positive or negative, the cone opens toward the direction of motion (positive β) or opposite to the direction of motion (negative β). The terminology applied to the tool inclination is lead (positive β) and lag (negative β). The geometry of Figure 14.12 is drawn with a lead angle. Of course, the cones formed by angles α and β must intersect or there can be no solution for orienting the tool.

14.2.4 AUTOPS With Tool Axis Tilt

The tool axis orientation for the AUTOPS discussions of Section 8.2.4 and 9.2.4 was parallel with the z-axis. We now use this statement when the tool axis is inclined to the z-axis.

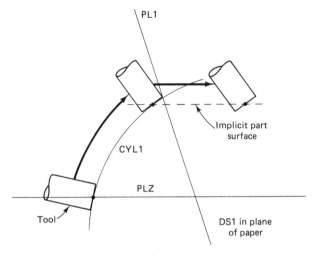

TLAXIS/NORMPS
INDIRV/V1
GO/TO, DS1, TO, CYL1, ON, PLZ
INDIRV/V2
TLRGT, GOFWD/DS1, TO, PL1
TLAXIS/1
AUTOPS
TLONPS, GOFWD/DS1, . . .

Figure 14.13 **Implicit part surface with**
tool axis tilt.

The AUTOPS statement always defines a part surface passing through the tool endpoint and parallel with the *xy*-plane. So, to resume contouring and to maintain the tool within tolerance of the surfaces, we must now include the TLONPS modifier as shown in Figure 14.13. Because the implicit part surface is declared in an ad hoc manner, this matters only if the tool axis is variable and should not penetrate what otherwise would have been its part surface under a TLOFPS condition.

14.3 VTLAXS: A SPECIAL VARIABLE ORIENTATION

The TLAXIS statements allowed us to tilt the tool axis relative to the part and either fix it in an absolute sense or in a relative sense contingent on tool-to-part relationships. Thus far, only the TLAXIS statement related to a ruled surface gave the appearance of tilting the tool throughout its cutting motion, but then only if the ruled surface was defined as a variably orienting surface. The special five-axis feature that does continually vary the tool axis tilt angle over a cutting sequence via a transformation is invoked with the VTLAXS (variable tool axis) statement. It is sometimes accompanied by a special inside cornering statement, WCORN (see Section 14.3.2).

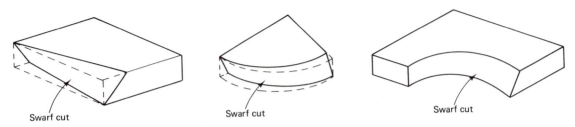

Figure 14.14   Swarf cuts produced with the VTLAXS feature.

14.3.1 The VTLAXS Statements

The variable tool axis feature is a special transformation controlled by a pair of VTLAXS statements between which the tool axis change is prorated linearly over the cut vectors as a function of distance over which the tool must travel. It is applied after cut vectors are computed, not during their computation. This feature is useful for beveling surfaces, for cutting relief angles on blanking dies, and for cutting draft angles on casting patterns. The result is a swarf cut as shown in Figure 14.14.

The motion statements for the path to be transformed are bounded by VTLAXS/ON and VTLAXS/OFF statements as follows:

VTLAXS/ON, . . .
$\qquad$: $\qquad$ } Motion commands for which the
$\qquad$: $\qquad$ } CL-file data are transformed
VTLAXS/OFF, . . .

The CL-file point immediately preceding the VTLAXS/ON statement is also transformed by this feature. Thus at least one such point must have been written to the CL-file before the VTLAXS feature is invoked. Unlike the TLAXIS statements, which are modal in their effect, the VTLAXS feature is self-canceling with the VTLAXS/OFF statement.

These restrictions apply to the VTLAXS feature:

1. The part described and the motion commands between the VTLAXS bounding pair must be two-dimensional—part surface parallel with the xy-plane and cutter axis vector [0, 0, 1].
2. The MULTAX condition must have been specified prior to the VTLAXS/ON statement.
3. The cutter path between the bounding statements must have no inflections.

The syntax of the bounding statements follows:

$$\text{VTLAXS/ON,} \begin{Bmatrix} \text{RIGHT} \\ \text{LEFT} \end{Bmatrix}, \alpha, d, r, \begin{Bmatrix} \text{RIGHT} \\ \text{LEFT} \end{Bmatrix} [\text{,ZSMALL}]$$

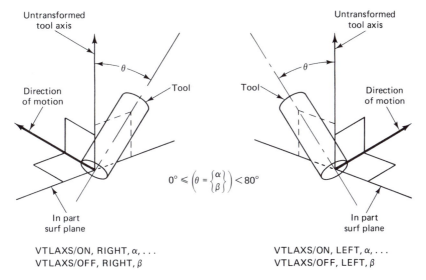

$$0° \leqslant \left(\theta = \begin{Bmatrix} \alpha \\ \beta \end{Bmatrix}\right) < 80°$$

VTLAXS/ON, RIGHT, α, . . .
VTLAXS/OFF, RIGHT, β

VTLAXS/ON, LEFT, α, . . .
VTLAXS/OFF, LEFT, β

Figure 14.15   Interpretation of the VTLAXS first RIGHT/LEFT modifier.

ON	Keyword for the upper bounding statement.
RIGHT or LEFT	First occurrence—a modifier that specifies the direction of cutter tilt at the beginning of the CL-data being transformed. It is obtained by viewing in the direction of motion and noting which side from vertical the tool axis is to be tilted (see Figure 14.15). It is used in conjunction with angle α.
α	The angle ($0° \leqslant \alpha < 80°$, in degrees) of the cutter tilt from vertical in the direction of the first RIGHT/LEFT modifier (see Figure 14.15).
d	The cutter diameter value, its meaning identical to the corresponding parameter of the CUTTER/d,r statement.
r	The cutter corner radius value, its meaning identical to the corresponding parameter of the CUTTER/d,r statement. It must be zero if the ZSMALL option is used, implying a flat-end tool. The cutter shapes of Section 2.2 are only those allowed.
RIGHT or LEFT	Second occurrence—a modifier that specifies the location of the cutter relative to the drive surface when viewed in the direction of motion before the transformation is applied. RIGHT is selected for a TLRGT tool position modifier, LEFT for the TLLFT modifier. The tool must not have been in a TLON condition.
ZSMALL	This parameter allows the tool end to penetrate the part surface as the tool is tilted. If not given, the tool is maintained in a TLOFPS condition (see Figure 14.16). It is allowed only for flat-end cutters ($r = 0$).

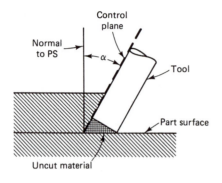

VTLAXS/ON, RIGHT, α, d, 0, RIGHT

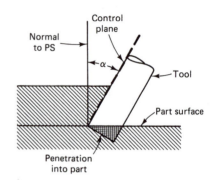

VTLAXS/ON, RIGHT, α, d, 0, RIGHT, ZSMALL

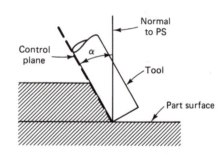

VTLAXS/ON, LEFT, α, d, 0, RIGHT
or
VTLAXS/ON, LEFT, α, d, 0, RIGHT, ZSMALL

Figure 14.16   Tool-to-part relationship of flat-end tool used with and without ZSMALL in VTLAXS statement.

$$\text{VTLAXS/OFF,} \begin{Bmatrix} \text{RIGHT} \\ \text{LEFT} \end{Bmatrix}, \beta$$

OFF   Keyword for the lower bounding statement.

RIGHT   A modifier that specifies the direction of cutter tilt at the end of the
or   CL-data being transformed. It is obtained by viewing in the direction
LEFT   of motion and noting which side from vertical the tool axis is to be
            tilted (see Figure 14.15). It is used in conjunction with angle β.

β   The angle ($0° \leq \beta < 80°$, in degrees) of the cutter tilt from vertical
            in the direction of the above RIGHT/LEFT modifier (see Figure 14.15).

To understand the action of this feature, first mentally position a control plane parallel with the direction of tool motion and tangent to the untransformed tool on the

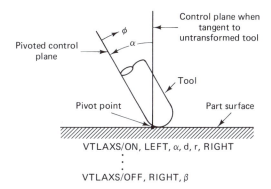

VTLAXS/ON, LEFT, α, d, r, RIGHT
.
.
.
VTLAXS/OFF, RIGHT, β

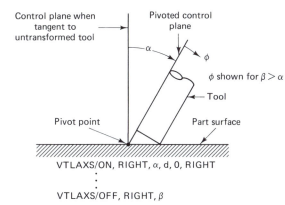

VTLAXS/ON, RIGHT, α, d, 0, RIGHT
.
.
.
VTLAXS/OFF, RIGHT, β

Figure 14.17   Pivoting action of the VTLAXS statements.

drive surface side of the tool. This plane is perpendicular to the part surface. Next, with a pivot point in the part surface plane, pivot the control plane through the angle α while maintaining the tool in contact with the part surface and in sliding contact with the control plane, unless the ZSMALL modifier is used in which case assume that the tool is fastened to the control plane. The result is tool axis transformation for the initial point, that immediately preceding the VTLAXS/ON statement, as shown in Figures 14.16 and 14.17. This transformation, the swarf effect, takes place after the two-dimensional cut vectors are computed.

Thereafter, the control plane is pivoted from its initial inclination (determined by α and the first RIGHT/LEFT modifier of the VTLAXS/ON statement) by an amount ϕ toward its final inclination (determined by β and the RIGHT/LEFT modifier of the VTLAXS/OFF statement) as the tool is driven on the path between the bounding statements. Angle ϕ is determined linearly with tool location on the path by the expression

$$\phi = \frac{l_p}{l_t} |\alpha \pm \beta|$$

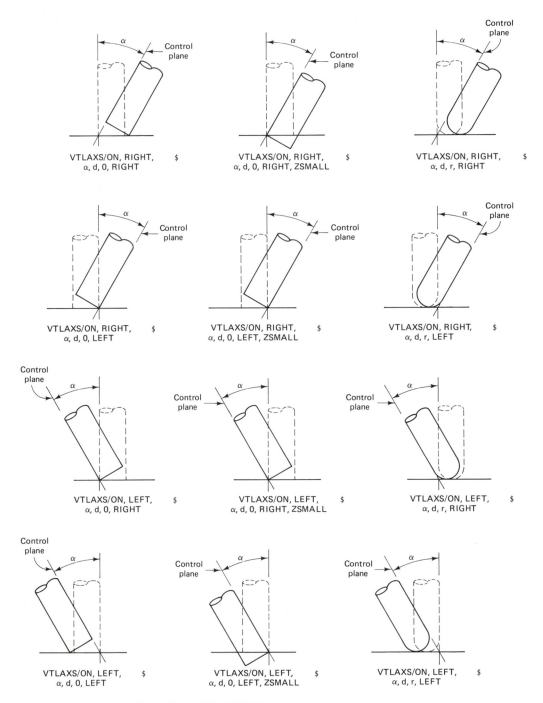

Figure 14.18 VTLAXS/ON tool axis orientations.

where l_p = distance of the tool endpoint from the initial point to its current position on the path between the bounding statements

l_t = total distance on the path from the initial point to the last point computed between the bounding statements

± = plus sign chosen if the α and β RIGHT/LEFT modifiers differ, else the negative sign is chosen

Figure 14.18 shows the tool in its initial position for various combinations of the VTLAXS/ON statement modifiers. The tool is at all times pivoted in a plane perpendicular to its direction of motion to create the swarf effect. This is in contrast with its orientation when using the TLAXIS statement with a ruled surface, where the tool axis is always parallel with the rulings. The rulings are not necessarily perpendicular to the direction of motion, nor do they cause a linear distribution of tool angular orientations along the surface.

Tool endpoints derived from a curve with an inflection point are shown in Figure 14.19, where the definition of an inflection point in terms of the cosine of the angle from the normal to a line joining point p_i with p_{i-1} and a line from point p_i to p_{i+1}. To avoid the inflection, motion commands generating points P_1 through P_4 should be between one VTLAXS bounding pair while those generating points P_5 through P_7 should be between another VTLAXS bounding pair.

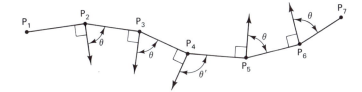

cos θ > 0.01 (no inflection)
cos θ' < 0.01 (inflection)

Figure 14.19   Inflection at tool endpoint P_4.

Here are other factors that the part programmer should be aware of when using the VTLAXS feature.

1. Tool axis orientation following the VTLAXS/OFF statement is returned to [0, 0, 1].
2. The drive surface can be any legitimate surface not necessarily oriented perpendicular to the part surface.
3. There can be no FROM statement between the bounding statements.
4. There must be at least one motion statement between the bounding statements.
5. The VTLAXS/ON statement contains its own cutter description parameters, which are interpreted as for two-dimensional contouring (see Section 2.2).

These additional limitations apply when the VTLAXS feature is used with the TRACUT and COPY features.

1. TRACUT and COPY may not appear between the VTLAXS bounding statements.

2. The INDEX statement cannot appear between the VTLAXS/ON statement and its preceding motion statement, nor between the VTLAXS bounding statements.

Example 14.2

The following motion commands illustrate the VTLAXS feature when applied to the geometry of Figure 14.20. The CL-printout from an IBM APT-AC system appears in Figure 14.21. Liberal tolerances were used to reduce the amount of printout.

MULTAX	1300
CUTTER/1,.5	1400
TLAXIS/0,0,1	1500
FROM/SETPT	1600
INDIRV/0,1,0	1700
GO/TO,C1	1800
VTLAXS/ON,LEFT,10,1,.5,RIGHT	1900
TLRGT,GORGT/C1,TO,C2	2000
GORGT/C2,ON,LX	2100
VTLAXS/OFF,RIGHT,5	2200
GOTO/SETPT	2300

Transformed cut vectors are contained within the rectangular outline of Figure 14.21. We see that the first point preceding the VTLAXS/ON statement is also transformed and that the tool axis is returned to [0, 0, 1] after the VTLAXS/OFF statement. The figure is annotated with the tool axis tilt from vertical. Two warnings of possible cutter path inflections

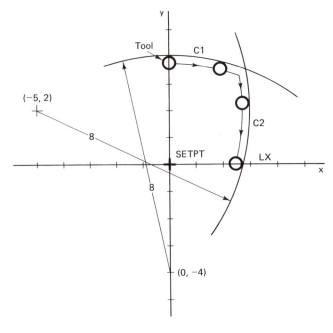

Figure 14.20 Geometry for Example 14.2.

```
                    ┌─────── Tool endpoint ────────┐      ┌────── Tool axis vector ──────┐
0014 CUTTER/   1.00000000   0.50000000                                              00017 00001400
0016                                                                                00018 00001600
0016 FROM/        SETPT                                                             00020 00001600
               0.0          0.0          0.0       0.0        0.0     1.00000000  0°  00021 00001800
0018                                                                                00021 00001800
0018 GOTO/        C1                                                                00023 00001800
               0.0          3.50000000   0.0       0.0        0.0     1.00000000  0°
0022 GOTO/        C1                                                                00023 00002200
              -0.00002737   3.49362615   0.00759612  0.00074574  0.17364658  0.98480775  10°
─ ─ ─ ─ ─ ─ ─ ─ ─ ─ ─ ─ ─ ─ ─ ─ ─ ─ ─ ─ ─ ─ ─ ─ ─ ─ ─ ─ ─ ─ ─ ─ ─ ─ ─ ─ ─ ─ ─ ─ ─ ─
0022 VTLAXS/      ON          LEFT     10.00000000  1.00000000                      00025 00002200
               0.50000000    RIGHT
0022 SURFACE      C1                    CIRCLE    DS(IMP-TO)                        00027 00002200
               0.0         -4.00000000   0.0
               0.0          0.0          1.00000000  8.00000000
0022 GOTO/        C1                                                                00028 00002200
               0.30797684   3.49323177   0.00639555  0.00068468  0.15943065  0.98720889  9.174°
0022 GOTO/        C1                                                                00029 00002200
               0.92161640   3.44440036   0.00430246  0.01073536  0.13046287  0.99139507  7.522°
0022 GOTO/        C1                                                                00030 00002200
               1.52925354   3.34512991   0.00262149  0.01671718  0.10089107  0.99475702  5.87°
0022 GOTO/        C1                                                                00031 00002200
               2.12674309   3.19604932   0.00135403  0.01793179  0.07132493  0.99729194  4.218°
0022 GOTO/        C1                                                                00032 00002200
               2.41875345   3.09839787   0.00087571 [0.01884018] 0.03607869  0.99824858  3.392°
0022 SURFACE      C2                    CIRCLE    DS(IMP-TO)                        00034 00002200
              -5.00000000   2.00000000   0.0
               0.0          0.0          1.00000000  8.00000000
0022 GOTO/        C2                                                                00035 00002200
               2.46289355   2.78938521   0.00049603 [0.04408300] 0.00631181  0.99900793  2.552°
** ERROR 3210  ISN 00022         SEQ. NO. 00002200    COND. CODE 04 *
** WARNING ... REVERSAL IN TOOL AXIS VECTOR COMPONENT DETECTED.  CHECK FOR INFLECTION IN CUTTER PATH. **
0022 GOTO/        C2                                                                00036 00002200
               2.50312979   2.16495175   0.00005818 [0.01522388][0.00097082][0.99988364] 0.874°
0022 GOTO/        C2                                                                00037 00002200
               2.49079722   1.53931253   0.00004926 -0.01403333 [0.00027678][0.99990149] 0.804°
** ERROR 3210  ISN 00022         SEQ. NO. 00002200    COND. CODE 04 *
** WARNING ... REVERSAL IN TOOL AXIS VECTOR COMPONENT DETECTED.  CHECK FOR INFLECTION IN CUTTER PATH. **
0022 GOTO/        C2                                                                00038 00002200
               2.42594207   0.91692093   0.00046926 -0.04308475 [0.00443892][0.99906147] 2.483°
0022 GOTO/        C2                                                                00039 00002200
               2.30898929   0.30222236   0.00131785 -0.07129795  0.01345554  0.99736430  4.161°
0022 GOTO/        C2                                                                00040 00002200
               2.22641228   0.00054394   0.00190265 -0.08411200  0.02283188  0.99619470  5°
0022 VTLAXS/      OFF          RIGHT     5.00000000                                 00042 00002200
─ ─ ─ ─ ─ ─ ─ ─ ─ ─ ─ ─ ─ ─ ─ ─ ─ ─ ─ ─ ─ ─ ─ ─ ─ ─ ─ ─ ─ ─ ─ ─ ─ ─ ─ ─ ─ ─ ─ ─ ─ ─
0023 GOTO/        SETPT                                                             00044 00002300
               0.0          0.0          0.0       0.0        0.0     1.00000000  0°
```

Figure 14.21   CL-printout for Example 14.2.

are given. In this case there were no inflections but reversal in tool axis vector components was detected. These reversals are boxed off as shown. Such reversals are expected when the tool swings from one side of vertical to the other.

14.3.2 Cornering with the WCORN Statement

Linear proration of the angular tool axis difference between the VTLAXS bounding statements may cause undesired cutting effects at inside corners as shown in Figure 14.22(a). What really may be desired is the tool axis tilt shown in Figure 14.22(b). The latter swarf effect may be achieved with the WCORN statement in conjunction with a VTLAXS bounding pair.

The WCORN statement affects tool axis tilt in the vicinity of a corner by the angles given in the WCORN statement, rather than allowing this tilt to be prorated by the

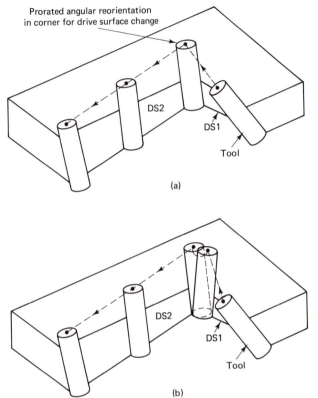

Figure 14.22 (a) Part after VTLAXS transformation; (b) inside corner clean-out with WCORN feature.

VTLAXS bounding pair. It causes cut vectors to be inserted very close together on both sides of the corner. The number of such cut vectors is predetermined and causes the total number of cut vectors between the VTLAXS bounding pair to be increased by this amount. In effect, the WCORN feature warps the linear proration of tilt on the cutter path as required by the VTLAXS statements, yet allows proration to be applied for nearly the entire path. This statement may not be available for all APT processors.

The syntax of the statement is as follows:

$$\text{WCORN/}\left\{\begin{matrix}\text{RIGHT}\\\text{LEFT}\end{matrix}\right\},\phi,\left\{\begin{matrix}\text{RIGHT}\\\text{LEFT}\end{matrix}\right\},\psi,h[,n]$$

RIGHT or LEFT	First occurrence—a modifier that specifies the direction of cutter tilt as it approaches the corner. Its geometric interpretation is identical to that described for the VTLAXS statement but applies to the tool as it approaches the last cut vector on the CL-file produced by the motion statement immediately preceding the WCORN statement.

φ   The approach angle (0° ≤ φ < 80°, in degrees) of the cutter tilt from
             vertical in the direction of the first-occurrence RIGHT/LEFT modifier.

RIGHT   Second occurrence—a modifier that specifies the direction of cutter tilt
or   as it leaves the corner. Its geometric interpretation is identical to that
LEFT   of the first-occurrence modifier but applies to the tool as it leaves the
             last cut vector on the CL-file produced by the motion statement im-
             mediately preceding the WCORN statement.

ψ   The departure angle (0° ≤ ψ < 80°, in degrees) of the cutter tilt from
             vertical in the direction of the second-occurrence RIGHT/LEFT mod-
             ifier.

h   The vertical height of the part at the corner subject to being cut by the
             tool in the units of measurement for the part. This parameter is used
             by the WCORN algorithm for cornering in conjunction with the ap-
             proach and departure path length.

n   The number of cut vectors to be inserted on each side of the corner
             cut vector (see RIGHT/LEFT modifier descriptions). If omitted, the
             default value is 6; if given the value 1 or 2, it is assigned the value 3;
             otherwise, it is assigned the given value. Sufficient path length must
             be allowed on each side of the corner cut vector to effect the angular
             reorientation of the tool axis in the corner. Tool endpoint movement
             for each cut vector is defined to be equal to or greater than 0.0005
             units.

The WCORN statement must immediately follow the motion command that places
the tool into the corner. It applies only to that one command and there can be only one
WCORN statement per VTLAXS bounding pair.

Tool axis tilt is affected by the angles of the VTLAXS and WCORN statements
with their respective RIGHT/LEFT modifiers as follows:

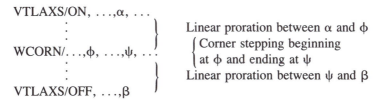

The effect of ZSMALL is sustained while corner stepping. Also, the corner may
not exceed 180°. That is, during a turn in one direction, the tool cannot be on the opposite
side of the drive surface.

Although the resultant corner depends upon the combination of RIGHT/LEFT modifiers and angles α, β, φ, and ψ, there is no restriction on the relative size of these angles.
While stepping, the tool-end movement may be small but the tool-axis angular change
may be large. Generally, a smaller-diameter tool is stepped through a smaller angular
swing compared to that for a larger-diameter tool for the same corner.

		Tool endpoint		Tool axis vector				
0036	GOTO/	C1						00063 00003600
		0.30798206	3.49444741	0.00486595	0.00059768	0.13917158	0.99026810	8°
0036	GOTO/	C1						00064 00003600
		0.92183257	3.44702742	0.00121797	0.00572069	0.06952142	0.99756406	4°

```
**×××××××××××××××××××××××××××××××××××××××××××××××××××××××××××××××××××××
** ERROR 3210   ISN 00036           SEQ. NO. 00003600   COND. CODE 04 *
** WARNING ... REVERSAL IN TOOL AXIS VECTOR COMPONENT DETECTED. CHECK FOR INFLECTION IN CUTTER PATH. **
**×××××××××××××××××××××××××××××××××××××××××××××××××××××××××××××××××××××
```

0036	GOTO/	C1						00065 00003600
		1.52964027	3.34746389	0.0	0.00000001	0.00000008	1.00000000	0°

```
**×××××××××××××××××××××××××××××××××××××××××××××××××××××××××××××××××××××
** ERROR 3210   ISN 00036           SEQ. NO. 00003600   COND. CODE 04 *
** WARNING ... REVERSAL IN TOOL AXIS VECTOR COMPONENT DETECTED. CHECK FOR INFLECTION IN CUTTER PATH. **
**×××××××××××××××××××××××××××××××××××××××××××××××××××××××××××××××××××××
```

0036	GOTO/	C1						00066 00003600
		2.12673133	3.19600253	0.00121797	-0.01700813	-0.06765098	0.99756406	4°
0036	GOTO/	C1						00067 00003600
		2.41804751	3.09629661	0.00273905	-0.03328892	-0.09908606	0.99452190	6°
0036	GOTO/	C1						00068 00003600
		2.41651096	3.09671106	0.00365687	0.02434010	-0.11824356	0.99268626	6.934°
0036	GOTO/	C1						00069 00003600
		2.41523991	3.09683011	0.00643444	0.08230973	-0.13710308	0.98713112	9.202°
0036	GOTO/	C1						00070 00003600
		2.41456879	3.09654255	0.01106141	0.13997095	-0.15544887	0.97767718	12.074°
0036	GOTO/	C1						00071 00003600
		2.41482717	3.09574566	0.01746568	0.19668502	-0.17308234	0.96506866	15.189°

```
**×××××××××××××××××××××××××××××××××××××××××××××××××××××
TOOL AXIS TILT ANGLE AT CORNER VERTEX IN X-Z PLANE= 11.5193560      DEGREES
TOOL AXIS TILT ANGLE AT CORNER VERTEX IN Y-Z PLANE= -10.1677393     DEGREES
**×××××××××××××××××××××××××××××××××××××××××××××××××××××
```

0036	WCORN /	RIGHT	6.00000000		LEFT	10.00000000		00073 00003600
		1.00000000	4.00000000					
0036	SURFACE	C2			CIRCLE	DS(IMP-TO)		00075 00003600
		-5.00000000	2.00000000	0.0				
		0.0	0.0	1.00000000	8.00000000			
0036	GOTO/	C2						00076 00003600
		2.42496799	3.01957003	0.01316523	0.19120749	-0.12412613	0.97366955	13.177°
0036	GOTO/	C2						00077 00003600
		2.43540112	2.94285190	0.01006973	0.18522054	-0.07460999	0.97986055	11.518°
0036	GOTO/	C2						00078 00003600
		2.44610520	2.86578099	0.00821180	0.17876571	-0.02490538	0.98357640	10.398°
0036	GOTO/	C2						00079 00003600
		2.45705361	2.78854905	0.00759613	0.17189514	0.02461197	0.98480775	10°
0036	GOTO/	C2						00080 00003600
		2.50018439	2.16476392	0.00338083	0.11585766	0.00738817	0.99323835	6.667°
0036	GOTO/	C2						00081 00003600
		2.49004922	1.53932728	0.00084593	0.05813368	-0.00114659	0.99830815	3.333°

```
**×××××××××××××××××××××××××××××××××××××××××××××××××××××××××××××××××××××
** ERROR 3210   ISN 00036           SEQ. NO. 00003600   COND. CODE 04 *
** WARNING ... REVERSAL IN TOOL AXIS VECTOR COMPONENT DETECTED. CHECK FOR INFLECTION IN CUTTER PATH. **
**×××××××××××××××××××××××××××××××××××××××××××××××××××××××××××××××××××××
```

0036	GOTO/	C2						00082 00003600
		2.42642952	0.91687048	0.0	0.00000023	-0.00000002	1.00000000	0°

```
**×××××××××××××××××××××××××××××××××××××××××××××××××××××××××××××××××××××
** ERROR 3210   ISN 00036           SEQ. NO. 00003600   COND. CODE 04 *
** WARNING ... REVERSAL IN TOOL AXIS VECTOR COMPONENT DETECTED. CHECK FOR INFLECTION IN CUTTER PATH. **
**×××××××××××××××××××××××××××××××××××××××××××××××××××××××××××××××××××××
```

0036	GOTO/	C2						00083 00003600
		2.30950084	0.30212582	0.00084391	-0.05713594	0.01078285	0.99830818	3.333°
0036	GOTO/	C2						00084 00003600
		2.22641228	0.00054394	0.00190265	-0.08411200	0.02283188	0.99619470	5°
0036	VTLAXS/	OFF	RIGHT	5.00000000				00086 00003600

Figure 14.23   Partial CL-printout for Example 14.3.

Example 14.3

The following motion commands illustrate the WCORN feature applied to the code of Example 14.2. The geometry of the part is shown in Figure 14.20. A partial CL-printout from an IBM APT-AC system appears in Figure 14.23. Liberal tolerances were used.

```
MULTAX                              2600
CUTTER/1,.5                         2700
```

TLAXIS/0,0,1	2800
FROM/SETPT	2900
INDIRV/0,1,0	3000
GO/TO,C1	3100
VTLAXS/ON,LEFT,10,1,.5,RIGHT	3200
TLRGT,GORGT/C1,TO,C2	3300
WCORN/RIGHT,6,LEFT,10,1,4	3400
GORGT/C2,ON,LX	3500
VTLAXS/OFF,RIGHT,5	3600
GOTO/SETPT	3700

Cut vectors added by the WCORN statement are contained within the rectangular outline of Figure 14.23 (there are $n = 4$ on each side of the corner). Not all cut vectors appear in Figure 14.23, but the tool axis is swung from an initial angle of $\alpha = 10°$ LEFT to $\phi = 6°$ RIGHT at the start of the corner. Through the corner it is stepped from $\phi = 6°$ RIGHT to $\psi = 10°$ LEFT. As it leaves the corner it swings from $\psi = 10°$ LEFT to the ending angle of $\beta = 5°$ RIGHT. Additional inflection warnings also appear in this printout. Figure 14.23 is annotated with the tool axis tilt from vertical.

14.4 EXAMPLES

The following examples of three-dimensional contouring illustrate some of the principles introduced above and in Chapter 13. The part programmer should expect to expend considerably more program development and debugging effort for such problems than was required for two-dimensional contouring applications. Even though APT is a mature product, its algorithms for computing cut vectors under multiaxis conditions may exhibit unstable tendencies, thus requiring workarounds from what may otherwise be an obvious way to program the part. For this reason, not all APT processors can be expected to produce identical results for a given problem with the same amount of effort.

These guidelines are offered to help overcome difficulties with program checkout.

1. Keep the geometric descriptions as simple as possible. Wherever tangent or normal relationships exist, use defining formats that involve such relationships.
2. Introduce auxiliary geometry when necessary to simplify motion commands.
3. Verify all geometric definitions via PRINT/3 statements and special contouring statements to further define the part geometry to avoid unexpected results from invisible check surfaces.
4. Write and check out the code in segments. Work in increments using MACROs to simplify the code.
5. Maintain explicit tool-to-surface relationships to avoid tool positioning problems while contouring.
6. Adjust the tool description for help in avoiding unexpected check conditions.
7. When advantageous, reposition the tool with the GO command; otherwise, bind the tool to drive and part surfaces to the greatest extent possible.

8. If necessary, use the DNTCUT-CUT bounding pair for code that repositions the tool under difficult positioning conditions.

9. Use transformations to generate additional cut vectors whenever possible to avoid computational and geometric definition problems.

Example 14.4

The geometry of the part is shown in Figure 14.24. For this example, we are concerned only with the recessed cut bounded by LX, L1, L2, and PL2. We expect to use PL1 and CY1 as part surfaces for cutting with a 1-in.-diameter flat-end tool.

The plan is to machine in a left/right direction on PL1 and CY1 with the tool normal to the part surface. Parallel lines will be defined dynamically via CANON for the drive surface. A back/forth movement is then required to remove the uncut material at the intersection of PL2 with CY1 as shown in Figure 14.24. A TLOFPS condition with PL2 as drive surface is reasonably expected to do the job. Tool movement and tool axis alignment for the plan reduces the problem to a four-axis case.

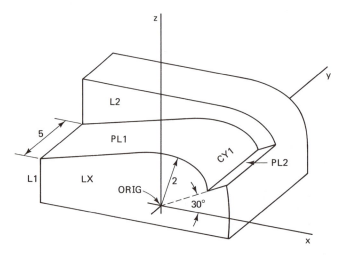

Figure 14.24   Geometry of Example 14.4.

The coordinate axes are shown in Figure 14.24, where line LX is the *zx*-plane. Preliminary to driving the tool we have the following program code (not all geometric definitions are given here):

VEC	= VECTOR/COSF(120),0,SINF(120)		100
PL2	= PLANE/ORIG,PERPTO,VEC		110
VPY	= VECTOR/0,1,0		120
VNY	= VECTOR/0,−1,0		130
SETPT	= POINT/−5,−2.5,3		140
	MULTAX		150
	D = 1	$$ TOOL DIAMETER	160
	YL = 5	$$ DIST L2 FROM LX	170

N = INTGF(YL/D) + 1	\$\$ NUM LEFT/RIGHT CUTS	180
DY = YL/N	\$\$ ACTUAL CUT STEP DIST	190
Y = 0		200
Y = Y + DY		210
LL = LINE/LX,CANON,,,,Y	\$\$ 1ST LEFT/RIGHT DS	220
RADCYL = 2	\$\$ RADIUS OF CY1	230
PHI = ATANF(D/2/RADCYL)		240
DPHI = 2	\$\$ STEP ANG FOR UNCUT MAT	250
ND = INTGF(PHI/DPHI) + 1	\$\$ NUM BACK/FORTH PASSES	260
NDPHI = PHI/ND	\$\$ ACTUAL CUT STEP ANGLE	270
TLPHI = 30 + PHI	\$\$ STARTING ANGLE	280
PHI90 = TLPHI + 90		290
VEC = VECTOR/VEC,CANON,COSF(PHI90),0,SINF(PHI90)		300
PL3 = PLANE/ORIG,PERPTO,VEC	\$\$ BACK/FORTH PLANE	310
CUTTER/D		320

Variables N and ND contain the number of left/right and back/forth passes, respectively. The same amount of material is cut during each pass for a given type of operation. Macros are used to reduce the amount of program code. The following code and macro perform the left/right cuts.

DRVLN = MACRO/TLPOS = TLRGT,TLDIR = GORGT,S1,S2,S3		400
INDIRV/0,1,0		410
TLON,GOFWD/S1,TO,LL		420
TLPOS,TLDIR/LL,PSTAN,S2		430
PSIS/S2		440
GOFWD/LL,ON,S3		450
Y = Y + DY		460
LL = LINE/LX,CANON,,,,Y		470
TERMAC		480
\$\$**		490
FROM/SETPT		500
TLAXIS/NORMPS		510
GO/ON,L1,TO,PL1,TO,LX		520
J = 0		530
LOOPST		540
LA1) CALL/DRVLN,S1 = L1,S2 = CY1,S3 = PL3		550
J = J + 1		560
IF (J − N) LA2, LA3, LA3		570
LA2) CALL/DRVLN,TLPOS = TLLFT,TLDIR = GOLFT, \$		580
S1 = PL3,S2 = PL1,S3 = L1		590
J = J + 1		600
IF (J − N) LA1, LA3, LA3		610
LA3) . . .		620

We use the macro to drive right, line 550, then left, line 580. However, we do not use PL2 as a check surface when driving right nor as a drive surface when driving to the next line because of APT system difficulty in computing the cut vectors. Instead, we define

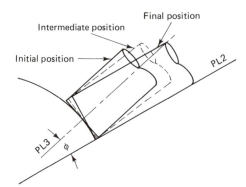

Figure 14.25 **Typical tool positions for removing uncut material.**

plane PL3 (see Figure 14.25) to check and drive the tool on with an ON modifier. The following code and macro perform the back/forth cuts.

```
DRVPL = MACRO/CK1,S1,CK2,S2,V                               800
        TLPHI = TLPHI − NDPHI                               810
        COPHI = COSF(TLPHI)                                 820
        SIPHI = SINF(TLPHI)                                 830
        VEC = VECTOR/VEC,CANON,COPHI,0,SIPHI                840
        TLAXIS/VEC                                          850
        GODLTA/0,0,.5                                       860
        PL3 = PLANE/PL3,CANON,COPHI,0,SIPHI,RADCYL          870
        GO/ON,PL3,TO,CY1,CK1,S1                             880
        PSIS/PL3                                            890
        INDIRV/V                                            900
        TLON,GOFWD/PL3,CK2,S2                               910
        TERMAC                                              920
$$***********************************************************  930
LA3)    M = N − 2*INTGF(N/2)                                620
        IF (M) LA4, LA4, LA5                               630
LA4)    GOTO/1.5,4.25,2.5                                   640
LA5)    J = 0                                               650
LA6)    CALL/DRVPL,CK1 = TO,S1 = L2,CK2 = ON,S2 = LX,V = VNY  660
        J = J + 1                                           670
        IF (J − ND) LA7, LA8, LA8                          680
LA7)    CALL/DRVPL,CK1 = ON,S1 = LX,CK2 = TO,S2 = L2,V = VPY  690
        J = J + 1                                           700
        IF (J − ND) LA6, LA8, LA8                          710
LA8)    LOOPND                                              720
        GOTO/2,1,2                                          730
        TLAXIS/0,0,1                                        740
        GOTO/SETPT                                          750
```

Variable M is used to determine if the tool, on the last left/right cut, ended up at the left or right end of its motion. Essentially, this is an odd/even test. Tool orientation is shown in Figure 14.25. The first cut is made with the tool located on PL3 (initial position).

Thereafter, it is angularly reoriented on a new plane tangent to CY1 (intermediate position) and TO PL2. Angular increments are given by variable NDPHI. A new tool axis is computed for each back/forth pass. Of course, the code segments given for this example would be grouped differently in the actual part program.

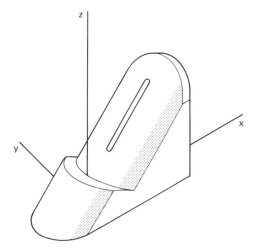

Figure 14.26   Geometry of Example 14.5.

Example 14.5

The purpose of this example is to illustrate solutions to problems in orienting the tool. The geometry of the part is shown in Figure 14.26, while some geometric details are given in Figure 14.27. We are addressing three features of the part: the uniform width in shaping the upper end of the semicylinder, the bevel near the lower end of the semicylinder, and the slot in the semicylinder.

To contour the semicylinder to a uniform width at its upper end requires that we follow cylinder CYL1 such that we remove material relative to plane PL1. The shape of this cut is not readily described for use as a drive surface so we choose to orient the tool axis normal to PL1 and use the CUTTER/OPTION statement with the ring spaced ½ in. from the plane. Plane PL6 serves to position the tool tangent to the semicylinder. Code for this operation is as follows:

```
PL1    = PLANE/1,0,0,4.5                          100
PL6    = PLANE/0,0,1,4                            110
CYL1   = CYLNDR/0,0,0,1,0,1,1.5                   120
SETPT1 = POINT/2, - 3,6                           130
SETPT2 = POINT/2,3,6                             140
         CUTTER/1                                 150
         FROM/SETPT1                              160
         TLAXIS/ - 1,0,0                          170
         CUTTER/OPTION,2,.5,.5    $$ RING CUTTER  180
         GO/TO,CYL1,TO,PL1,ON,PL6                 190
         INDIRV/0,0,1                             200
         TLRGT,GOFWD/CYL1,ON,PL6                  210
```

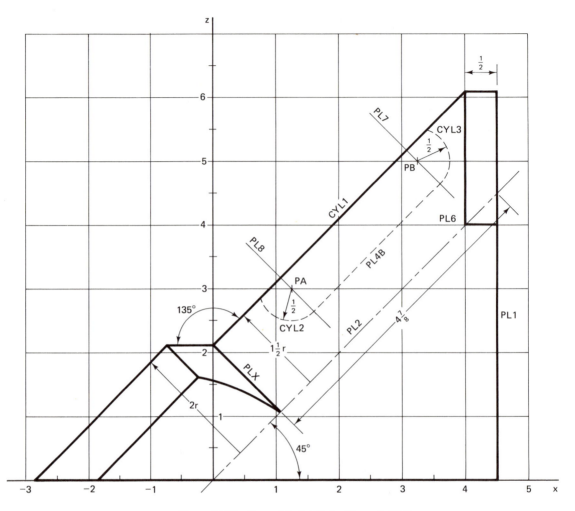

Figure 14.27   Geometric details for Example 14.5.

CUTTER/OPTION,2,OFF	220
GOTO/SETPT2	230

The 45° bevel at the lower end of the semicylinder is handled by orienting the tool axis at 45° relative to the cylinder surface normal. Plane PL5, the controlling surface, is defined parallel with and above plane PLX and used as the drive surface. CYL1 is the part surface. Code for this operation is as follows:

SC45 = COSF(45)	300
PL2 = PLANE/ − SC45,0,SC45,0	310
DPL5 = 4.5*SQRTF(2) − 4.875 + SC45	320
PL5 = PLANE/SC45,0,SC45,DPL5	330

```
        CUTTER/1                                                   340
        FROM/SETPT1                                                350
        TLAXIS/ATANGL,2,45          $$ ALPHA = 45 DEG              360
        GO/PAST,PL5,TO,CYL1,ON,PL2                                 370
        INDIRV/ − 1,0,1                                            380
        TLLFT,GOFWD/PL5,ON,PL2                                     390
        GOTO/SETPT2                                                400
```

The ½-in. slot in the semicylinder, made with a ½-in.-diameter tool, requires manipulation of the tool relative to auxiliary surfaces and suppression of superfluous cut vectors with the DNTCUT/CUT pair. These superfluous cut vectors arise because we reestablish the tool position relative to the constraining surfaces after reorienting the tool axis. Code for this operation is as follows:

```
LX    = LINE/XAXIS            $$ ZX-PLANE                500
PA    = POINT/1.25,0,3                                   510
PB    = POINT/3.25,0,5                                   520
PC    = POINT/1.25,1,3                                   530
PL4A  = PLANE/PA,PB,PC                                   540
PL4B  = PLANE/PARLEL,PL4A,ZSMALL,.5                      550
PL7   = PLANE/PB,PARLEL,PL5                              560
PL8   = PLANE/PA,PARLEL,PL5                              570
CYL2  = CYLNDR/PA,0,1,0,.5                               580
CYL3  = CYLNDR/PB,0,1,0,.5                               590
        CUTTER/.5,.25, 0,0,0,0, .25    $$ .5 DIA BALL-END  600
        FROM/SETPT1, − 1,0,0    $$ SET TOOL AXIS         610
        GO/ON,LX,ON,PL7,TO,CYL1                          620
        PSIS/PL7                                         630
        TLON,TLONPS,GODOWN/LX,PAST,CYL3                  640
        PSIS/CYL3                                        650
        TLON,TLOFPS,GOFWD/LX,PSTAN,PL4B                  660
        DNTCUT                                           670
        GOTO/PB, − 1,0,1        $$ NEW TOOL AXIS         680
        GO/ON,LX,TO,PL4B,ON,PL7  $$ REPOSITION TOOL      690
        CUT                                              700
        PSIS/PL4B                                        710
        INDIRV/ − 1,0,1                                  720
        GOFWD/LX,ON,PL8                                  730
        DNTCUT                                           740
        GOTO/PA,0,0,1           $$ NEW TOOL AXIS         750
        GO/ON,LX,TO,PL4B,TO,PL8  $$ REPOSITION TOOL      760
        CUT                                              770
        PSIS/PL4B                                        780
        INDIRV/ − 1,0, − 1                               790
        GOFWD/LX,PSTAN,CYL2                              800
        PSIS/CYL2                                        810
        GOFWD/LX,ON,PL8                                  820
        GODLTA/0,0,.5                                    830
        GOTO/SETPT1                                      840
```

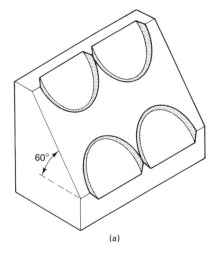

(a)

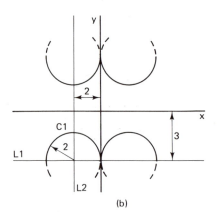

(b)

Figure 14.28 Geometry of Example 14.6.

Example 14.6

The purpose of this example is to illustrate the VTLAXS feature under COPY-TRACUT transformations. The geometry of the part is shown in Figure 14.28 with symbolic labeling of entities given in Figure 14.28(b). The plan is to tilt the tool axis with the VTLAXS feature along circle C1. Two sets of VTLAXS bounding statements are needed because of the inflection while tilting the tool axis. The set of cut vectors produced is then transformed via nested COPY statements, which, in turn, are then transformed by the TRACUT feature to translate and incline them at the 60° slope of Figure 14.28(a). The part program is as follows:

L1	= LINE/XAXIS, − 3	100
L2	= LINE/YAXIS, − 2	110
C1	= CIRCLE/ − 2, − 3,0,2	120

PL2 = PLANE/0,0,1,2	130
PLZ = PLANE/0,0,1,0	140
MAT1 = MATRIX/YZROT,60	150
MAT2 = MATRIX/TRANSL,0,0,(3*SINF(60))	160
MAT3 = MATRIX/MAT2,MAT1	170
SETPT = POINT/−6,−5,3	180
MULTAX	190
FROM/SETPT,0,0,1	200
TRACUT/LAST,MAT3	210
INDEX/5	220
GO/TO,L1,TO,PL2,TO,C1	230
GODLTA/0,0,−2	240
INDIRV/1,0,0	250
PSIS/PLZ	260
INDEX/3	270
TLRGT,GOFWD/L1,PAST,C1	280
VTLAXS/ON,LEFT,20,1,0,RIGHT	290
GOLFT/C1,ON,L2	300
VTLAXS/OFF,RIGHT,0	310
VTLAXS/ON,RIGHT,0,1,0,RIGHT	320
GOFWD/C1,PAST,L1	330
VTLAXS/OFF,LEFT,20	340
GODLTA/0,0,0,0,0,1	350
COPY/3,TRANSL,4,0,0,1	360
GODLTA/4,0,2,0,0,1	370
COPY/5,XYROT,180,1	380
TRACUT/LAST,NOMORE	390
GOTO/SETPT,0,0,1	400

PROBLEMS

14.1. In Problem 8.6 we imagined that our tool was a welding tip that we directed along the intersection of plates welded together to form the object of Figure P8.6. There, our tool axis was parallel with the z-axis. Repeat the welding operation but now position the tool with its axis perpendicular to the weld line and in a plane that bisects the exterior angle formed by the two plates of each weld line.

14.2. Write an APT part program to machine the object of Figure P8.6 from a solid block of material. Select a suitable shape and size of cutter.

14.3. Figure P14.3(c) shows the proportions of an "ideal brilliant" diamond. Write an APT part program to machine the crown only, including all facets whose formation can be determined from the geometry of Figure P14.3(a) and (b). Let the diameter of the girdle be 100% = 3 in. Select a suitable shape and size of cutter.

14.4. Figure P14.4(a) shows a hollow tube of square cross section into the sides of which are machined geometric shapes from Figure P14.4(b). Write an APT part program to machine

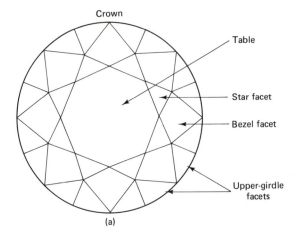

(a)

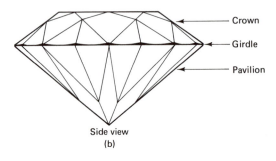

Side view
(b)

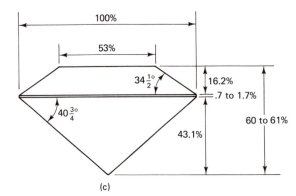

(c)

Figure P14.3

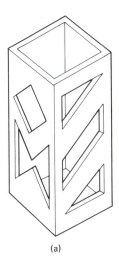

(a)

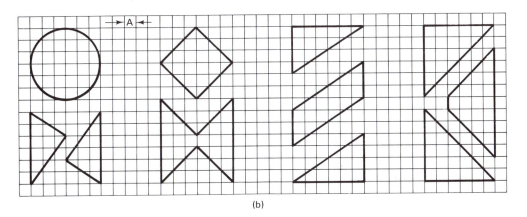

(b)

Figure P14.4

each of the shapes on a side of the tube while mounted in a vertical position. Determine dimensions from the grid of Figure P14.4(b) with $A = \frac{1}{4}$ in. Select a suitable shape and size of cutter.

14.5. Write an APT part program to machine the exterior surface of the hemisphere shown in Figure P14.5, including the conical cutouts. Assume that the hemisphere is a shell of the thickness given. The interior is not to be machined. Select a suitable shape and size of cutter.

14.6. Write an APT part program to machine the object of Example 13.10. Select a suitable shape and size of cutter.

14.7. Write an APT part program to machine the object of Figure P12.3. Select a suitable shape and size of cutter.

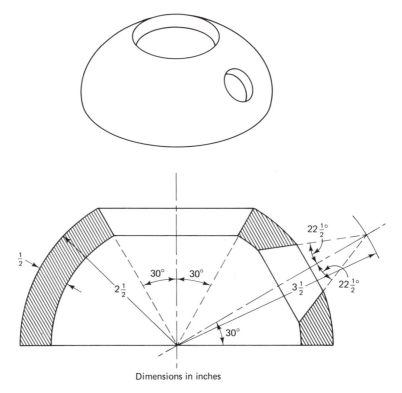

Dimensions in inches

Figure P14.5

14.8. Repeat Problem 14.7 but for Figure P12.4.

14.9. Repeat Problem 14.7 but for Figure P12.7. Omit the inside elliptic cylinder.

14.10. Repeat Problem 14.7 but for Figure P12.8.

14.11. Repeat Problem 14.7 but for Figure P12.10.

14.12. Repeat Problem 14.7 but for Figure P12.11.

14.13. Repeat Problem 14.7 but for Figure P12.13. Write the program to produce a four-bladed propeller with $R = 4$ in. and $\psi = 15°$.

14.14. Repeat Problem 14.7 but for Figure P12.14(a). Write the program to produce a three-bladed propeller with $R = 4$ in. and $\psi = 15°$.

14.15. Repeat Problem 14.14 for Figure P12.14(b).

14.16. Repeat Problem 14.14 for Figure P12.14(c).

APT Reserved Words

Listed below are 532 reserved words from various APT systems. These words are often referred to as vocabulary words. They have special meaning to a given APT system and should not be used for symbol and variable names. APT major and minor words are included here as well as many postprocessor words, except for postprocessor names for machine tools. There are variations on postprocessor words for different APT systems because postprocessor programs are not standardized. Often, words for APT system maintenance purposes are also reserved, but not many such words will be found in this list.

Sometimes, a list of reserved words is supplied in the reference manual for a given APT system. Such lists should probably be augmented with more postprocessor words. The APT documentation for early systems was in two volumes: an encyclopedia and a dictionary. For such documentation, there may not be a reserved word list and it will be necessary to consult the dictionary for the words. In any case, spelling errors may appear in any word list, so care must be used when interpreting diagnostics.

There are advantages to having later APT systems accept part programs written for earlier APT versions produced by the same software developer. Consequently, reserved word lists may be lengthy to maintain upward compatibility of the systems. Extensions to the APT language probably appear in every APT system. These also help to lengthen the reserved word lists. At least one such system allows reserved words to be used for symbol names and variables provided that the reserved word feature is not available in that system.

AAXIS	CENTER	DASH	EXEC	INCR	LOOPND
ABSF	CHANGE	DATREF	EXPF	INDEX	LOOPST
ABSLTE	CHECK	DCOORD	FACE	INDIRP	LOW
ADD	CHUCK	DEBUG	FACEML	INDIRV	LPRINT
ADJUST	CIRCLE	DECR	FAM	INSERT	LTV
AIR	CIRCUL	DEEP	FEDRAT	INTCOD	MACH
ALL	CIRLIN	DEEPHL	FEDTAB	INTENS	MACHIN
ANGLF	CLAMP	DELAY	FEED	INTERC	MACRO
ANTSPI	CLDIST	DELET	FEEDRT	INTGF	MAGTAP
APROMT	CLEARP	DELETE	FEET	INTGRV	MAIN
ARC	CLEARV	DELTA	FILE	INTOF	MAJOR
ARCSLP	CLPRNT	DEPTHV	FINI	INTOL	MANUAL
ASLOPE	CLRSRF	DHOLE	FINISH	INVERS	MATL
AT	CLTV	DIAG1	FLOOD	INVOC	MATRIX
ATANF	CLW	DIAG2	FMATL	INVX	MAXDP
ATANGL	CM	DIAG3	FOURPT	INVY	MAXDPM
ATAN2F	CMIT	DIAMTR	FROM	IPM	MAXIPM
ATTACH	CNSINK	DISPLY	FRONT	IPR	MAXRPM
AUTO	CODEL	DISTF	FULL	ISTOP	MAXVEL
AUTOPOL	COLLET	DITTO	FUNOFY	JUMPTO	MAX1F
AUTOPS	COMDMP	DMILL	GAPLES	KEYBOR	MCHFIN
AUXFUN	CONE	DNTCUT	GCONIC	LARGE	MCHTOL
AVOID	CONSEC	DNTLR	GO	LAST	MDEND
BAXIS	CONST	DNTLRP	GOBACK	LATER	MDWRIT
BCD	CONT	DNTR	GOCLER	LCONIC	MED
BCHIP	CONTUR	DOTF	GODLTA	LEADER	MEDIUM
BEVEL	COOLNT	DOTTED	GODOWN	LEFT	MESH
BEVELS	COPY	DOWN	GOFWD	LENGTH	MILL
BINARY	CORNFD	DRAFT	GOHOME	LETTER	MINOR
BLACK	COSF	DRAWLI	GOLFT	LIBRY	MINUS
BLANK	COUPLE	DRESS	GORGT	LIFTOF	MIN1F
BLUE	CROSS	DRILL	GOTO	LIGHT	MIRROR
BORE	CRSSPL	DS	GOUGCK	LINCIR	MIST
BOREOS	CRTLIN	DSTAN	GOUP	LINE	MIT
BOTH	CS	DWELL	GREEN	LINEAR	MM
BRKCHP	CSINK	DWELLV	GRID	LIST	MMPM
CALL	CTREAC	DWL	GROOVE	LITE	MMPR
CAM	CUT	DYNDMP	HEAD	LNTHF	MODE
CAMERA	CUTANG	EDIT	HIGH	LOADTL	MODIFY
CANON	CUTCOM	EDITND	HOLDER	LOCAL	MOTION
CATLOG	CUTTER	ELLIPS	HYPERB	LOCK	MOVETO
CAXIS	CYCLE	ELMSRF	IF	LOCKX	MULTAX
CBORE	CYLNDR	END	IFRO	LOFT	MULTRD
CBRTF	DAC	ENDARC	IN	LOGF	MXMMPM
CCLW	DARK	ERCOND	INCHES	LOG10F	NAUDIT

NCDB	PATERN	PTSLOP	SEC1	TERMAC	UNIT
NDTEST	PBS	PULBOR	SEC2	TEST	UNITMM
NEGX	PEN	PULFAC	SELCTL	THETAR	UNITS
NEGY	PENDWN	PUNCH	SEQNO	THICK	UNLIKE
NEGZ	PENUP	QADRIC	SETANG	THREAD	UNLOAD
NEXT	PERPTO	RADIUS	SETOOL	THRU	UP
NIXIE	PERSP	RAIL	SFM	TIME	VECTOR
NOCS	PICKUP	RANDOM	SIDE	TIMES	VTLAXS
NOMORE	PILOTD	RANGE	SIGNF	TITLES	WCORN
NOPLOT	PITCH	RAPID	SINF	TLAXIS	WDEFAC
NOPOST	PIVOTZ	READ	SLOPE	TLLFT	WEIGHT
NOPS	PLABEL	REAM	SLOWDN	TLNDON	XAXIS
NORMAL	PLANE	REAMA	SMALL	TLOFPS	XCOORD
NORMDS	PLOT	REAR	SOLID	TLON	XLARGE
NORMPS	PLUNGE	RED	SOURCE	TLONPS	XREF
NOSLT	PLUS	REFSYS	SPDRL	TLRGT	XSMALL
NOW	PNCHID	REGBRK	SPDTAB	TMARK	XYOF
NOX	PNTSON	REMARK	SPECDP	TO	XYPLAN
NOY	PNTVEC	REPLAC	SPECFR	TOLER	XYROT
NOZ	POCKET	RESERV	SPEED	TOOL	XYVIEW
NUMBR	POINT	RESET	SPHERE	TOOLNO	XYZ
NUMF	POLCON	RETAIN	SPINDL	TOOLST	YAXIS
NUMPTS	POLYGN	RETRCT	SPINSP	TORS	YCOORD
OBTAIN	POSTN	REV	SPLINE	TORUS	YLARGE
OFF	POSX	REVERS	SPMIL	TP	YSMALL
OFFSET	POSY	REWIND	SQRTF	TPI	YZPLAN
OMIT	POSZ	RIGHT	SRFREV	TPMM	YZROT
ON	POWER	RLDSRF	SRFVCT	TRACUT	YZVIEW
OPEN	PPLOT	ROOT	SSURF	TRANPT	ZAXIS
OPSKIP	PPRINT	ROTABL	START	TRANS	ZCOORD
OPSTOP	PPWORD	ROTHED	STEP	TRANSL	ZERO
OPTION	PREFUN	ROTREF	STOP	TRANTO	ZIGZAG
ORIGIN	PRINT	ROUGH	SWITCH	TRAV	ZLARGE
OUT	PROBX	ROUND	SYN	TRFORM	ZSMALL
OUTTOL	PROBY	RPM	SYSLIB	TRYBOR	ZSURF
OVCONT	PROC	RTHETA	TABCYL	TRYBOS	ZXPLAN
OVPLOT	PROCND	SADDLE	TANCRV	TUNEUP	ZXROT
PARAB	PROPF	SAFETY	TANDS	TURN	ZXVIEW
PARLEL	PS	SAME	TANF	TURRET	2DCALC
PART	PSIS	SCALE	TANON	TWOPT	3DCALC
PARTNO	PSTAN	SCRIBE	TANSPL	TYPE	3PT2SL
PASS	PTFORM	SCRUCT	TANTO	TYPEF	4PT1SL
PAST	PTNORM	SCULPT	TAP	ULOCKX	5PT
PATCH	PTONLY	SECTN3	TAPKUL		

Bibliography

AMERICAN NATIONAL STANDARDS INSTITUTE, INC., 1430 Broadway, New York, NY 10018. The documentation center for various NC standards developed by the International Standards Organization.

ASSOCIATION FOR INTEGRATED MANUFACTURING TECHNOLOGY, P.O. Box 1234, Beloit, WI 53511. Formerly the Numerical Control Society. Publisher of NC conference proceedings and NC, CAM, and manufacturing publications.

CHASEN, SYLVAN H., *Geometric Principles and Procedures for Computer Graphic Applications*. Englewood Cliffs, NJ: Prentice-Hall, Inc., 1978. Although not evident from its title, this book contains a thorough discussion of the mathematics of geometric entities found in NC applications.

CHILDS, JAMES J., *Numerical Control Part Programming*, New York: Industrial Press, Inc., 1973. This is a widely used textbook that contains a general discussion of machine tools and introductory discussions of manual part programming and the APT language. It contains many photographs of NC machine tools.

IBM CORP., STAFF, *System/370 APT-AC Numerical Control Processor Program Reference Manual*, Vol. 1, Form SH20-1414. White Plains, NY: IBM Corporation, 1973. This is the reference manual for the APT system used to run all examples in this book.

IIT RESEARCH INSTITUTE, APT LONG RANGE PROGRAM STAFF, *APT Part Programming*. New York: McGraw-Hill Book Company, 1967. This is the first book on the APT language. Its layout resembles the encyclopedia/dictionary format of many early APT language reference manuals.

LESLIE, W. H. P., ed., *Numerical Control Users' Handbook*. London: McGraw-Hill Book Company, 1970. Numerous authors contributed to this book with the intent of recording the advances made in the utilization of NC. The subject matter covers NC machine tools, NC economic factors, NC standards, various high-level NC languages, and postprocessors.

MODERN MACHINE SHOP, *NC/CAM Guidebook*. New York: Gardner Publications, Inc. (published annually). This book contains comprehensive NC equipment and services directories. It also has a brief discussion on each of a large number of NC programming languages.

OLESTEN, NILS O., *Numerical Control*. New York: Wiley-Interscience, 1970. This book is well illustrated with drawings and photographs of NC machines and systems that are described in detail for much of the book. It also contains one chapter on the mathematics of NC and a small amount on manual and computer-aided part programming.

ROBERTS, ARTHUR D., AND Richard C. PRENTICE, *Programming for Numerical Control Machines* (2nd ed.). New York: McGraw-Hill Book Company, 1978. This is a popular introductory textbook on manual NC part programming. It is abundantly illustrated with photographs of various types of NC machine tools.

Index